PHYSIOLOGICAL ECOLOGY OF ESTUARINE ORGANISMS

THE BELLE W. BARUCH LIBRARY IN MARINE SCIENCE

THE BELLE W. BARUCH LIBRARY IN MARINE SCIENCE NUMBER 3

Physiological Ecology
of Estuarine
Organisms

Edited by F. John Vernberg

Published for the Belle W. Baruch Institute for Marine Biology and

Coastal Research by the

UNIVERSITY OF SOUTH CAROLINA PRESS

COLUMBIA, SOUTH CAROLINA

Library of Congress Cataloging in Publication Data
Main entry under title:

Physiological ecology of estuarine organisms.

   (The Belle W. Baruch library in Marine Science, no. 3)
   Papers of a symposium held Apr. 11–14, 1973.
   Includes bibliographies.
   1. Estuarine ecology—Congresses. 2. Adaptation (Biology)—Congresses. 3. (Physiology)—
Congresses. I. Vernberg, F. John, 1925–   ed. II. Series.
QH541.5.E8P49   576'.15   74-14572
ISBN 0-87249-320-2

This volume is dedicated to Professor Dr. Carl Schlieper who is one of the first important investigators to blend the physiological and ecological viewpoints. Since the late twenties he has published over 100 papers and books. In the book "Biologie des Brackwassers" which was co-authored with Dr. Remane, Dr. Schlieper discussed in detail the physiological features of life in brackish waters. A revised English edition, "Biology of Brackish Waters," was published in 1971.

Born on August 20, 1903 in Fritzlar/Hessen, Germany he eventually studied natural sciences at the Universities of Rostock, Jena and Kiel. He received the degree Dr. phil. in 1926. The famous physiologist von Buddenbrech encouraged Carl Schlieper to study physiology. Soon his work on problems of osmoregulation of marine and estuarine animals was known internationally. After various research and teaching positions Professor Schlieper joined the faculty of the University of Kiel in 1952 as leader of the Department of Marine Zoology. He remained there as full professor until his retirement in 1971.

The central theme of his research has been physiological adaptation of aquatic organisms to environmental stress. He and his students have vigorously attacked this problem at the organismic, tissue and enzyme levels, examining problems of osmotic and ionic regulation, metabolic adaptation, and survival under extreme conditions of high hydrostatic pressure, osmotic and ionic stress, temperature, and oxygen.

Not only has Professor Schlieper been recognized as an international scientific figure through his publications and participation in scientific meetings, but also his honest, quiet but forceful personality has been a positive force to all who have had the honor of knowing and working with him.

CONTRIBUTORS

J. W. ANDERSON, *Department of Biology, Texas A & M University, College Station, Texas.*

B. BAYNE, *Institute for Marine Environmental Research, Citadel Road, Plymouth PLl 3AX, England.*

L. E. BURNETT, *Department of Biology, College of William and Mary, Williamsburg, Virginia.*

J. D. COSTLOW, *Duke University Marine Laboratory, Beaufort, North Carolina.*

R. M. DARNELL, *Departments of Oceanography and Biology, Texas A & M University, College Station, Texas.*

J. M. DEAN, *The Belle W. Baruch Institute for Marine Biology and Coastal Research and the Department of Biology, University of South Carolina, Columbia, South Carolina.*

J. L. DUPUY, *Virginia Institute of Marine Science, Gloucester Point, Virginia.*

R. L. FERGUSON, *National Marine Fisheries Service, Atlantic Estuarine Fisheries Center, Beaufort, North Carolina.*

R. B. FORWARD, JR., *Zoology Department, Duke University and Duke University Marine Laboratory, Beaufort, North Carolina.*

H. J. FYHN, *Duke University Marine Laboratory, Beaufort, North Carolina.*

S. R. HOPKINS, *The Belle W. Baruch Institute for Marine Biology and Coastal Research and the Department of Biology, University of South Carolina, Columbia, South Carolina.*

G. W. HYATT, *Department of Biological Sciences, Box 4348, University of Illinois at Chicago Circle, Chicago, Illinois.*

K. JOHANSEN, *Friday Harbor Laboratories, University of Washington, Seattle, Washington, and Department of Zoophysiology, University of Aarhus, Aarhus, Denmark.*

C. P. MANGUM, *Department of Biology, College of William and Mary, Williamsburg, Virginia.*

Contributors (cont.)

T. G. MILLER, *The Belle W. Baruch Institute for Marine Biology and Coastal Research and the Department of Biology, University of South Carolina, Columbia, South Carolina.*

G. S. MOREIRA, *Instituto de Biologia Marinha, Universidade de São Paulo, São Paulo, Brazil.*

R. C. NEWELL, *Department of Zoology and Comparative Physiology, Queen Mary College, London, England.*

L. C. OGLESBY, *Zoology Department, Pomona College, Claremont, California.*

J. A. PETERSEN, *Friday Harbor Laboratories, University of Washington, Seattle, Washington, and Department of Zoophysiology, University of Aarhus, Aarhus, Denmark.*

W. H. QUEEN, *Chesapeake Research Consortium, University of Maryland, College Park, Maryland.*

T. R. RICE, *National Marine Fisheries Service, Atlantic Estuarine Fisheries Center, Beaufort, North Carolina.*

A. N. SASTRY, *Graduate School of Oceanography, University of Rhode Island, Kingston, Rhode Island.*

H. THEEDE, *Institut für Meereskunde an der Universität Kiel, Kiel, Germany.*

B. L. UMMINGER, *Department of Biological Sciences, University of Cincinnati, Cincinnati, Ohio.*

F. J. VERNBERG, *The Belle W. Baruch Institute for Marine Biology and Coastal Research and the Department of Biology, University of South Carolina, Columbia, South Carolina.*

W. B. VERNBERG, *The Belle W. Baruch Institute for Marine Biology and Coastal Research and the Department of Biology, University of South Carolina, Columbia, South Carolina.*

T. E. WISSING, *Department of Zoology, Miami University, Oxford, Ohio.*

R. G. ZINGMARK, *The Belle W. Baruch Institute for Marine Biology and Coastal Research and the Department of Biology, University of South Carolina, Columbia, South Carolina.*

PREFACE

The study of the functional capability of marine organisms to successfully survive environmental fluctuation is basic in understanding the nature of the marine environment. This blending of physiological and environmental considerations is called physiological ecology. Because of the increasing awareness of the importance of the marine environment to human society and because of the increased scientific activity by physiological ecologists, a symposium entitled "Physiological Ecology of Estuarine Organisms" was held April 11-14, 1973 for the purpose of reviewing the current status of existing research areas and reporting on new research direction. This volume represents papers presented at the symposium.

The principal research areas which were covered are: resistance adaptations, respiration and energetics, water and ions, reproduction and development, and perception of the environment.

This is Volume 3 in a continuing series in the Belle W. Baruch Library in Marine Science sponsored by the Belle W. Baruch Institute for Marine Biology and Coastal Research, University of South Carolina and published by the University of South Carolina Press. The editor wishes to acknowledge the expert assistance of the various scientists who reviewed the papers and to those who offered editorial assistance, especially Ms. Dorothy Knight, Ms. Bettye Dudley, Ms.Barbara Caldwell, Ms. Hilda Merritt,and Ms. Susan Counts. The staff of the Institute assisted in numerous ways both in the organizing and conducting of the symposium and in the preparation of this volume. Dr. Winona B. Vernberg deserved special recognition for service beyond the call of

CONTENTS

Contents (cont.)

Response of estuarine phytoplankton to environmental conditions

T. R. Rice and R. L. Ferguson

Approximately three-fourths of the earth's surface is covered with water and is inhabited by planktonic algal flora, the base of the aquatic foodweb. To be able to live in the many different types of environments around the world, various species of phytoplankton have become adapted to the extremes of environmental factors such as temperature and salinity. For example, some species grow in polar regions at $-2°C$ while others grow in hot springs and have become adapted to temperatures as high as 70°C (Ukeles, 1961). Phytoplankton live in fresh, brackish, and sea water and some species can divide at salinities ranging from about 5 to 45 o/oo (Braarud,1961). Although temperature and salinity in estuaries do not fluctuate over the entire range tolerated by some species of phytoplankton, environmental factors in estuaries do fluctuate over wider ranges and more frequently than in either fresh water or sea water. Many species of phytoplankton inhabit estuaries where fluctuating conditions are permanent features of the environment (Hedgpeth, 1957). There are seasonal and diurnal changes in temperature, salinity, nutrient concentrations, light intensity and duration, turbidity, and other factors. Most of these factors also fluctuate with tides and over wide ranges within short periods of time.

In addition to natural fluctuations of environmental conditions, phytoplankton encounter toxic materials, heated water, and growth-promoting substances introduced by man into estuaries along our industrialized coasts. Water runoff from land carries nutrients from fertilized farmlands as well as many toxic substances including pesticides into the estuary. Municipal wastes cause an enrichment of the water and also act as a source of many toxic substances. Many substances carried into estuaries by rivers flocculate and settle to the bottom. Contaminants can be adsorbed on particles and become trapped in the sediments. By contrast, water is a dilutant for some contaminants and may disperse them in the estuary. The extent of either process also depends upon the behavior of the particular substance. The relative volumes of fresh water and sea water entering and mixing within the estuary determine the concentrations of many elements and the salinity of the water. All of these phenomena can be important in determining the species of phytoplankton present, the population size and its rate of increase, and the succession of species. Phytoplankton must adjust not only to the effects of natural fluctuations, but also to perturbations brought about by man's activities. The requirements of phytoplankton growing in estuaries characterized by these fluctuations need to be better understood in relation to both short- and long-term effects.

The interactions of man's effects and natural factors upon populations of phytoplankton in estuarine systems have been summarized from the literature and unpublished data. Results of studies carried out in the Newport River estuary, North Carolina, by the authors and other scientists of the Atlantic Estuarine Fisheries Center have been used to illustrate the extent of fluctuations in environmental conditions occurring in shallow drowned-river estuaries of the southeastern United States. This review has considered mainly the abundance, growth, and physiological responses of phytoplankton within a changing estuarine environment.

THE ESTUARINE ENVIRONMENT

The level of environmental factors and their rates of change in the estuary are determined by many natural and man-imposed conditions. To a large extent, man's activities have resulted in an increase or decrease in both the levels of natural environmental factors and the rates of change of these levels. These changes are the result of environmental pollution defined as the addition of non-nutrient, nutrient, and toxic substances into estuarine systems or redistribution of these substances within the estuary. The latter occurs through physical modification of exposed and submerged land which affects drainage and circulation characteristics of coastal areas. The most unusual man-imposed change is the introduction of synthetic toxicants.

Environmental factors which affect phytoplankton distribution are referred to as limiting, controlling, or lethal (McCombie, 1960). Limiting factors are light energy and nutrients which may be present at such low levels that no cell division occurs. A controlling factor, such as temperature, affects the rate at which phytoplankton utilize available energy supplies and nutrients. Phytoplankton tolerate a range of concentrations or intensities of all environmental

factors. Above or below these ranges or as a result of extreme
changes over short periods of time, most factors produce physiological stress, or may become lethal to phytoplankton cells.

Natural Factors

In estuaries both natural phenomena and man's impingement cause
environmental factors such as salinity, temperature, light, nutrients, metabolites, and toxic substances to fluctuate over wide ranges
(Table 1). In an estuary, sea water is measurably diluted with

TABLE 1.
Natural and man-imposed conditions which determine levels and rates
of change of factors affecting abundance and succession of estuarine
phytoplankton

Factors	Natural conditions	Man-imposed conditions
Salinity	Precipitation, runoff, evaporation, circulation of water	Water impoundment, channelization, dredge and fill, mosquito ditching
Temperature	Latitude, season, weather, time of day, circulation of water	Heated effluent, dams, canals and waterways, stream channelization
Light intensity		
At surface	Latitude, season, weather, time of day	Air pollution--smog
Below surface	Reflection, absorption, scattering	Dredging, waste dumping, erosion
Nutrients	Drainage, runoff, circulation of water, sediments	Sewage and industrial wastes, urban and agricultural drainage, erosion
Metabolites	Living and dead plants and animals	Sewage, urban and agricultural drainage, erosion
Toxic substances		
Petroleum	Deposits	Leaks and spills during drilling, transport, storage, use or disposal
Radionuclides	Primordial deposits, cosmic-ray produced	Fallout, nuclear power reactors, other releases
Heavy metals	Terrestrial deposits, sediments, land drainage	Industrial and domestic wastes, mining, erosion
Synthetic toxicants		Industrial, agricultural, and domestic use

fresh water derived from land drainage (Pritchard, 1967). Salinity
and temperature of the water and the rates of change of these factors are determined by the geomorphology of the estuary, by local
weather, and by wind and tidal mixing of the water. In the Newport
River estuary, North Carolina, salinity and temperature were determined on an hourly basis from October 1969 to October 1970 at five
stations (Table 2). The range of mean salinity at these stations

TABLE 2
Fluctuations in salinity and temperature at five stations in the
Newport River estuary, North Carolina, from October 1969 to
October 1970[*]

Station	Salinity (o/oo)				Temperature (°C)			
	Mean	Range High	Range Low	Maximum hourly gradient	Mean	Range High	Range Low	Maximum hourly gradient
1	27.8	35.5	15.4	7.6	19.0	31.0	2.1	4.0
2	23.2	36.5	6.0	12.8				
3	21.0	32.0	7.0	6.7				
4	17.2	30.0	0.0	12.0				
5	3.8	26.3	0.0	20.0	18.0	32.5	0.0	6.2

[*]Data of the Atlantic Estuarine Fisheries Center, unpublished.

was 3.8 to 27.8 o/oo and the range of mean temperature was 18.0 to
19.0°C. Precipitation, local drainage, wind-driven water, and tidal
flow have a marked effect on temperature and salinity because of the
shallowness of the estuary. Fluctuations as large as 20 o/oo and
6.2 C within a period of 1 hr have occurred in the freshwater-
dominated portion of the estuary. The maximum change within a period
of 1 hr in the most saline station in the estuary was 7.6 o/oo and
4.0°C. Salinity and temperature can be controlling or lethal factors to populations of phytoplankton. In order to survive, phytoplankton living in the estuary must be euryhaline and eurythermal
freshwater or marine forms able to tolerate this fluctuating osmotic
and thermal environment.

Light reaching the surface of estuarine waters varies seasonally. The extremes of average meridian light intensity reaching the
surface of the earth in the southeastern United States are approximately 87 to 170 klux and the duration of light ranges from 10 to
14 hrs/day (Ferguson, Collier, and Meeter, unpublished manuscript)
(Fig. 1). The relation between the intensity of light at depths in
the water and the turbidity of the water indicates that extremes of
turbidity common to the Newport River estuary can have a greater
effect on light as a limiting factor than seasonal changes in intensity at the surface (Ferguson, 1972). The shallowness of the water
subjects the submerged sediments to frequent resuspension dependent

Fig. 1. Effect of turbidity
on light penetration into
Newport River estuarine water
in winter and summer. Depths
of water at which cell division
is dependent upon light intensity varies with turbidity and
with season (Ferguson, Collier
and Meeter, unpublished manuscript).

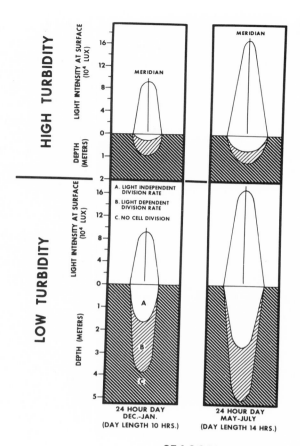

upon wind and tides. The seasonal variation in day length will also
affect phytoplankton (Castenholz, 1964; Ferguson, 1972). Ultraviolet
light and high intensity light or the lack of light over extended
periods can be lethal (Curl and McLeod, 1961; Goldman, Mason, and
Wood, 1963; Halldal, 1967; McLeod and McLachlan, 1959; Steemann
Nielsen, 1952; and Yentsch and Reichert, 1963).

Although salinity, temperature, and light intensity fluctuate
over wide ranges in short periods of time, estuaries are frequently
very productive because of relatively high nutrient concentrations.
Estuarine waters receive land drainage enriched through leaching of
plant nutrients from soil and farmland. Also, microbial populations
in the sediments of the estuary regenerate nutrients. Sea water drawn
from below the euphotic zone by subsurface countercurrents also may
enrich some estuaries (Ketchum, 1967), but not shallow enclosed
estuaries such as the Newport River estuary. The concentration of
a nutrient can be a limiting or a lethal factor to phytoplankton.
In general, only one nutrient will be limiting to cell division at
any given time and this deficiency of a single essential nutrient
will prevent cell division. The concentration of a nutrient becomes
lethal when it is low enough to preclude uptake at adequate rates
for cell maintenance or high enough to be toxic.

In estuaries, nutrients such as inorganic carbon, trace metals, or silicon seldom limit the growth of phytoplankton populations (Williams, 1973). Even though the concentrations of inorganic nitrogen and phosphorus tend to be higher in estuarine waters than in the open ocean, they are most frequently the limiting nutrients. Estuaries also tend to have different relative amounts of nitrogen and phosphorus than the open ocean. The ratio of inorganic nitrogen to phosphorus on a molar basis is 15:1 in the open ocean (Redfield, Ketchum, and Richards, 1963), but this ratio is not typical of estuarine water, in which it is normally much less (Williams, 1973). In the Newport River estuary, the ratio usually varies from 2:1 to 3:1 although at times there may be no measurable nitrate, nitrite, or ammonium nitrogen present (Thayer, 1971). Some species of phytoplankton can use the nitrogen present in organic compounds (Ferguson, 1972; Guillard, 1963; and Schaffner, 1970). Even though nitrogen occurs in many inorganic and organic compounds in the water, its availability probably most often limits cell division in estuaries.

The role of metabolites in the aquatic environment and their influence upon the size of populations and species composition has been reviewed by a number of investigators (Hartman, 1960; Lucas, 1938; Saunders, 1957). In estuaries, these metabolites can either originate on land and be carried into the estuary by drainage or they can originate within the estuary from living or dead organisms. It has been demonstrated in laboratory cultures that a number of species of phytoplankton can produce metabolites which can serve as a direct nutrient or accessory growth factor for other species of phytoplankton (Jørgensen, 1956; Lefevre and Jakob, 1949; McVeigh and Brown, 1954). A number of species also produce and release metabolites which inhibit cell division (Jørgensen, 1956; Lefevre and Nisbet, 1948; Proctor, 1957; Rice, 1949). Metabolites are interchanged most effectively in the aquatic environment and probably are quite significant in the ecology of an estuary.

Although abiotic and biotic factors can vary widely in the Newport River estuary, it is considerably more productive and contains a larger phytoplankton population based on chlorophyll a measurements than either its freshwater source or the adjacent seawater-dominated area, Lookout Bight (Thayer, 1969, 1971)(Table 3). However, the Newport River estuary, similar to other estuaries and coastal zone waters, receives pollution and is modified through physical alteration resulting from man's activities. Thus, it is in the estuary that man's impact upon the marine environment is most pronounced.

TABLE 3

Hydrographic and productivity data from the Newport River estuary, North Carolina, its major freshwater source and the ocean-water-dominated Lookout Bight[1]*

Data	Newport River		Newport Estuary[2]		Lookout Bight	
	May–Sept.	Sept.–May	May–Sept.	Sept.–May	May–Sept.	Sept.–May
Salinity (o/oo)	0	0	29	32	34	36
Temperature (°C)	24	11	26	12	26	13
Phosphate (μM/l)	0.99	0.87	0.32	0.32	0.20	0.18
Inorganic N (μM/l)[3]	1.15	1.23	0.72	0.85	0.75	0.50
Chlorophyll $\underline{a}$ (mg/m^3)	2.45	2.25	9.08	4.09	2.94	3.30
Productivity (mg C/m^3/day)	374	131	583	189	148	89
Secchi depth (m)	0.40	0.41	0.89	1.2	2.4	1.7
Depth (m)[4]	0.5	0.5	1.7	1.7	2.6	2.6
Tidal range (m)	0	0	<0.8	<0.8	0.8	0.8

*After Thayer 1969, 1971.
[1]Mean of samples collected at high tide ± 2 hours.
[2]Mean of 4 stations.
[3]Total NO_3^-, NO_2^-, and NH_4^+.
[4]Mean low water.

Man-imposed Factors

Pollution from man's activities is carried to estuaries by rivers and land drainage, but some wastes may be piped or dumped directly into estuaries. In addition, lead oxides from automobile exhausts and radioactive fallout are washed out of the atmosphere by precipitation and reach estuaries either directly or from land drainage (Goldberg, 1968, 1971). Industrial and municipal wastes are the largest sources of pollution in the estuaries. There are many other important sources of pollution, however, such as pesticides and fertilizer runoff from land; oil spills and losses from commercial and pleasure craft; radioactive elements from nuclear power plants, nuclear-powered submarines, and ships; and sedimentation from dredging and filling.

Practically all wastes released into estuaries are considered to have adverse effects except possibly secondary-treated sewage which contains nutrients that may increase primary production. Many of the industrial wastes contain contaminants of a highly toxic nature, such as cyanides, heavy metals, and polychlorinated biphenyls. Some elements, such as copper and mercury, released as wastes are already present in estuaries, but only in trace amounts. The addition of more of these elements in wastes can raise concentrations to toxic levels. Finally, heat from electric power generating plants also can enhance or be detrimental to primary production. In this era of rapidly increasing power demand, nuclear power plants and their tremendous volumes of warm water discharges require that they be properly sited in areas of adequate dilution capacity (Ragotzkie, Teal, and Bader, 1972).

It would be possible to evaluate the toxicity of a given pollutant in the natural environment if all other environmental conditions remained relatively constant. It already has been pointed out, however, that seldom, if ever, are environmental conditions static in the estuary. Furthermore, most pollutants probably very seldom, if ever, reach the estuary at a constant rate. Their entry into the estuary is most frequently on a periodical basis. Thus, concentrations of pollutants can fluctuate over wide ranges from high concentrations near the source of entry to very low or undetectable amounts at increasing distances from the source of entry. Also, concentrations will vary with time at any given distance from the source of entry if discharge is not continuous.

PHYTOPLANKTON RESPONSE

Many hypotheses have been advanced over the past seventy-five years to explain the observed fluctuations in abundance of phytoplankton (Rice, 1949; Ferguson, 1971). The abundance of each species, the size of the total population, and the succession of species during the season are controlled by changes in physical factors and in the availability of nutrients, or by a combination of these two influences. The action of filter-feeding animals in reducing the numbers of phytoplankton also has been noted. The further suggestion has been made that the intensive growth or bloom of one species might affect the growth of another species through excretion of metabolically active compounds, and thus exert an influence on the seasonal succession of species.

Phytoplankton occurring in an estuary may have a high or a low species diversity depending upon the conditions existing at a given time (Patrick, 1967; Riley, 1967). High species diversity, defined as a large number of species with moderate population sizes, occurs in estuaries because of the variability of environmental conditions which restricts the growth of species and because of the increased numbers of species brought together by the confluence of fresh water and seawater. Low species diversity occurs as a result of competition during periods of high productivity or as a result of extremely unfavorable conditions which restrict growth to a few resistant species, thus resulting in large populations of these species (Odum, 1959).

Phytoplankton cells can either respond or have no response to fluctuations in environmental factors. If a response occurs, it can be extensive and readily observed within a short period of time or it can be subtle, progressing over an extended period of time and be unobserved. The response can be so slight that it cannot be measured or so severe that it results in the death of the cell. A response can occur as a result of a change in the physiology of the cell and can appear as a cytological or morphological change in the cells (Holmes, 1966). A problem related to measuring the physiological response of these microscopic organisms is that a large number of cells are needed to produce measurable changes. This, of course, precludes determination of the extremes of the response for individual cells. While this is a problem that cannot be solved for measuring physiological responses, changes in the appearance of individual cells or a change in the numerical abundance of species can be observed under the microscope.

At any given time, a standing crop of phytoplankton in the natural environment is composed of species and individual cells in all states of physiological condition, from those in good condition and rapidly dividing, to those in poor condition and unable to divide, or to those which are dying. The same is true in cultures in the laboratory held in a limited volume of medium (Fogg, 1965). This is particularly so during the lag phase and when the population is approaching a maximum population size. During the logarithmic growth phase, however, most, if not all, cells are in good physiological condition. Also, in a synchronous culture all the cells divide at about the same time so that their physiological condition will be similar, even though physiological changes occur rapidly within the cell during division (Edmunds, 1965; Hastings and Sweeney, 1964; Hoogenhout, 1963; Pirson and Lorenzen, 1966; Tamiya, 1966).

Measurement

Measurements of phytoplankton responses can be made in the fiel
or in the laboratory. By measuring the environmental conditions whe
a species occurs, it should be possible to identify the requirements
for growth (Gran and Braarud, 1935; Hopkins, 1966; Marshall, 1966,
1969; Patten, Mulford, and Warinner, 1963; Pratt, 1965; Riley, 1947;
Smayda, 1958; Thomas and Simmons, 1960; Welch, 1968). If cells were
being fed upon at a rate somewhat equal to the rate of cell division
however, a potentially large population of cells could be limited by
grazing. Conversely, in the absence of grazing, a very large popula
tion might occur in the presence of an adverse condition which slowe
but did not prevent cell division. Thus, there is always the possibility that a change in population size cannot be directly related t
measurements of the factor responsible for the increase or decrease.

Although data on the responses of phytoplankton to environmenta
factors can be gathered in the laboratory using controlled experimen
there are difficulties in relating this information to the response
phytoplankton to changes in these factors in the natural environment
(Braarud, 1961; Fogg, 1965). For instance, in most laboratory exper
ments the number of cells per unit volume of water is much larger th
that found in the natural environment. Furthermore, concentrations
of nutrients are much higher at the beginning of an experiment and
may change more drastically than concentrations occurring in nature.
In confined volumes of water under laboratory conditions the wide
fluctuations in those factors affected by the growth and metabolism
of phytoplankton have made it difficult to culture species other tha
those which are the most hardy (Fogg, 1965). With the application o
continuous culture techniques to laboratory experiments, however, da
can be obtained with nutrient levels and cell densities more nearly
approaching those occurring in the natural environment (Herbert, 195
Monod, 1950; Phillips and Myers, 1954; Sorokin and Krauss, 1959).

Generally, the presence of detritus and dead organic material i
the natural environment precludes the direct measurement of phytoplankton biomass by weight determination. The change in phytoplankt
biomass, that is, the change in ash-free dry weight with time, is th
most direct measure of phytoplankton response. None of the measurements of standing crop of phytoplankton, however, is completely sati
factory for determining phytoplankton biomass in the field (Strickla
1960). The standing crops of natural populations of phytoplankton
in the estuary frequently are determined by counting the cells. By
using a microscope for cell counts, it is possible to identify individual species and determine the dominant forms. Also, when determi
ing the number of cells, changes can be observed in the standing crc
in species composition, and in the morphology of cells. Microscopic
observation further enables living cells to be distinguished from
detritus. The disadvantage of estimating biomass by cell counts is
that frequently it requires concentrating or preserving the cells,
resulting in fragmentation of some of the fragile forms. In additic
to convert cell numbers to amounts of biomass present, the total

volume of cells must be estimated (Strickland, 1960).

The determination of carbon in cells is probably the best measure of phytoplankton standing crop (Strickland, 1960), since standing crop in terms of carbon is directly related to biomass. Carbon is one of the least variable constituents of phytoplankton cells, accounting for 53 ± 5% of the ash-free dry weight (Ryther, 1956). The percentage of carbon per cell does vary under certain extreme conditions, such as when photosynthesis continues in an environment devoid of nitrogen which can be used by the cells (Fogg, 1959; Syrett, 1962). Changes in standing crop with time in relation to carbon can be related to photosynthetic rate. Estimates of biomass based on measurement of carbon, like the direct measurement of ash-free dry weight will be inaccurate in field samples to the extent that detrital carbon is present.

The standing crop in field studies has been followed by measuring other constituents of cells such as chlorophyll a, deoxyribonucleic acid (DNA), and adenosine triphosphate (ATP) (Hobbie et al., 1972). Chlorophyll a, DNA, and ATP are synthesized by living phytoplankton cells and will degrade upon death of the cells. Considerable amounts of both chlorophyll a and DNA, however, can be detected for some time after death of the organism and can result in an overestimate of phytoplankton biomass. Although the ratio of chlorophyll a to biomass varies with species, light intensity, availability of nutrients, and physiological condition of the cells, chlorophyll a is used as an estimate of cellular carbon by multiplying by a factor of 30 when the natural population is known to have no nutrient deficiencies and by a factor of 60 when populations are grown at high light intensities or in warm nutrient-deficient water (Strickland, 1960). Measurements of many other cellular constituents such as silicon, nitrogen, or phosphorus could be used to estimate biomass if phytoplankton maintained a constant relationship between the amount of these elements and total biomass. The elemental composition of phytoplankton cells, however, is subject to large variations (Vinogradov, 1953; Strickland, 1960).

Measurements of ATP also can result in an overestimate of phytoplankton biomass even though it is not associated with nonliving detrital material (Hamilton and Holm-Hansen, 1967). This is because ATP, like DNA, is also a constituent of all living cells including those heterotrophic cells, such as bacteria, which will be present in phytoplankton samples. Values for ATP can be converted to carbon in microorganisms and while the ATP cellular carbon conversion factor does vary with physiological condition of the cells, it is probably less variable than the chlorophyll a carbon factor (Holm-Hansen and Booth, 1966).

Determining changes in phytoplankton biomass with time requires confinement of samples in containers in order to eliminate such effects as sinking of cells and population dilution or concentration by water movement. The adsorption of particulate and dissolved matter on the surfaces of the container will affect the concentration of dissolved compounds and the bacterial activity in the enclosed water during the relatively long periods of time required to detect changes in phytoplankton (ZoBell, 1946). The productivity of natural populations has been routinely determined by measuring the uptake of inorganic carbon 14, isotope-labeled nutrients, or changes in levels of oxygen in confined volumes of water (Steemann Nielsen, 1960; Strickland, 1960). These indirect measurements of changes in phytoplankton biomass have provided considerable data on the response of phytoplankton exposed to different environmental conditions, since measurable responses occur within periods of time short enough to eliminate the effects due to the container.

Responses to Natural Factors

Physical and chemical factors influencing phytoplankton can be divided into two categories similar to those of Blackman (1905) for photosynthesis by leaves. These two categories are (1) factors which limit the supply of material or energy and (2) factors which affect only the rates of cellular processes. As examples, light intensity and nutrients belong to the first category, while temperature belongs to the second. It should be pointed out, however, that while serving as a source of energy or material, factors such as light intensity and nutrients can also affect the rate of a process when they approach a limiting or lethal level or concentration.

Phytoplankton respond to changes in environmental factors primarily by changing their rate of division and their cellular composition. Cell division rate is a summary response and is controlled

TABLE 4
Phytoplankton responses selected for illustrating effects of environmental conditions

---

Cell Division Rate - changes in any one of the following physiological processes of a cell can affect this summary response.

  Photosynthesis
  Respiration
  Nutrient uptake
  Assimilation
  Excretion
  Osmotic balance

Assimilation Number - rate of photosynthetic carbon production per weight of chlorophyll a can vary with physiological condition.

Cell Composition - the following components of a cell can vary over wide ranges, depending upon the physiological condition of the cell.

  Pigments
  Proteins
  Carbohydrates
  Lipids
  Nutrients
  Non-nutrients

---

by the rates of the physiological processes of the cells, which in
turn are affected by environmental conditions (Eppley and Strickland,
1968)(Table 4). The rate of cell division cannot always be obtained
by direct methods of measurement in field studies. In these instances,
indirect measurements are made and then used to estimate division
rates. Thus, the assimilation number, or the photosynthetic rate per
unit of chlorophyll a, is frequently determined since it is a response
which can be measured for natural populations and related to the population doubling rate.

Phytoplankton respond to fluctuations in environmental factors
by changing their cellular composition (Vinogradov, 1953; Strickland,
1960). The cellular composition of phytoplankton varies, depending
on the environmental conditions and the species. Environmental
changes, however, can affect the organic composition of cells more
drastically than do species differences when environmental conditions
are favorable for growth (Parsons, Stephens, and Strickland, 1961).
Both cell division (which controls abundance of phytoplankton) and
cellular composition (which controls the value of cells as food) have
the greatest impact on higher trophic levels and therefore are the
responses which will be discussed.

Salinity. The rate of cell division at different salinities varies
from species to species. For instance, the division rate of
Thalassiosira pseudonana clone 3H is unaffected by salinities between
0.5 to 36 o/oo (Guillard and Ryther, 1962) while the division rate of
Isochrysis glabana cells is unaffected by salinities between 15 to
40 o/oo and is reduced by 25% at a salinity of 10 o/oo (Kain and
Fogg, 1958b). The division rate of *Prorocentrum micans* is maximum at
a salinity of 25 o/oo and decreases by 25% when the salinity is
increased to 40 o/oo or decreases to 20 o/oo and ceases at a salinity
of 15 o/oo (Kain and Fogg, 1960). *Astrionella japonica* cells divide
at a maximum rate between 30 and 35 o/oo. This rate decreases at 20 o/oo
and cell division ceases when the salinity decreases to 15 o/oo
(Kain and Fogg, 1958a).

The range of salinity over which phytoplankton species divide
when placed in water with a different salinity can be correlated with
the range of salinity in the water from which they were isolated
(Guillard, 1963). Species of phytoplankton isolated from oceanic
waters divide at over half maximum rates above salinities ranging
from 18 to 24 o/oo (Table 5). Minimum salinities at which estuarine
forms divide at over half maximum rates range from 0.5 to 8 o/oo.
The response of neritic species to salinity is intermediate to that
of oceanic and estuarine forms. Cell division can continue in fresh
water for some estuarine species but most freshwater species will
divide only at very low salinities of less than 2 o/oo (Chu, 1942).
A notable exception is the freshwater species, *Chlamydomonas moewusii*,
which will divide at salinities up to 29 o/oo (Guillard, 1960).

Phytoplankton cells physiologically adapt to salinity by adjusting their internal salt concentration to a level somewhat higher than
that of the growth medium (Guillard, 1962). This provides the cells

TABLE 5
Salinity tolerance of phytoplankton from different
marine environments[*]

Marine environment	Number of species tested	Minimum salinity (o/oo) allowing half-maximum division rate[1]
Oceanic	10	18 - 24
Neritic	6	0.5 - 20
Estuarine	6	0.5 - 8

[*]After Guillard, 1963.
[1]Range for species tested.

with an osmotic pressure which favors the flow of water into the
cell and maintains turgor. Therefore, the range of salinity at whic
cell division proceeds and the rate of cell division depend upon the
metabolic rates as affected by altered internal salt concentration.
Adaptive adjustment of the internal salt concentration must proceed
rapidly, since cell division rates of euryhaline diatoms at differen
test salinities do not change as a result of conditioning to these
salinities (Williams, 1964).

The genetic adaptation of metabolic rates to internal salt concentrations has been investigated only to a limited extent and avail
able data are not in agreement. For example, two different clones
of the species *Peridinium trochoideum*, when isolated from water with
salinities of 20 o/oo for one and 37 o/oo for the other, responded
in a similar manner to a variation in salinity (Braarud, 1961).
Oceanic, neritic, and estuarine clones of *T. pseudonana*, however,
showed at least half maximum division rates between 20 and 36, 4 and
36, and 0.5 and 36 o/oo, respectively (Guillard and Ryther, 1962;
Guillard, 1963).

Temperature. Phytoplankton cells respond to temperature by changing
their rate of division (Eppley, 1972) and their assimilation number
(Ichimura, 1968; Talling, 1955; Williams and Murdoch, 1966). Cell
division rates of marine phytoplankton in laboratory cultures generally increase by a factor of 2 to 4 times with a 10 degree increas
in temperature, if these temperatures are within the range of temperatures favorable to growth (Fogg, 1965). Cells of *T. pseudonana*
estuarine clone 3H divide twice as fast at 20°C as at 10°C (Ferguson
1971). An approximation to the temperature-division rate relation
for phytoplankton species in general is:

$$\log_{10} \mu = 0.0275\ T - 0.070$$

where μ is the cell division rate in doublings per day and T is
temperature in degrees Celsius (Eppley, 1972). This is an estimate

of the maximum division rate which might occur in those environments where division rate is not restricted by other factors. The Q_{10}, the ratio of the rates observed with a 10°C increase in temperature, of this equation is 1.88. The positive correlation of assimilation number with temperature for natural phytoplankton populations in the Newport River estuary was established by Williams and Murdoch (1966) as follows (Fig. 2):

$$\log_{10} A = 0.0353 \ T + 0.138$$

where A equals assimilation number in mg C/mg chlorophyll $\underline{a}$/hour. The Q_{10} of this relation was 2.25, which is a typical assimilation number Q_{10} value for natural phytoplankton populations (Ichimura, 1968; Talling, 1955). Both division rate and assimilation number increase with temperature until unfavorably high temperatures are reaches. This is because metabolic rates, including the dark reaction rates of photosynthesis, are temperature-dependent.

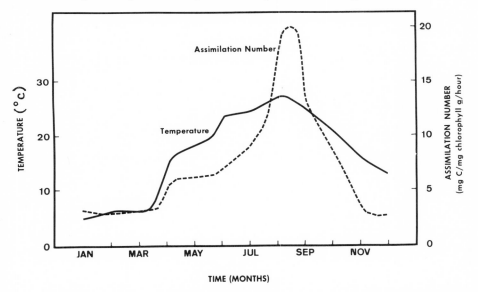

Fig. 2. Correlation of temperature with assimilation number of phytoplankton in the Newport River estuary (after Williams and Murdoch, 1966).

Phytoplankton respond to temperature with a change in the composition of their cells. *Skeletonema costatum* cells increase photosynthetic enzymes and organic matter at low temperatures and double their carbon content per cell as temperature decreases from 20 to 7°C (Jørgensen, 1968). The increase in carbon per cell and carbon per unit chlorophyll $\underline{a}$ with decreasing temperature appears to be a characteristic of marine phytoplankton (Eppley and Sloan, 1966; Jørgensen, 1968). Since cell division and dark reaction rates of photosynthesis depend upon rates of enzymatic processes, an increase in amount of cellular enzymes per cell offsets to some extent the decrease of enzyme activity with decrease in temperature.

Light. Phytoplankton cells respond to different light intensities by changing their rate of division, their assimilation number, and their chemical composition. The rate of cell division is dependent upon the supply of photosynthetically produced carbon and is therefore light-dependent. At very low light intensities, cellular carbon is used faster than it is produced. The rate of production is equal to its rate of use at the compensation intensity. Further increases in light intensity increase division rate until unfavorably high light intensities are reached or until some other factor becomes limiting. Division rate of the estuarine diatom, *T. pseudonana* clone 3H, increases to a maximum of 2.08 doublings/day at 6.5 klux (Ferguson, Collier, and Meeter, unpublished manuscript) (Fig. 3). Further increases in light intensity to 16 klux did not affect the division rate.

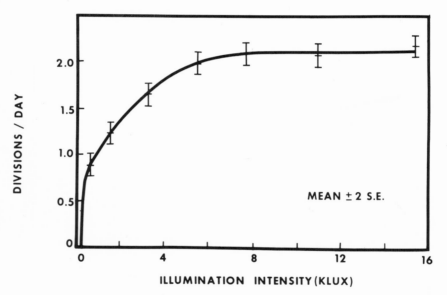

Fig. 3. Division of *Thalassiosira* as a function of light intensity. Cells were exposed to light for 12 hrs per day (Ferguson, Collier, and Meeter, unpublished manuscript).

Phytoplankton adapt to changes in light intensity by changing the amounts of pigments or the amounts of photosynthetic enzymes in the cells. Steemann Nielsen and Jørgensen (1968) have shown that *Chlorella pyrenoidosa* responds by decreasing the amount of pigment per cell as the light intensity increases (Fig. 4). By contrast *Cyclotella meneghiniana* adapts to increasing light intensity by increasing the maximum rates of the enzymatic reactions (Jørgensen, 1964a). Other species of algae adapt to light intensity in a manner similar to *Chlorella* or to *Cyclotella* (Jørgensen, 1964a, 1964b).

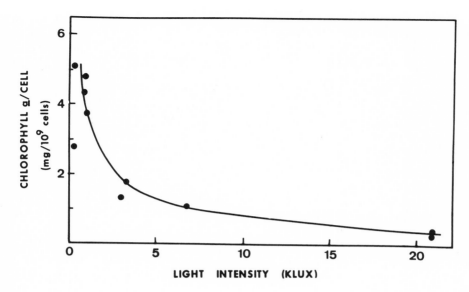

Fig. 4. Chlorophyll a per cell in *Chlorella* adapted to various
light intensities (Steemann Nielsen and Jørgensen, 1968).

Decreasing the chlorophyll a content of the cell increases the cell's
resistance to extreme light intensities. If the dark reactions are
not able to keep pace with photochemical reactions, photo-oxidations
may be induced which destroy enzymes, reduce overall photosynthesis,
or kill the cells (Steemann Nielsen and Jørgensen, 1968). Carotenoid
pigments, however, may function as photochemical stabilizing agents
(Krinsky, 1964, 1966). For instance, the diatom *C. meneghiniana*
shows no decrease in the rate of photosynthesis when grown at 3 klux
and then exposed to 100 klux even though the level of chlorophyll a
is maintained (Jørgensen, 1964a). As light intensity decreases,
chlorophyll a concentration increases, and assimilation number and
compensation level are reduced. These adaptive changes allow cells
to utilize light of lower intensity than cells which have not been
adapted. In the sun the compensation level is 600 to 1200 lux while
in the shade it is 200 to 450 lux (Steemann Nielsen and Hansen, 1959).

Nutrients. Phytoplankton respond to changes in nutrient concentration through changes in the rate of cell division, assimilation number, and the composition of cells. Determination of the relation
between nutrient concentration and division rate, however, is not
possible for most nutrients using standard culture techniques, i.e.,
cultures in confined volumes. In this type of experiment, the nutrient concentration is high initially and the cell population is low.
Uptake of nutrients and cell division proceed rapidly and tend to end
quite abruptly. Thus, the effect of nutrient concentration on division rate normally cannot be quantified except that depletion of a
nutrient will terminate cell division. At concentrations of a
nutrient low enough to affect division rate, nutrient uptake

rapidly decreases the concentration of nutrient in the medium. In continuous cultures, nutrient concentrations do not vary drastically but even in these cultures the effect of nutrient concentration can be difficult to determine.

While cell division can be limited by the uptake and assimilation of nutrients at low concentrations, the uptake of some nutrients may exceed requirements even at very low concentrations in the medium. For instance, it has been found that the rate of uptake of phosphorus by *Phaeodactylum* at low phosphate concentrations, 0.01 to 0.17 µg atoms P/1, is greater than that needed for maximal division rate (Kuenzler and Ketchum, 1962). The concentration of nitrogen in natural waters can have an effect upon cell division rate and, as already pointed out, it is most frequently the limiting nutrient in estuaries (Williams, 1973). The concentration of nitrate-nitrogen, which reduces uptake of this nutrient by a factor of 1/2, (Ks), has been measured for a number of phytoplankton species (Table 6). Values of Ks are related to levels of nitrate concentrations characteristic of a species' habitat. For example, clones of the same species isolated from different waters may show markedly different Ks values (Carpenter and Guillard, 1970). These investigators also have shown

TABLE 6
Nitrate uptake constants for phytoplankton from different marine environments*

Marine environment	Number of species tested	Mean half-saturation uptake constant for nitrate (µg at N/1)
Oceanic	6	0.3
Neritic	15	2.3
Estuarine	4	5.1

*After Carpenter and Guillard, 1970; Eppley and Coatsworth,1968; Eppley, Rogers and McCarthey, 1969.

that clones isolated from Sargasso Sea water with a low nitrate concentration of $\geq$ 0.5 µg atoms N/1 retained their relatively low Ks values even after having been cultured for 12 years in a medium with high nitrate. Under conditions of nitrogen limitation, in continuous culture, *Isochrysis galbana* divides more rapidly with an increase flow rate of the medium (Caperon, 1968). This observation indicates that the rate of supply of nitrogen can also determine the rate of cell division.

Nitrogen deficiency is reflected in a reduction in the assimilation number. With progressive nitrogen deficiency, the chlorophyll content of cells will decrease more rapidly than the nitrogen content and this chlorophyll is relatively less active than that in nondeficient cells (Bonger, 1956; Fogg, 1959). Thus, the assimilation number of nitrogen-deficient cells will be less than that of non-deficient cells. With the addition of nitrogen to the growth medium, cells will recover from their deficiency and their chlorophyll content will increase with higher assimilation numbers again occurring (Syrett, 1962).

Nutrient availability also will affect the amounts of other organic compounds in the cells, as for example in *Chlorella* cells where the proportions of carbohydrate, protein, and lipid vary in relation to nutrient availability and the length of time the cells have grown in nutrient-depleted medium (Milner, 1948). Carbohydrate content varied from 15 to 37.5%, protein from 7.9 to 46.4%, and lipid from 20.2 to 77.1%. This variability results from a shift in metabolic end products in response to nutrient deficiency. With progressive nitrogen deficiency, for example, the cellular content of nitrogen decreases from about 8 or 10% to 2% (Bonger, 1956; Fogg, 1959; Thomas and Krauss, 1955). During this period the products of photosynthesis are not converted to protein and the carbohydrate production is quantitatively most important (van Oorschot, 1955). This metabolic shift has an adaptive advantage since nitrogen-deficient cells are able to take up ammonium nitrogen at rates of four to five times those of normal cells (Bongers, 1956; Harvey, 1953). This uptake of ammonium nitrogen is at the expense of the large carbohydrate reserves and occurs with a four- to five-fold increase in respiration rate (Syrett, 1953).

Interaction of Environmental Factors

Most laboratory experiments have been designed to examine the effect of changes in only one factor at a time so that the interactions of two or more factors have not been observed. In these experiments most, if not all, of the critical factors are held at an optimum level and are more or less constant during the testing. An interaction occurs when the response of phytoplankton to changes in the level of one factor is affected by the level of another factor in a way not predictable from observed responses when the factors have been varied singly. Since population growth of phytoplankton in the natural environment may be influenced by the levels of two or more factors at any given time, it is important that those factors which interact and the mechanism of their interaction be more thoroughly investigated. The interaction of a number of environmental factors upon phytoplankton response has been demonstrated (Curl and McLeod, 1961; Droop, 1956; Ferguson, 1971; Halldal and French, 1958; Jitts et al., 1964; Jørgensen, 1960; McCombie, 1960; Talling, 1957; Thomas, 1966).

Salinity has a temperature-dependent effect upon the division rate of *Dunaliella tertiolecta*. In seawater the upper limit of temperature for cell division is 39 °C, while the upper tolerance limit is 36°C in sea water diluted to one-quarter its normal salt

content (Ukeles, 1961). In artificial seawater, with salinity two
to four times that of normal seawater, the maximum temperature for
cell division is 41 and 42°C, respectively. The mechanism of the
interaction is unknown but may involve an effect of salinity on the
thermal stability of enzymes.

An interaction between temperature and light arises because
the enzyme activity of the dark reactions is temperature-dependent
while the light reactions of photosynthesis are light-dependent. The
division rate of *T. pseudonana* increases with an increase in temperature to a maximum at 21°C, but this rate is dependent upon light
as well as the particular combination of the two factors (Ferguson,
Collier, and Meeter, unpublished manuscript) (Fig. 5). At the
lowest light intensity, the division rate was temperature-dependent,
with the maximum rate at 21°C and reduced rates at 17 and 25°C. At
twice this intensity, the division rate was not affected over the
entire experimental temperature range. This indicates that at low
light intensity, the division rate is adversely affected by temperatures which do not affect the division rate at somewhat higher
intensities. At four times the lowest light intensity, division rat
was reduced at 17 C°but not at 25°C.

Also reported was another manifestation of the interaction between temperature and light which concerned the response of cultures
of *Monochrysis lutheri* and *Thalassiosira nordenskioldii* to accidenta
chemical poisoning (Jitts et al., 1964). Although this poisoning
was not sufficiently severe to kill the cultures, in both cases
growth was restricted to the light and temperature optima of the
species. The addition of this chemical stress was sufficient to

Fig. 5. Division rate of
Thalassiosira cells at
different combinations
of light intensity and
temperature (Ferguson,
Collier, and Meeter,
unpublished manuscript).

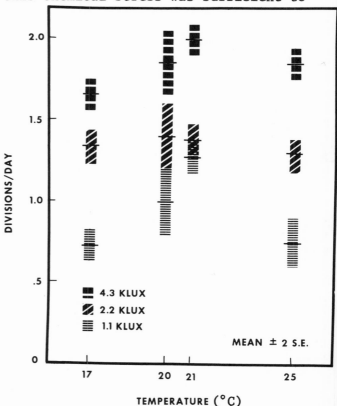

prevent growth in those cultures exposed to suboptimal conditions of temperature and light which otherwise would have produced some limited growth by these species.

Phytoplankton division rate is also affected by an interaction between temperature and nutrients. *T. pseudonana* divides more rapidly at 25°C than at 17°C if the concentration of inorganic nitrogen is very low or if the nitrogen source is urea (Ferguson, Collier, and Meeter, unpublished manuscript). If the nitrogen source is ammonium chloride or potassium nitrate, the division rate is not significantly different at the two temperatures (Fig. 6). The freshwater alga *Chlamydomonas reinhardi* is affected also by an interaction of temperature and nutrient concentration (McCombie, 1960). With double the normal concentration of nutrients the optimum temperature for cell division decreases from 28°C to 18°C. The maximum division rate increases but the minimum temperature for cell division also increases from 6°C to 12°C. When exposed to increased nutrient concentrations in continuous culture, *Nitzschia closterium* and *Tetraselmis* sp. divide at higher temperatures than normally tolerated (Maddux and Jones, 1964). With *Skeletonema costatum*, however, increased nutrients lower the temperature from 20°C to 15°C for maximum photosynthesis (Curl and McLeod, 1961).

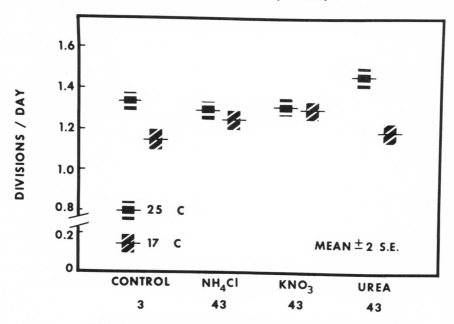

Fig. 6. Division rate of *Thalassiosira* cells at different combinations of temperature and nitrogen sources. Control cultures contained 3 μg atoms N/liter as ammonia and nitrate introduced as impurities from reagents used to prepare the medium (Ferguson, Collier, and Meeter, unpublished manuscript).

Specific information concerning the underlying causes of the interactions of environmental factors on the response of phytoplankton has not been obtained but the general mechanisms of their respon do give some insight as to why interactions of some environmental factors have been observed. At this point in our understanding of phytoplankton survival and growth, it is not as urgent to determine the exact mechanisms by which different factors interact as it is to recognize the potential significance of interactions on phytoplankton response. Contrary to the conclusions of Maddux and Jones (1964), however, the existence of interactions in the response of phytoplankton to some environmental factors does not negate the valu of the concepts of limiting and controlling factors. This information does indicate the necessity to incorporate modifications in experimental design and/or further caution in the interpretation and application of experimental results when studying the response of phytoplankton to potentially interacting environmental factors.

Metabolites. Metabolites can reach the estuary in fresh water from land drainage, and in seawater, or they can originate within the estuary from living and dead organisms. The excretion of soluble, extracellular substances by phytoplankton has received the attention of a number of investigators over the past 35 years (Hartman, 1960; Lucas, 1938; Saunders, 1957; Seiburth, 1968). Studies have been concerned with the chemical nature of the excreted products, conditions of the environment and of the cells which control the production of these substances, and their ecological effects.

This discussion will be limited to the effects of metabolites in the estuary upon the growth rate of phytoplankton populations. Studying these growth modifying substances has required eliminating any possible effects of other organisms. For this reason most of the studies have been carried out with bacteria-free cultures of phytoplankton. Evidence for the production of extracellular substances by phytoplankton which may influence their growth has been obtained by either one or two methods (Hartman, 1960). The first method has been to grow a test species of phytoplankton in filtered medium in which that species or another species had previously grown thus, metabolites could have accumulated in this medium. The second method has been to grow two species together in a mixed culture using a medium in which either species could grow alone. The use of filtrates eliminates the competition for space since only one species at a time is grown in the filtered medium. Growing two species together more nearly approaches the natural situation of competition among species.

A metabolite may act as an accessory growth substance, as an inhibitory substance, or it may have no effect upon the growth of a given species of phytoplankton. In laboratory cultures a number of species of phytoplankton produce metabolites which can serve as direct nutrient or accessory growth substances for other species. The growth rates of *Scenedesmus obliquus* and *Dictyosphaerum ehrenbergianum* have approximately doubled when grown in medium

containing water extracts prepared from large quantities of phytoplankton collected from Pymatuning Reservoir at the end of a summer
bloom (Hartman, 1960). Jørgensen (1956) has found a stimulation of
the growth of *Scenedesmus quadricauda* when grown in filtrates from
cultures of *Chlorella pyrenoidosa* and *Nitzschia palea*. Finally, the
filtrates of *Chlamydomonas* promote the growth of *Chlorella* when it
would not otherwise develop (Allen, 1955).

Certain species of phytoplankton also can produce metabolites
which inhibit their own division rate as well as that of other
species. Pratt and Fong (1940) grew cultures of *Chlorella vulgaris*
in inorganic medium for different periods of time and until different
population sizes had been obtained. The cells were then removed by
filtering the culture medium. This filtered culture medium, which
shall be referred to as "conditioned medium," was used after pH
adjustment in different proportions with fresh medium to prepare
new cultures. The growth of *Chlorella* in these media was found to
be slower than in culture medium to which no conditioned medium had
been added. Their conclusions are as follows: (1) that the growth
of *Chlorella* is inhibited by the presence of the conditioned medium
in the culture medium; (2) that the depression of growth increases
as the percentage of conditioned medium increases; and (3) that for
a given concentration of conditioned medium in the culture medium,
the depression of growth varies inversely with the size of the
Chlorella population in the conditioned medium prior to filtering.
An apparent case of phytoplankton antagonism, similar to that
observed by Pratt and Fong, was reported by Worthington (1943), who
found that the water of Lake Windermere, when an *Asterionella* bloom
was disappearing, was unsuitable for the preparation of culture
medium. This suggests the possibility that *Asterionella* produces
a substance similar to chlorellin produced by *Chlorella*. Experiments also have shown that a number of other species produce substances which inhibit their own growth, e.g., *Skeletonema* (Levring,
1945) and *Nitzschia palea* (Jørgensen, 1957). A number of other
species have been found to produce such substances, but no effort
will be made here to list all of them.

Phytoplankton also produce metabolites which when released
into the water will inhibit the growth of other species (Johnston,
1955; Jørgensen, 1957; Lefevre, Jakob, and Nisbet, 1952; Proctor,
1957). The growth rates of both *Chlorella vulgaris* and *Nitzschia
frustulum* are less when the species are grown together in mixed
cultures than when they were grown alone, depending upon the size
of the populations used (Rice, 1954). An increase in the inhibition
of growth rate of *Chlorella* populations occurred with an increase in
the initial concentration of *Nitzschia*. Similarly, an increase in
inhibition of the growth rate of small populations of *Nitzschia* was
brought about by increasing the initial concentration of *Chlorella*.

In estuaries one species of phytoplankton seldom,if ever,exists
alone; but blooms, or unusually large populations of phytoplankton,
are generally dominated by one species. There are undoubtedly many
factors contributing to the dominance of a bloom by one species, but

it is not inconceivable to postulate that seasonal fluctuations in total phytoplankton numbers and in the numbers of each species, as well as a succession of species, may in part be dependent upon the phytoplankton itself.

Responses to Man's Activities

Many estuarine areas are being indiscriminately damaged as suitable places for aquatic life by pollution resulting from man's activities. Environmental pollution is more inclusive that the definition of "any substance added to the environment as a result of man's activities which has a measurable and generally detrimental effect upon the environment" (Ketchum, 1967). Also involved are other man-induced alterations of the aquatic environment such as dredging, land filling, and thermal discharges from fossil fuel and nuclear electric power stations. Dredging and filling alter the shape and reduce the total area of an estuary, destroy or remove the shallow water benthic environments of estuaries, increase turbidity, smother sedentary bottom forms through siltation, and change water flow patterns. However, we will restrict our discussion here to the potentially harmful effects of thermal, chemical, and radioactive pollution on the phytoplankton of estuaries.

Pollutants have a threshold level below which their presence has no detectable effects. This level for a specific pollutant varies from species to species of phytoplankton so that a given level can affect some species while having no direct effect on others. Since phytoplankton concentrate many substances to thousands of times their concentrations in water, chronic exposure to a pollutant may result in deleterious internal concentrations when acute exposures would not. Similarly, exposures to elevated temperature can be classed as chronic and acute. Duration of exposure, magnitude of temperature change, rate of temperature change, and initial acclimation temperature all influence the response of the phytoplankton.

Thermal pollution. The potential harmful effects of thermal pollution on the phytoplankton of estuaries range from elimination of this important base of the estuarine food web to eutrophication of the estuary. Less drastic effects, which may occur and be of long-term significance, are moderate increases or decreases in phytoplankton productivity and a decrease in the diversity of the phytoplankton community (Cairns, 1971; Coutant and Goodyear, 1972). As already discussed, the phytoplankton division rate and the assimilation number will increase with increases in temperature within the range of temperature favorable to growth. Temperatures above or below this range or extreme changes in temperature over short periods of time can produce physiological stress or be lethal to phytoplankton. This discussion will include the effects of thermal stress on the survival, growth, photosynthesis, and species diversity of phytoplankton. The study of thermal stress on phytoplankton has been extensive in both laboratory and field studies. There is a need for on-site investigations of thermal stress which includes the complicating factors of the passage of natural populations of phytoplankton through the pumps, cooling coils, and effluent canals of the generator plant. These conditions would be difficult or impossible to duplicate in the laboratory (Adams, 1969; Coutant, 1970; Patrick, 1969).

The effect of temperature increase on survival and growth is dependent upon the temperature to which the phytoplankton are acclimated and how near this temperature is to the upper lethal limit (Gurtz and Weiss, 1972). With acclimation temperatures below optimum temperature for growth, rapid temperature increases do not produce the physiological stress and lethal effects of temperature increases approaching or exceeding upper lethal limits. Increases to moderate temperatures may enhance growth. Cairns and Lanza (1972) subjected pure cultures of *Navicula seminulum* var. *hustedtii* Patrick which were acclimated to 18, 23, or 29°C to thermal stresses of (1) plus 10 to 12°C for 20 seconds with a return to ambient temperature in less than 60 seconds; (2) plus 7°C in 1 hour and maintained for about 96 hours; and (3) a combination of stresses (1) and (2). These stresses produced measurable decreases in the survival of cells and in their division rates only when the cells had been acclimated to the highest temperature, 29°C. A number of studies utilizing phytoplankton on slides exposed to thermal stress in the laboratory or in thermal effluents of power plants support these data. Growth of populations of phytoplankton collected on these slides tends to be adversely affected during summer months but stimulated in the late fall and winter (Cairns and Lanza, 1972; Gurtz and Weiss, 1972).

The study of the effect of thermal stress on the photosynthesis and the species diversity of natural populations of phytoplankton which have passed through the condenser coil of a generator plant is complicated by associated stresses of entrainment, mechanical damage by the impellers of pumps, and chlorination of the coolant water. These stresses, however, are inherent in the passage of water through these coolant systems and are also important aspects of the overall effect of these plants (Gurtz and Weiss, 1972). Studies on the photosynthesis of natural populations of phytoplankton present in coolant water emphasize the importance of the temperature of the incurrent water upon whether the increase in temperature will enhance or reduce the rate of photosynthesis and the magnitude of the change in the photosynthetic rate. Increases or decreases in the rate of photosynthesis have been observed in water passing through power plants on the York River at Yorktown, Virginia (Warinner and Brehmer, 1966) and on the Patuxent River estuary in Maryland (Morgan and Stross, 1969). In these investigations, the critical temperatures of incurrent water above which a reduction in photosynthesis always occurred were 14.5 and 16°C. Although temperature increase seems to favor the dominance of blue-green algae over diatoms and over green algae (Cairns, 1971; Merriman, 1970; Trembley,

1960), chlorination of the cooling water reduces the survival and productivity of all algae and makes the thermal effect less apparent (Carpenter, Peck, and Anderson, 1972). For example, Gebelein (1971) found no consistent difference in the species of phytoplankton present in the intake and effluent of a steam-electric generating plant at Northport, New York, except in September, during the highest temperatures observed in the effluent, when diatom species were noticeably reduced relative to other species.

Chemical and radioactive pollution. Phytoplankton cells have the capacity to take up many nutrient, non-nutrient, and toxic substances at rates which exceed dilution within the cell due to division of the cell. This occurs when concentrations of these substances in the water are high relative to those which would limit the rate of uptake. For a nutrient such as zinc, phosphorus, or nitrogen, the division rate of cells will be independent of nutrient concentration when uptake exceeds dilution within the cell by division (Fig. 7).

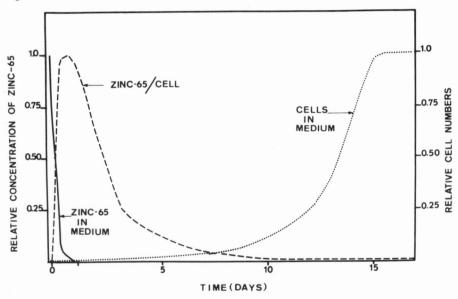

Fig. 7. Uptake of zinc and population growth of *Nitzschia* in a limited volume of medium (Rice, unpublished manuscript).

When the rate of uptake of a nutrient is reduced, the nutrient content of the cell will fall and may equilibrate at a reduced cell division rate. With a cessation of uptake, cell division proceeds at a decreasing rate until it is stopped at some lower level of nutrient content in the cell. With non-nutrient substances, a high initial uptake will increase the concentration within the cells to high levels but with the loss of this substance from the medium it can be reduced to levels that are not measurable. It is probable that the initially high uptake rate and high internal concentration of a toxic substance will be most likely to affect cell division.

If cell divison proceeds, even at a reduced rate, the cellular concentration of the toxicant will decrease unless the concentration of the toxicant is maintained in the water. With a decrease in cellular concentration there would be a tendency for the substance to have less of an effect upon cell division rate. High initial concentrations of toxicants within cells will be maintained only if the rate of supply of the toxicant to the water is equal to or exceeds the rate of uptake due to the increase of total cell volume.

It is of fundamental importance to determine if the response of phytoplankton to a particular concentration of a toxicant is the result of differential uptake, differential sensitivity, or some combination of the two. If a cell does not accumulate a toxicant, no deleterious effects may occur. If the toxicant is accumulated by the phytoplankton cell but the cell is not affected by the presence of the toxicant inside the cell, there will be no response and, therefore, no deleterious effect on the cell. However, with any given species exposed at any concentration, it is possible that the experimental approach may alter the response to the toxicant.

Several experimental approaches can be used to test the effect of toxic substances on phytoplankton. In the past, species of phytoplankton in cultures of limited volumes confined in flasks have been used. Although this approach may be acceptable for testing acute effects, it is not suitable for testing long-term effects. The use of a continuous-flow culture technique is required to obtain long-term effects at relatively low concentrations. As previously discussed with essential nutrients, the observed concentration of a substance in phytoplankton cells will vary with time in cultures in confined volumes, while in flow-type cultures a steady state will be established and the concentration of the toxicant in the medium can be held more or less constant.

The conditions of culture will affect the accumulation of toxic substances by the cells. The presence of organic substances in the culture medium which chelate heavy metal ions such as copper may reduce the availability of the metal to the cells and therefore reduce the apparent toxicity (Steemann Nielsen and Wium-Andersen, 1970). These organic chelators may be included in the formulation of synthetic media (Provasoli, McLaughlin, and Droop, 1957) or excreted by the phytoplankton growing in the media (Steemann Nielsen and Wium-Andersen, 1971). The number of cells present in a culture to which a toxic substance is added also can influence the effect of the substance on the cells. If the deleterious effect of a toxic substance results from the accumulation of a given amount of the substance, it is obvious that with more cells in the culture a smaller amount will be available to each cell. The reduction in the photosynthesis of *Skeletonema* exposed to DDT was greatest in cultures with fewer cells, while at the high cell concentrations tested, photosynthesis was not much different from that of the control (Wurster, 1968) (Fig. 8). Thus, in confined volumes of water, non-essential materials, which are taken up at rates exceeding the dilution due to cell division, will reach maximum internal concentrations and then drop in concentration if cell division continues after the material is depleted from the water.

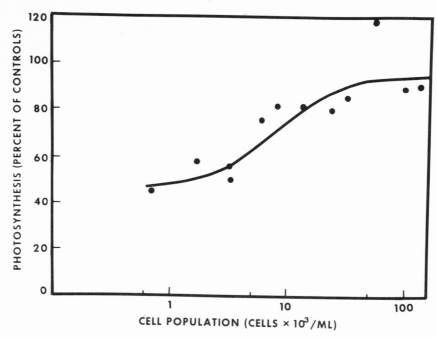

Fig. 8. Comparison of photosynthesis of *Skeletonema* cells at different population densities when exposed to 10 ppb DDT (Wurster, 1968).

The differential uptake and sensitivity of a number of species of marine phytoplankton to copper has been shown by Mandelli (1969). The dinoflagellate, *Glenodinium foliaceum*, appeared to be as sensitive to low concentrations of copper as the blue-green alga, *Coccochloris elabans*. At an initial concentration of 0.03 μg Cu/ml, a concentration detrimental to both species, *Glenodinium* accumulated more copper from the medium than *Coccochloris*. *Glenodinium* concentrated copper 1,500 times over levels in the medium or at 0.03 μg Cu/ml the cells contained 45 μg Cu/mg dry wt, while *Coccochloris* with a concentration factor of 1,100 concentrated copper to only 33 μg Cu/mg. Not only does this show a differential uptake of copper by these species but also it demonstrates that *Coccochloris* was affected by a smaller amount of copper in the cells and was therefore more sensitive to copper inside the cells since these cells concentrated copper to a lower level.

The concentration of some toxicants required to inhibit the growth of phytoplankton has been found to vary inversely with the nutrient concentration (Hannan and Patouillet, 1972a). An example of this relation between nutrient concentration and a surface-active agent [Deriphat CH_3 $(CH_2)_{10}$ CH_2-NH CH_2COONa] is shown in Figure 9. *Chlorella* was grown in a culture medium containing 10^3, 10^4, or 10^5

Fig. 9. Effect of a surface-
active agent (detergent) on
the growth of *Chlorella* cells
cultured in media with different concentrations of
nutrients (Hannan and
Patouillet, 1972a).

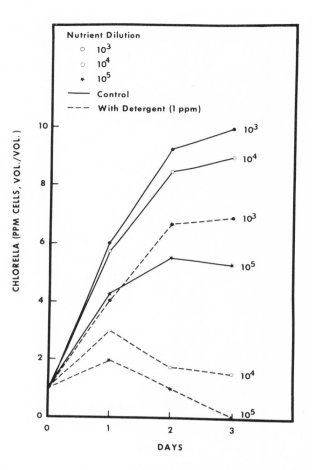

dilution of the stock medium. To the test cultures was added 1 ppm
of the surface-active agent and the total volume of the cells was
determined daily. Growth in the presence of the surface-active agent
increased with an increase in nutrient concentration, indicating that
inhibition decreased with an increase of nutrient concentration.

A nutrient can reduce the growth of phytoplankton when present
in such small concentrations that cells cannot obtain sufficient
amounts for division or when present at levels so high as to inhibit
cell division. The effects of relatively high levels of zinc on
the division rate of *Nitzschia closterium* were tested (Chipman,
Rice, and Price, 1958; Boroughs, Chipman, and Rice, 1957). Zinc uptake has been followed by adding small quantities of zinc 65 so that
the uptake of total zinc could be followed (Fig. 10). At the lowest
concentration of zinc almost all was removed in about 4 days. The
lowest concentration used is about 10 times higher than the average
zinc concentration of seawater. At 1.0 mg Zn/l a little more than
60% was removed in about 2 days and the curve representing the uptake of zinc at this concentration was beginning to level off, indicating that the cells either could not take up any more zinc or they
were beginning to respond adversely to this higher level of zinc in
the water. At 5 mg Zn/l the cells continued to take up zinc for

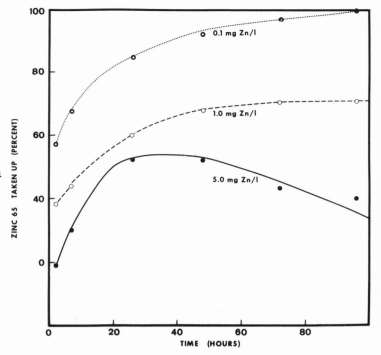

Fig. 10. Uptake and retention of zinc 65 by *Nitzschia* cells in medium containing different concentrations of zinc (Boroughs, Chipman, and Rice, 1957).

about 30 hours. After this time uptake leveled off and then the amount in the cells began to decrease, indicating that the cells had been adversely affected by this highest concentration of zinc tested. This same phenomenon has been found for some other nutrient and nonnutrient elements (Ferguson, 1971; Rice, 1949).

In most instances, testing the effects of toxic substances has been followed for only relatively short periods of time. In an experiment testing the radiation effects of cesium 137 on the division rate of *Nitzschia closterium*, cultures were subcultured weekly for 56 weeks to maintain the cells in a state of division (Rice et al., 1972). Since rapidly dividing cells are more sensitive than nondividing cells to the effects of ionizing radiation, studies were carried out on the effects of chronic irradiation on division rate. Each subculture was started with 2 million cells and 14.3 µCi/1 of cesium 137 in the medium. This concentration is 715 times the maximum concentration of cesium 137 legally permitted in drinking water (Code of Federal Regulations). After 26 weeks the cesium 137 was increased ten-fold to 143 µCi/1 and subculturing was continued for an additional 30 weeks. During the 56-week period, cells grown in the radioactive culture medium divided 426 times, and, following the same general pattern, cells grown in the control culture medium divided 427 times (Fig. 11). It is obvious that the *Nitzschia* cells are quite resistant to long-term exposure at the levels tested.

From a review of the literature, the toxicity of eight different types of substances on a number of species of phytoplankton has been arranged in Table 7. The toxic effect varies with (1) the species, which will affect species composition in mixed populations (Mosser, Fisher, and Wurster, 1972), and (2) the particular toxic substance,

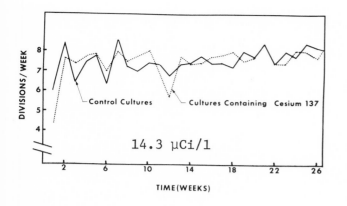

Fig. 11. Division rate of *Nitzschia* cells during long-term exposure to cesium 137 (after Rice et al., 1972).

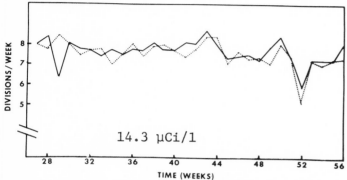

as can be seen by the range of the concentrations causing a deleterious effect on the cells. Phytoplankton cells used in these experiments, in most instances, were subcultures taken from cultures containing cells in good physiological condition. Also, the testing of the different species was usually for relatively short periods of time. By comparing the range of concentrations causing deleterious effects, it can be seen that some of the pesticides were most toxic while nitrogen and phosphorus were least toxic. Of the pesticides, the herbicides are generally most toxic (Ware and Roan, 1970). Copper was the most toxic heavy metal ion but the toxicity of zinc and mercury was increased when combined in certain organic compounds used as pesticides.

SUMMARY

Fluctuations in the biomass of phytoplankton and in species composition in estuaries occur when phytoplankton respond to changes in environmental factors. Even though environmental factors fluctuate rapidly and over wide ranges, estuaries are highly productive due to a large extent to nutrient enrichment from land drainage. This also makes estuaries directly susceptible to man's activities. For the most part, man's impact is to increase or decrease both the

TABLE 7
Concentrations of substances toxic to phytoplankton

Substance tested Type	Number	Species tested (number)	Range of Minimum Concentrations Producing toxicity[1] (ppm)			Duration of experiment (days)
			Low	High	Geometric mean	
[2]Chlorine	1	2	1.5	> 20	--	< 0.1
[3]Detergents	21	32	0.1	>1000	14.0	1 - 210
[4]Heavy metals (Hg, Cu, Zn, Ag)	5	53	0.001	16	0.34	< 1 - 26
[5]Hydrocarbons	2	14	0.01	100	0.52	2 - 5
[6]Nitrogen (NO_3, NO_2, NH_4)	3	4	1.1	28,000	120.0	< 1 - 42
[7]Pesticides	50	46	0.000015	> 200	0.70	1 - 30
[8]Phosphorus (PO_4)	1	2	--	70,000	--	7
[9]Polychlorinated biphenyls	1	5	0.01	> 1	0.08	4

[1]Measured response varied from a change in photosynthetic rate to 100% death of cells.

[2]Carpenter, Peck and Anderson,1972; Hirayama and Hirano,1970.

[3]Boney,1970; Davis and Gloyna,1969; Hannan and Patouillet,1972b; Ukeles,1965.

[4]Erickson, Lackie and Maloney,1970; Gross, Pugno and Dugger,1970; Hannan and Patouillet,1972a; Maloney and Palmer,1956; Marvin, Lansford and Wheeler,1961; Steemann Nielsen and Wium-Andersen,1970; Young and Lisk, 1972.

[5]Aksenova and Trufanova,1971; Hannon and Patouillet,1972b; Mironov,1972.

[6]Ferguson,1971; Natarajan,1970; ZoBell,1935.

[7]Aksenova and Trufanova,1971; Bowes,1972; Hannan and Patouillet,1972b; Harriss, White and Macfarlane,1970; Maloney and Palmer,1956; Menzel, Anderson and Randtke,1970; Mosser et al.,1970;Stroganov, Khobot'ev and Kolosova,1970; Ukeles,1962; Vance and Drummond,1969; Vance and Smith,1969; Wurster,1968.

[8]Rice,1949.

[9]Mosser et al.,1972.

levels and rates of fluctuation of environmental factors, such as salinity, temperature, light (through turbidity of water), nutrients, metabolites, and toxic substances such as pesticides and heavy metals. Practically all wastes when released into estuaries have adverse effects upon phytoplankton, except perhaps secondary-treated sewage and thermal effluent which may in certain instances increase primary production.

The precise determination of the response of phytoplankton to changes in environmental conditions and the identification of adverse environmental conditions in the natural environment is extremely difficult. Most wastes enter the estuary on a periodic or occasional basis which superimposes their fluctuations upon those environmental fluctuations of natural origin. Further, more wastes enter the estuary from point sources which produce gradients in the estuary in both space and time. The response of phytoplankton to these simultaneous fluctuations and cross gradients is complicated by water movements which may disperse or mix populations and by herbivore predation upon the phytoplankton.

The precise determination of the response of phytoplankton to environmental change, however, can be made in the laboratory. Care must be taken not only in the design of laboratory experiments but also in the use of these data in the interpretation of observations made in the natural environment. The observed responses to the manipulation of one or two environmental factors are a function of the species of phytoplankton and are related to the physiological condition of the cells at the beginning of the experiment. Also, it is a function of the levels at which other factors are held constant. Unnatural extremes of nutrient concentration, population density, and light quality and duration which characterize many laboratory studies may also limit the applicability of data. Finally, most controlled experiments are performed over relatively short periods of time and may miss crucial effects resulting from chronic exposures.

The response of phytoplankton is most appropriately determined by measurements of cell division rate, assimilation number, and cellular composition, including accumulated nutrient and toxic materials, in the laboratory and when possible in the field. Changes in species composition and in the succession of dominant species also have yielded significant data. From our research and that available in the literature we have been able to formulate the following conclusions concerning the response of estuarine phytoplankton to the adverse environmental impacts of man.

1. The effect of environmental factors on the uptake, stability, and potency of toxicants must be considered in the evaluation of results since physical and chemical conditions of the water may increase or decrease susceptibility to toxicity (Hannan and Patouillet, 1972b; Portmann, 1968).

2. The variation of sensitivity of different species of phytoplankton to toxicants will be important ecologically in influencing the species present.

3. Since the species present in the phytoplankton show seasonal variations and since they show differential sensitivity to the various types of pollutants, pollution at different times of the year can result in different levels of effect.

4. If acute exposure to a pollutant does not kill all the cells, the population will replace itself. If the death rate due to the pollutant is equal to the division rate, the population will maintain itself unchanged.

5. The effects of a pollutant on phytoplankton are related to the concentration of the pollutant and of the cells in the water and the length of time the phytoplankton cells are exposed to the pollutant.

6. The effects of a pollutant on phytoplankton cells are related to the amount of the pollutant in the cells but it is not known whether the uptake of a pollutant is dependent upon the physiological condition of the cell.

7. The presence of pollutants concentrated in phytoplankton cells may be of greater ecological significance to those species which feed upon phytoplankton and to species in higher trophic levels than to the phytoplankton themselves.

ACKNOWLEDGMENTS

The authors express their appreciation to Mrs. Peggy M. Keney for her assistance in the literature search and compiling Table 7 and the references.

LITERATURE CITED

Adams, J. R. 1969. Ecological investigations around some thermal power stations in California tidal waters. Chesapeake Sci. 10: 145-54.

Aksenova, Ye. I. and Z. A. Trufanova. 1971. Effect of chlorophos and petroleum products of Protococcaceae and blue-green algae (experimental data). Hydrobiol. J. (Engl. transl. Godrobio. Zh.) 7:74-76.

Allen, M. B. 1955. General features of algal growth in sewage oxidation ponds. Calif. State Water Pollut. Control Board, Publ. 13.

Blackman, F. F. 1905. Optima and limiting factors. Ann. Biol. 19:281-98.

Boney, A. D. 1970. Toxicity studies with an oil-spill emulsifier and the green alga Prasinocladus marinus. J. Mar. Biol. Assoc. U. K. 50:461-73.

Bongers, C. H. J. 1956. Aspects of nitrogen assimilation by cultures of green algae (*Chlorella vulgaris*, strain A and *Scenedesmus*). Meded. Landbouwhogesch. Wageningen 56:1-52.

Boroughs, H., W. A. Chipman, and T. R. Rice. 1957. Laboratory experiments on the uptake, accumulation and loss of radionuclides by marine organisms. In: The Effects of Atomic Radiation on Oceanography and Fishes, pp. 80-87. Washington: National Academy of Sciences.

Bowes, G. W. 1972. Uptake and metabolism of 2,2-bis-(p-Chlorophenyl)-1, 1,1-trichloroethane (DDT) by marine phytoplankton and its effect on growth and chloroplast electron transport. Plant Physiol. 49:172-76.

Braarud, T. 1961. Cultivation of marine organisms as a means of understanding environmental influences on populations. In: Oceanography (M. Sears, ed.), pp. 271-98. Washington: American Association for the Advancement of Science, Publ. 67.

Cairns, J. Jr. 1971. Thermal pollution - a cause for concern. J. Water Pollut. Control Fed. 43:55-66.

Cairns, J. Jr. and G. R. Lanza. 1972. The effects of heated waste waters on some microorganisms. Blacksburg: Virginia Polytechnic Institute, Water Resources Research Center, Bull. 48. 101 pp.

Caperon, J. 1968. Population growth response of *Isochrysis galbana* to nitrate variation at limiting concentrations. Ecology 49: 866-72.

Carpenter, E. J. and R. R. L. Guillard. 1970. Intraspecific differences in nitrate half-saturation constants for three species of marine phytoplankton. Ecology 52:183-85.

Carpenter, E. J., B. B. Peck, and S. J. Anderson. 1972. Cooling water chlorination and productivity of entrained phytoplankton. Mar. Biol. (Berl.) 16:37-40.

Castenholz, R. W. 1964. The effect of daylength and light intensity on the growth of littoral marine diatoms in culture. Physiol. Plant. 17:951-63.

Chipman, W. A., T. R. Rice, and T. J. Price. 1958. Uptake and accumulation of radioactive zinc by marine plankton, fish, and shellfish. U. S. Fish Wildl. Serv. Fish. Bull. 58:279-92.

Chu, S. P. 1942. The influence of the mineral composition of the medium on the growth of planktonic algae. Part 1. Methods and culture media. J. Ecol. 30:284-325.

Code of Federal Regulations. 1967, with supplements through 1970, Atomic Energy, Title 10. Standard for protection against radiation, pt. 20, pp. 63-80b. Washington: U. S. Government Printing Office.

Coutant, C. C. 1970. Biological aspects of thermal pollution. I. Entrainment and discharge canal effects. CRC Critical Reviews in Environmental Control 1:241-381.

_____ and C. P. Goodyear. 1972. Thermal effects. J. Water Pollut. Control Fed. 44:1250-94.

Curl, H. Jr. and G. C. McLeod. 1961. The physiological ecology of a marine diatom, *Skeletonema costatum* (Grev.) Cleve. J. Mar. Res. 19:70-88.

Davis, E. M. and E. F. Gloyna. 1969. The role of algae in degrading detergent surface active agents. J. Water Pollut. Control Fed. 41:1494-1504.

Droop, M. R. 1956. Optimum relative and actual ionic concentrations for growth of some euryhaline algae. Int. Assoc. Theor. Appl. Limnol. Verh. 13:722-30.

Edmunds, L. N. 1965. Studies on the synchronously dividing cultures of *Euglena gracilis* Klebs (strain Z). II. Patterns of biosynthesis during the cell cycle. J. Cell. Comp. Physiol. 66:159-81.

Eppley, R. W. 1972. Temperature and phytoplankton growth in the sea. U. S. Natl. Mar. Fish. Serv. Fish. Bull. 70:1063-85.

_____ and J. L. Coatsworth. 1968. Uptake of nitrate and nitrite by *Ditylum brightwellii*. Kinetics and mechanisms. J. Phycol. 4:151-56.

_____, J. N. Rogers, and J. J. McCarthy. 1969. Half-saturation constants for uptake of nitrate and ammonium by marine phytoplankton. Limnol. Oceanogr. 14:912-20.

_____ and P. K. Sloan. 1966. Growth rates of marine phytoplankton: Correlation with light adsorption by cell chlorophyll a. Physiol. Plant. 19:47-59.

_____ and J. D. H. Strickland. 1968. Kinetics of marine phytoplankton growth. In: Advances in Microbiology of the Sea (M. R. Droop and E. J. F. Wood, eds.), pp. 23-62. New York: Academic Press.

Erickson, S. J, N. Lackie, and T. E. Maloney. 1970. A screening technique for estimating copper toxicity to estuarine phytoplankton. J. Water Pollut. Control Fed. 42:R270-78.

Ferguson, R. L. 1971. Growth kinetics of an estuarine diatom: Factorial study of physical factors and nitrogen sources. Ph.D. dissertation, Florida State University. 131 p.

_____. 1972. Population density limitation of an estuarine diatom by illumination and nitrogen source. J. Elisha Mitchell Sci. Soc. 88:188.

_____, A. Collier, and D. A. Meeter. Unpublished manuscript. Kinetic response of an estuarine diatom to illumination, temperature, and nitrogen source. Atlantic Estuarine Fisheries Center, Beaufort, N. C.

Fogg, G. E. 1959. Nitrogen nutrition and metabolic patterns in algae. Symposia Soc. Exp. Biol. 13:106-25.

_____. 1965. Algal Cultures and Phytoplankton Ecology. Madison: University of Wisconsin Press.

Gebelein, N. 1971. Analysis of phytoplankton and periphyton. In: Studies on the Effects of a Steam-electric Generating Plant on the Marine Environment at Northport, New York (G. G. Williams, ed.), pp. 31-59. Stony Brook: State University of New York, Marine Sciences Research Center, Technical Report 9.

Goldberg, E. D. 1968. The chemical invasion of the ocean by man. In: Global Effects of Environmental Pollution (S. F. Singer, ed.), pp. 178-85. New York: Springer-Verlag.

_____. 1971. Atmospheric transport. In: Impingement of Man on the Oceans (D. W. Hood, ed.), pp. 75–88. New York: Wiley-Interscience.

Goldman, C. R., D. T. Mason, and B. J. B. Wood. 1963. Light injury and inhibition in Atlantic freshwater phytoplankton. Limnol. Oceanogr. 8:313–22.

Gran, H. H. and T. Braarud. 1935. A quantitative study of the phytoplankton in the Bay of Fundy and the Gulf of Maine (including observations on hydrography, chemistry and turbidity). J. Biol. Board Can. 1:279–467.

Gross, R. E., P. Pugno, and W. M. Dugger. 1970. Observations on the mechanism of copper damage in Chlorella. Plant Physiol. 46: 183–85.

Guillard, R. R. L. 1960. A mutant of Chlamydomonas moewusii lacking contractile vacuoles. J. Protozool. 7:262–68.

_____. 1962. Salt and osmotic balance. In: Physiology and Biochemistry of Algae (R. A. Lewin, ed.), pp. 529–40. New York: Academic Press.

_____. 1963. Organic sources of nitrogen for marine centric diatoms. In: Symposium on Marine Microbiology (C. H. Oppenheimer, ed.), pp. 93–104. Springfield: Charles C. Thomas Publ.

_____ and J. H. Ryther. 1962. Studies of marine planktonic diatoms. I. Cyclotella nana Hustedt, and Detonula confervacea (Cleve) Gran. Can. J. Microbiol. 8:229–39.

Gurtz, M. E. and C. M. Weiss. 1972. Field investigations of the response of phytoplankton to thermal stress. Chapel Hill: University of North Carolina, Dept. of Environmental Sciences and Engineering, School of Public Health, ESE Publ. No. 321.152 p.

Halldal, P. 1967. Ultraviolet action spectra in algology: A review. Photochem. Photobiol. 6:445–60.

_____ and C. S. French. 1958. Algal growth in crossed gradients of light intensity and temperature. Plant Physiol. 3:249–52.

Hamilton, R. D. and O. Holm-Hansen. 1967. Adenosine triphosphate content of marine bacteria. Limnol. Oceanogr. 12:319–24.

Hannan, P. J. and C. Patouillet. 1972a. Effects of pollutants on growth of algae. Washington, D. C.: U.S. Naval Research Laboratory, February 1972 Progress Report.

_____. 1972b. Nutrient and pollutant concentrations as determinants in algal growth rates. In: Marine Pollution and Sea Life (M. Ruivo, ed.), pp. 340–42. Surrey, England: Fishing News (Books) Ltd.

Harriss, R. C., D. B. White, and R. B. Macfarlane. 1970. Mercury compounds reduce photosynthesis by plankton. Science 170: 736–37.

Hartman, R. T. 1960. Algae and metabolites of natural waters. In: The Ecology of Algae (C. A. Tryon, Jr. and R. T. Hartman, eds.), pp. 38–55. Pittsburgh: Pymatuning Laboratory of Field Biology, Spec. Publ. No. 2. University of Pittsburgh.

Harvey, H. W. 1953. Synthesis of organic nitrogen and chlorophyll by Nitzschia closterium. J. Mar. Biol. Assoc. U. K. 31:477–87.

Hastings, J. W. and B. M. Sweeney. 1964. Phased cell division in the marine dinoflagellates. In: Synchrony in Cell Division and Growth (E. Zenthen, ed.), pp. 307-21. New York: Interscience.

Hedgpeth, J. W. 1957. Estuaries and lagoons. II. Biological aspects. In: Treatise on Marine Ecology and Paleoecology (J. W. Hedgpeth, ed.), pp. 693-729. Baltimore: Waverly Press.

Herbert, D. 1959. Some principles of continuous culture. In: Recent Progress in Microbiology (G. Tunevall, ed.), pp. 381-96. Stockholm: Almqvist and Wiksell.

Hirayama, K. and R. Hirano. 1970. Influences of high temperature and residual chlorine on marine phytoplankton. Mar. Biol. (Berl.) 7:205-13.

Hobbie, J. E., O. Holm-Hansen, T. T. Packard, L. R. Pomeroy, R. W. Sheldon, J. P. Thomas, and W. J. Wiebe. 1972. A study of the distribution and activity of microorganisms in ocean water. Limnol. Oceanogr. 17:544-55.

Holm-Hansen, O. and C. R. Booth. 1966. The measurement of adenosine triphosphate in the ocean and its ecological significance. Limnol. Oceanogr. 11:510-19.

Holmes, R. W. 1966. Light microscope observations on cytological manifestations of nitrate, phosphate, and silicate deficiency in four marine centric diatoms. J. Phycol. 2:136-40.

Hoogenhout, H. 1963. Synchronous cultures of algae. Phycologia 2: 135-47.

Hopkins, T. L. 1966. The plankton of the St. Andrew Bay system, Florida. Publ. Inst. Mar. Sci., Univ. Texas 11:12-64.

Ichimura, S. 1968. Phytoplankton photosynthesis. In: Algae, Man, and the Environment (D. F. Jackson, ed.), pp. 103-20. Syracuse: Syracuse University Press.

Jitts, H. R., C. D. McAllister, K. Stephens, and J. D. H. Strickland. 1964. The cell division rates of some marine phytoplankters as a function of light and temperature. J. Fish. Res. Board Can. 21: 139-57.

Johnston, R. 1955. Biologically active compounds in the sea. J. Mar. Biol. Assoc. U. K. 34:185-95.

Jørgensen, E.G. 1956. Growth inhibiting substances formed by algae. Physiol. Plant. 9:712-26.

_____. 1957. Diatom periodicity and silicon assimilation. Experimental and ecological investigations. Dan. Bot. Ark. 18:1-54.

_____. 1960. The effects of salinity, temperature, and light intensity on growth and chlorophyll formation of Nitzschia ovalis. Carnegie Inst. Wash. Year Book 59:348-49.

_____. 1964a. Adaptation to different light intensities in the diatom Cyclotella meneghiniana Kiitz. Physiol. Plant.17:136-45.

_____. 1964b. Chlorophyll content and rate of photosynthesis in relation to cell size of the diatom Cyclotella meneghiniana. Physiol. Plant. 17:407-13.

_____. 1968. The adaptation of plankton algae. II. Aspects of the temperature adaptation of Skeletonema costatum. Physiol. Plant. 21:423-27.

Kain, J. M. and G. W. Fogg. 1958a. Studies on the growth of marine phytoplankton. I. *Asterionella japonica* Gran. J. Mar. Biol. Assoc. U. K. 37:397–413.

_____. 1958b. Studies on the growth of marine phytoplankton. II. *Isochrysis galbana* Parke. J. Mar. Biol. Assoc. U. K. 37:781–88.

_____. 1960. Studies on the growth of marine phytoplankton. III. *Prorocentrum micans* Ehrenberg. J. Mar. Biol. Assoc. U. K. 39: 33–50.

Ketchum, B. H. 1967. Man's resources in the marine environment. In: Pollution and Marine Ecology (T. A. Olson and F. J. Burgess, eds.), pp. 1–11. New York: Interscience.

Krinsky, N. I. 1964. Carotenoid de-expoxidations in algae. I. Photochemical transformation of antheraxanthin to zeaxanthin. Biochem. Biophys. Acta 88:487–91.

_____. 1966. The role of carotenoid pigment as protective agents against photosensitized oxidation in chloroplasts. In: Biochemistry of Chloroplasts (T. W. Goodwin, Ed.), pp. 423–30. New York: Academic Press.

Kuenzler, E. J. and B. W. Ketchum. 1962. Rate of phosphorus uptake by *Phaeodactylum tricornutum*. Biol. Bull. 5:134–45.

Lefevre, M. and H. Jakob. 1949. Sur quelques propriétés des substances activée tirées des cultures d'Algues d'eau douce. C. R. Acad. Sci. Paris 229:234–36.

_____, _____, and M. Nisbet. 1952. Auto- et hétéroantagonisme chez les algues d'eau douce. Ann. Stn. Cent. Hydrobiol. Appl. 4:5.

_____ and M. Nisbet. 1948. Sur la sécrétion, par certaines espèces d'algues, de substances inhibitrices d'autres espèces d'algues. C. R. Acad. Sci. Paris 226:107–09.

Levring, T. 1945. Some culture experiments with marine plankton diatoms. Medd. fran Oceanogr. Inst. Goteborg 3:3–18.

Lewin, J. C. and R. R. L. Guillard. 1963. Diatoms. Annu. Rev. Microbiol. 17:373–414.

Lucas, C. E. 1938. Some aspects of integration in plankton communities. J. Cons. Cons. Int. Explor. Mer. 13:309–22.

McCombie, A. M. 1960. Actions and interactions of temperature, light intensity and nutrient concentration on the growth of the green algae, *Chlamydomonas reinhardi* Dangeard. J. Fish. Res. Board Can. 17:871–94.

McLeod, G. C. and J. McLachlan. 1959. The sensitivity of several algae to ultraviolet radiation of 2537 A. Physiol. Plant. 12: 306–09.

McVeigh, I. and W. H. Brown. 1954. In vitro growth of *Chlamydomonas chlamydogama* Bold and *Haematococcus plavialis* Flotow em. Wille in mixed cultures. Bull. Torrey Bot. Club 81:218–33.

Maddux, W. S. and R. F. Jones. 1964. Some interactions of temperature, light intensity, and nutrient concentration during the continuous culture of *Nitzschia closterium* and *Tetraselmis* sp. Limnol. Oceanogr. 9:79–86.

Maloney, T. E. and C. M. Palmer. 1956. Toxicity of six chemical compounds to thirty cultures of algae. Water Sewage Works 103: 509–13.

Mandelli, E. F. 1969. The inhibitory effects of copper on marine phytoplankton. Contrib. Mar. Sci., Univ. Texas 14:47-57.

Marshall, H. G. 1966. The distribution of phytoplankton along a 140 mile transect in the Chesapeake Bay. Va. J. Sci. 17:105-19.

_____. 1969. Observations on the distribution of phytoplankton in the Elizabeth River, Virginia. Va. J. Sci. 20:37-39.

Marvin, K. T., L. M. Lansford, and R. S. Wheeler. 1961. Effects of copper ore on the ecology of a lagoon. U. S. Fish Wildl. Serv. Fish. Bull. 61:153-60.

Menzel, D. W., J. Anderson, and A. Randtke. 1970. Marine phytoplankton vary in their response to chlorinated hydrocarbons. Science 167:1724-26.

Merriman, D. 1970. The calefaction of a river. Sci. Am. 222:42-52.

Milner, H. W. 1948. The fatty acids of Chlorella. J. Biol. Chem. 177:813-17.

Mironov, O. G. 1972. Effect of oil pollution on flora and fauna of the Black Sea. In: Marine Pollution and Sea Life (M. Ruivo, Ed.), pp. 222-24. Surrey, England: Fishing News (Books) Ltd.

Monod, J. 1950. La technique de culture continue; theorie et applications. Ann. Inst. Pasteur (Paris) 79:390-410.

Morgan, R. P., II and R. G. Stross. 1969. Destruction of phytoplankton in the cooling water supply of a steam electric station. Chesapeake Sci. 10:165-71.

Mosser, J. L., N. S. Fisher, T. C. Teng, and C. F. Wurster. 1972. Polychlorinated biphenyls: Toxicity to certain phytoplankters. Science 175:191-92.

_____, _____, and C. F. Wurster. 1972. Polychlorinated biphenyls and DDT alter species composition in mixed cultures of algae. Science 176:533-35.

Natarajan, K. V. 1970. Toxicity of ammonia to marine diatoms. J. Water Pollut. Control Fed. 42:R184-90.

Odum, E. P. 1959. Fundamentals of Ecology. Philadelphia: Saunders.

Parsons, T. R., K. Stephens, and J. D. H. Strickland. 1961. On the chemical composition of eleven species of marine phytoplankters. J. Fish. Res. Board Can. 18:1001-16.

Patrick, R. 1967. Diatom communities in estuaries. In: Fundamentals of Ecology (E. P. Odum, Ed.), pp. 311-15. Philadelphia: Saunders.

_____. 1969. Some effects of temperature on freshwater algae. In: Biological Aspects of Thermal Pollution (P. A. Krenkel and F. C. Parker, Eds.), pp. 161-85. Nashville: Vanderbilt University Press.

Patten, B. C., R. A. Mulford, and J. E. Warinner. 1963. An annual phytoplankton cycle in the lower Chesapeake Bay. Chesapeake Sci. 4:1-20.

Phillips, J. N. and J. Myers. 1954. Measurement of algal growth under controlled steady-state conditions. Plant Physiol. 29: 148-61.

Pirson, A. and H. Lorenzen. 1966. Synchronized cultures of algae. Ann. Rev. Plant Physiol. 17:439-58.

Portmann, J. E. 1968. Progress report on a programme of insecticide analysis and toxicity-testing in relation to the marine environment. Helgol. wiss. Meeresunters. 17:247-56.

Pratt, D. M. 1965. The winter-spring diatom flowering in Narrangansett Bay. Limnol. Oceanogr. 10:173-84.

Pratt, R. J. and J. Fong. 1940. Studies on *Chlorella vulgaris*. II. Further evidence that *Chlorella* cells form a growth-inhibiting substance. Am. J. Bot. 27:431-36.

Pritchard, D. W. 1967. What is an estuary: Physical viewpoint. In: Estuaries (G. H. Lauff, Ed.), pp. 3-5. Washington: American Association for the Advancement of Science, Publ. 83.

Proctor, V. W. 1957. Studies of algal antibiosis using *Hematococcus* and *Chlamydomonas*. Limnol. Oceanogr. 2:125-39.

Provasoli, L., J. J. A. McLaughlin, and M. R. Droop. 1957. The development of artificial media for marine algae. Archiv für Mikrobiologie 25:392-428.

Ragotzkie, R. A., J. M. Teal, and R. G. Bader. 1972. Urbanization and industrial development. In: The Water's Edge: Critical Problems of the Coastal Zone (B. H. Ketchum, Ed.), pp. 103-24. Cambridge: MIT Press.

Redfield, A. C., B. H. Ketchum, and F. A. Richards. 1963. The influence of organisms on the composition of sea-water. In: The Sea (M. N. Hill, Ed.), vol. 2, pp. 26-77. New York: Interscience.

Rice, T. R. 1949. The effect of nutrients and metabolites on populations of planktonic algae. Ph. D. dissertation, Harvard University, Boston, 131 p.

_____ 1954. Biotic influence affecting population growth of planktonic algae. U. S. Fish Wildl. Serv. Fish. Bull. 54:227-45.

_____, J. P. Baptist, F. A. Cross, and T. W. Duke. 1972. Potential hazards from radioactive pollution of the estuary. In: Marine Pollution and Sea Life (M. Ruivo, Ed.), pp. 272-76. Surrey, England: Fishing News (Books) Ltd.

Riley, G. A. 1947. Seasonal fluctuations of the phytoplankton population in New England coastal waters. J. Mar. Res. 6:114-25.

_____. 1967. The plankton of estuaries. In: Estuaries (G. H. Lauff, Ed.), pp. 316-26. Washington: American Association for the Advancement of Science, Publ. 83.

Ryther, J. H. 1956. The measurement of primary production. Limnol. Oceanogr. 1:72-84.

Saunders, G. W. 1957. Interrelations of dissolved organic matter and phytoplankton. Bot. Rev. 23:389-490.

Schaffner, W. R. 1970. A comparative study of nitrogen utilization by an estuarine species of phytoplankton. Ph. D. dissertation, Cornell University, Ithaca, N. Y. 62 p.

Sieburth, J. M. 1968. The influence of algal antibiosis on the ecology of marine microorganisms. In: Advances in Microbiology of the Sea (M. R. Droop and E. J. F. Wood, Eds.), pp. 63-94. New York: Academic Press.

Smayda, T. J. 1958. Biogeographical studies of marine phytoplankton. Oikos 9:158-91.

Sorokin, C. and R. W. Krauss. 1959. Maximum growth rates of *Chlorella* in steady-state and in synchronous cultures. Proc. Natl. Acad. Sci. 45:1740-44.

_____ and _____. 1965. The dependence of cell division in *Chlorella* on temperature and light intensity. Am. J. Bot. 52:331-39.

Steemann Nielsen, E. 1952. On the detrimental effects of high light intensities on the phytosynthetic mechanism. Physiol. Plant. 5: 334-44.

_____. 1960. Productivity of the oceans. Annu. Rev. Plant Physiol. 11:341-62.

_____ and V. K. Hansen. 1959. Light adaptation in marine phytoplankton populations and its interrelation with temperature. Physiol. Plant. 12:353-70.

_____ and E. G. Jørgensen. 1968. The adaptation of plankton algae. I. General part. Physiol. Plant. 21:401-13.

_____ and S. Wium-Andersen. 1970. Copper ions as poison in the sea and in freshwater. Mar. Biol.6:93-97.

_____ and _____. 1971. The influence of copper on photosynthesis and growth in diatoms. Physiol. Plant. 24:480-84.

Strickland, J. D. H. 1960. Measuring the production of marine phytoplankton. Bull. Fish. Res. Board Can. 122: 172 p.

Stroganov, N. S., V. G. Khobot'ev, and L. V. Kolosova. 1970. Study of the connection of the chemical composition of organometallic compounds with their toxicity for hydrobionts. Voprosy Vodnoi Toksikologii, pp. 66-74. (Trans. from Russian by R. M. Howland.) Washington: U. S. Dept. Interior.

Syrett, P. J. 1953. The assimilation of ammonia by nitrogen-starved cells of *Chlorella vulgaris*. I. The correlation of assimilation with respiration. Ann. Bot. 17:1-19.

_____. 1962. Nitrogen assimilation. In: Physiology and Biochemistry of Algae (R. A. Lewin, ed.), pp. 171-88. New York: Academic Press.

Talling, S. F. 1955. The relative growth rate of three plankton diatoms in relation to underwater radiation and temperature. Ann. Bot. 19:329-41.

Talling, J. F. 1957. The growth of two plankton diatoms in mixed culture. Physiol. Plant. 10:215-23.

Tamiya, H. 1966. Synchronous cultures of algae. Annu. Rev. Plant Physiol. 17:1-26.

Thayer, G. W. 1969. Phytoplankton production and factors influencing production in the shallow estuaries near Beaufort, N. C. Ph. D. dissertation, N. C. State University, Raleigh. 170 p.

_____. 1971. Phytoplankton production and the distribution of nutrients in a shallow unstratified estuarine system near Beaufort, N. C. Chesapeake Sci. 12:240-53.

Thomas, W. H. 1966. Effects of temperature and illuminance on cell division rate of tropical oceanic phytoplankton. J. Physiol. 2: 17-22.

_____ and R. W. Krauss. 1955. Nitrogen metabolism in *Scenedesmus* as affected by environmental changes. Plant Physiol. 30:113-22.

_____ and E. G. Simmons. 1960. Phytoplankton production in the Mississippi Delta. In: Recent Sediments, Northwest Gulf of Mexico, 1951-1958, pp. 103-381. Tulsa: American Association of Petroleum Geologists.

Trembley, F. J. 1960. Research Project on Effects of Condenser Discharge Water on Aquatic Life. Bethlehem, Pa.: Lehigh University, The Institute of Research, Progress Report, 1956-1959.

Ukeles, R. 1961. The effect of temperature on the growth and survival of several marine algal species. Biol. Bull. 120:255-64.

_____. 1962. Growth of pure cultures of marine phytoplankton in the presence of toxicants. Appl. Microbiol. 10:532-37.

_____. 1965. Inhibition of unicellular algae by synthetic surface-active agents. J. Phycol. 1:102-10.

Vance, B. D. and W. Drummond. 1969. Biological concentration of pesticides by algae. J. Am. Water Works Assoc. 61:360-62.

_____ and D. L. Smith. 1969. Effects of five herbicides on three green algae. Tex. J. Sci. 20:329-37.

van Oorschot, J. L. P. 1955. Conversion of light energy in algal culture. Meded. Landbouwhogesch. Wageningen 55:225-76.

Vinogradov, A. P. 1953. The Elementary Chemical Composition of Marine Organisms (Trans. from Russian by J. Efron and J. K. Setlow). Sears Foundation for Marine Research, Memoir 2. New Haven: Yale University Press.

Ware, G. W.,and C.C. Roan. 1970. Interaction of pesticides with aquatic microorganisms and plankton. Residue Rev. 33:15-45.

Warinner, J. E. and M. L. Brehmer. 1966. The effects of thermal effluents on marine organisms. Air Water Pollut. 10:277-89.

Welch, E. B. 1968. Phytoplankton and related water-quality conditions in an enriched estuary. J. Water Pollut. Control Fed. 40: 1711-27.

Williams, R. B. 1964. Division rates of salt marsh diatoms in relation to salinity and cell size. Ecology 45:877-80.

_____. 1973. Nutrient level and phytoplankton productivity in the estuary. In: Proceedings of the Coastal Marsh and Estuary Symposium (R. H. Chabreck, ed.), pp. 59-89. Louisiana State Univ.

_____ and M. B. Murdoch. 1966. Phytoplankton production and chlorophyll concentration in the Beaufort Channel, North Carolina. Limnol. Oceanogr. 11:73-82.

Worthington, E. B. 1943. Eleventh Annual Report of the Director, Year Ending March 31, 1943. Freshwater Biological Association of the British Empire.

Wurster, C. F., Jr. 1968. DDT reduces photosynthesis by marine phytoplankton. Science 159:1474-75.

Yentsch, C. S. and C. A. Reichert. 1963. The effects of prolonged darkness on photosynthesis, respiration and chlorophyll in the marine flagellate Dunaliella euchlora. Limnol. Oceanogr. 8: 338-42.

Young, R. G. and D. J. Lisk. 1972. Effect of copper and silver ions on algae. J. Water Pollut. Control Fed. 44:1643-47.

ZoBell, C. E. 1935. The assimilation of ammonium nitrogen by Nitzschia closterium and other marine phytoplankton. Proc. Natl. Acad. Sci. 21:517-22.

_____. 1946. Marine Microbiology. Waltham, Mass.: Chronica Botanica Company.

The effects of mercury on the photosynthesis and growth of estuarine and oceanic phytoplankton

R. G. Zingmark and T. G. Miller

The death of scores of people near Minimata Bay, Japan, due to mercury poisoning, focused international attention on a possibly widespread environmental hazard. The need to study the biological effects of mercury on the various trophic levels of ecosystems is critical. Mercury is introduced naturally into the environment through weathering processes, geothermal (volcanic) activity, and, increasingly, through man-caused pollution. The usual dissolved mercury concentrations in aquatic and marine ecosystems range between 0.06 to 0.3 parts per billion (ppb) (Anonymous, 1970; Klein and Goldberg, 1970; Harriss, 1971; Windom, 1973), except in the vicinity of a mercury discharge where it can be much higher. Sources of man-caused mercury pollution are industrial wastes, agricultural uses of organomercurial fungicides and bactericides, combustion of petroleum, and indirectly by increased erosion through land clearance (Peakall and Lovett, 1972). In the waters of Minimata Bay a range of 0.16 to 3.6 ppb total mercury was measured (Klein and Goldberg, 1970). The source of the mercury was traced to a Japanese chemical plant. These amounts of mercury were not in themselves alarming (5 ppb is the maximum allowable concentration of mercury in United States drinking water). However, the toxic effects of the mercury were presumed to be magnified as the mercury was concentrated through the various trophic levels in the bay.

Highly productive coastal waters are subject to the greatest part of the mercury entering the marine environment from freshwater runoff (Windom, 1973). Therefore, the necessity of learning how it may influence the marine primary producers is greatly increased. Despite their key ecological role, relatively few investigations of the effects of mercury on these organisms have been reported. Glooschenko (1969) showed that cells of the diatom *Chaetoceros costatum* would accumulate mercury. Others have shown that various organic and inorganic forms of mercury depress primary production and growth of freshwater and marine phytoplankton and some marine benthic algae (Boney and Corner, 1959; Boney, Corner, and Sparrow, 1959; Ukeles, 1962; Harriss, White, and MacFarlane, 1970; Boney, 1971; Harriss, 1971; Kamp-Nielsen, 1971; Hannan and Patouillet, 1972; Nuzzi, 1972; Shieh and Barber, 1973). Nuzzi (1972) also noted an increased frequency of morphological abnormalities in cultures exposed to sublethal concentrations of mercury. Most of the litera- ture indicates organic forms of mercurials to be more toxic than the inorganic forms (Boney, Corner, and Sparrow, 1959; Harriss, White, and MacFarlane, 1970; Boney, 1971). Hannan and Patouillet (1972), however, reported that $HgCl_2$, an inorganic form of mercury was more toxic to various algae than the organic form dimethyl mercury.

Mercury thus has been shown to be detrimental to freshwater and marine algae, the primary producers of aquatic ecosystems. Any factor which can affect primary producers would also affect other trophic levels in a food web. Therefore, the presence of mercury in the environment due to pollution should be of great concern. The degree of mercury toxicity in cultured phytoplankton has been shown to be influenced by the concentration of the organisms. Thus, it would seem important to determine experimentally the effects of mercury on phytoplankton concentrations that are ecologically sig- nificant. The purposes of this investigation were to determine the effects of $HgCl_2$ on the photosynthetic rates of natural phytoplank- ton communities and the growth and reproductive rates of selected cultured species of estuarine and marine phytoplankton, as influ- enced by the concentration of the mercury in the media, the concen- tration of cells and stage of the growth cycle upon exposure to the mercury, and the duration of exposure to mercury.

METHODS AND MATERIALS

Oceanic samples were taken from the R/V EASTWARD in the western Sargasso Sea (76° 0' W × 30° 3' N) and the Gulf Stream (78° 0' W × 32° 2' N). Estuarine samples were taken at Clambank Creek, Hobcaw Barony, South Carolina (79° 2' W × 33° 3' N). Samples used for measurements of primary production were taken from the depth that received 50% of the surface solar insolation (calculated from Secchi disk readings), as this depth was consistently found to con- tain the greatest concentration of chlorophyll. Primary production was measured by the standard light/dark bottle, ^{14}C uptake method as outlined by Strickland and Parsons (1968). Water samples in

125 ml BOD bottles in triplicate were incubated with ^{14}C (as 5 to 20 µci NaH^{14}CO$_3$) for 4 hrs in a Menzel "simulated in situ" productivity incubator under conditions of 50% surface insolation and ambient temperature. Uptake of radioactive carbon was measured by liquid scintillation spectrometry at an average counting efficiency of 88% (Pugh, 1970). Mercury (as HgCl$_2$) was added to experimental bottles at Hg^{++} concentrations of 0, 1, 10, 50, 100, and 500 ppb.

Unialgal cultures of the dinoflagellate *Amphidinium carterae* Hulbert (Guillard isolation, obtained from L. Provasoli, Haskins Laboratory, New Haven, Connecticut) and the centric diatom *Skeletonema costatum* (Grev.) Cleve (Strain N5J, obtained from P. E. Hargraves, University of Rhode Island) were maintained in "F"/2 enriched seawater medium (Guillard and Ryther, 1962). To the cultures of *S. costatum* were added 1.5 mg/l silicon (as SiO$_3$). Cultures were kept at 23 ± 1C and were constantly illuminated from above by cool-white, fluorescent lights at approximately 6,500 lux. Cultures in log phase were diluted with media to a concentration of 5 to 7 × 10^4 cells/l. The various concentrations of mercury from 0 to 500 ppb were added. Primary production measurements were made as above (except with 1 µci ^{14}C added) after 4 hrs or 24 hrs of incubation with mercury.

Cultures of *A. carterae* were subjected to limited and chronic exposures of Hg^{++} at different phases of their growth cycles. To some, the initial addition of mercury was made concurrently with inoculation of the cells; in other experiments, the initial mercury addition was made during the log phase of cell growth. Growth of cultures was monitored by daily cell counts, using a Neubauer hemocytometer and/or by discrete in vivo fluorescence measurements (Kiefer, 1973) with a Turner, Model 111 Fluorometer, modified for chlorophyll analysis (Lorenzen, 1966).

At the end of the growth experiments the culture vessels were measured for their mercury content. To measure mercury taken up by the cells, aliquots of cultures were filtered through 1.2 µ Millipore ® filters. The filters were digested in 35% nitric acid and the resulting solution analyzed. Mercury adsorbed onto the walls of the culture flasks was measured by washing the flasks with concentrated nitric acid and measuring the mercury dissolved in the acid wash. Mercury was also measured in the filtered media. All mercury analyses were performed on a Coleman Model MAS-50 Mercury Analyzing System.

RESULTS

The field data indicated that inorganic mercury inhibited primary production (net photosynthesis) of natural plankton communities. The degree of inhibition increased as the concentration of mercury was increased. As little as 1 ppb mercury inhibited primary production 17 to 19%, while 500 ppb depressed production 83 to 96%

(Fig. 1). At the lower mercury concentrations of 1 and 10 ppb the estuarine samples were not affected as much as were the oceanic samples, while at the higher concentrations (50, 100, and 500 ppb) the degree of inhibition was very similar in all samples (Fig. 1).

Fig. 1. The effect of mercury on the primary production (net photosynthesis) of natural estuarine and marine plankton communities.

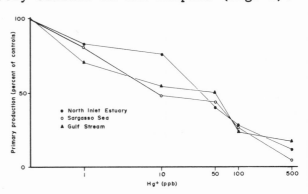

Populations of *Skeletonema costatum* appeared to be more sensitive to mercury than did *Amphidinium carterae* cultures after a 4-hr exposure to mercury. Concentrations of mercury of 1, 5, and 10 ppb had very little, if any, effect on primary production of *A. carterae* after 4 hrs, while these same concentrations seemed to inhibit primary production in *S. costatum* cultures equally at an average of 17% (Fig. 2). When measurements of primary production of *A. carterae* were delayed until 24 hrs after the exposure to mercury, a greater depression of production was seen with a given amount of mercury than in the 4-hr incubation experiments (Fig. 2). Primary production in *A. carterae* cultures exposed to 100 ppb for 4 hrs was depressed 32%, while after 24 hrs it was depressed 92%. Primary production in cultures exposed to 500 ppb mercury for 4 hrs was depressed by 80%, while after 24 hrs it was totally inhibited.

Fig. 2. The effect of 4-hr and 24-hr mercury exposures on the primary production (net photo-synthesis) of cultured marine phytoplankton.

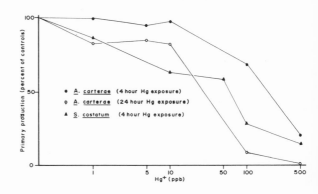

The curves resulting from both the field and the laboratory data resembled a typical bimodal, toxic dose response curve. The curves sloped gently at mercury concentrations between 1 and 10 through 50 ppb, but dropped much more steeply at concentrations above 50 ppb (Figs. 1 and 2). This suggested the possibility that

two biochemical pathways were being affected by mercury; one
sensitive to low mercury concentrations and another showing a
higher tolerance threshold to mercury.

Cell growth in *A. carterae* cultures normally involved a lag
phase of 5 to 7 days, followed by a log phase of 4 to 8 days. Cell
concentration reached 5 to 9 × 10^5 cells ml^{-1} after 12 to 19 days.
When cultures were exposed to single additions of 1, 5, or 10 ppb Hg,
made concurrently with inoculation of the organisms, growth was
indistinguishable from the nonmercury controls (Fig. 3). Cultures
with 10 or 500 ppb Hg appeared to be dead until, unexpectedly, after
3 to 5 wks, some recovery was noticed and measured. This prolonged
lag phase nurtured several possible explanations; perhaps a temporary metabolic adjustment (acclimation) or a permanent adjustment
(mutation) occurred in the cultures, or perhaps a change occurred
in the concentration of the mercury available to the cells. The
mercury remaining in the cultures upon termination of the experiment
represented in Figure 3 was measured. Most of the mercury that had
been added could not be accounted for. The possibility it had been
adsorbed onto the culture flasks was tested. It was found that in
the cultures containing the lower mercury concentrations of 1, 5,
and 10 ppb, the amount adsorbed was high relative to the initial
concentration, but at the higher concentration of 100 and 500 ppb,
less than 1% of the initial amount of mercury was retained by the
glass (Table 1). This loss of mercury resulted in a time-limited
exposure of the *A. carterae* cells to the mercury initially added.

Fig. 3. Growth of cultures
of *Amphidinium carterae*
exposed to single addi-
tions of inorganic mercury
made concurrently with
cell inoculation.

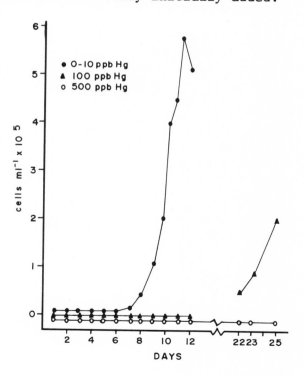

TABLE 1
Loss of mercury 18 days following a single addition of mercury at the time of inoculation of *Amphidinium carterae* cells

Initial Hg added to culture		Hg remaining in culture		Hg adsorbed on the glass		% Hg recovered from initial Hg added	% Hg unaccounted for
(ppb)	μg	μg	% total	μg	%		
500	195	6.3	3.2	2.6	1.3	4.5	95
100	39	1.3	3.3	1.1	2.8	6.1	94
10	3.9	0.35	8.9	2.6	67	76	24
5	2.0	0.25	13	1.6	80	93	7.0
1	0.39	0.10	26	0.33	85	111*	0

*Discrepancy due to variability in mercury analyzer at low Hg concentrations.

Chronic levels of mercury could provide a clearer idea of how a particular concentration of mercury could affect the organisms. Chronic levels were sustained by daily filtering aliquots from the cultures through 1.2 μ Millipore ® filters and measuring the mercury concentration in the filtrate. An amount equal to the calculated amount of "missing" mercury was then added to the respective cultures, bringing the mercury concentration up to the initial concentration.

Chronic mercury levels of 100 ppb or higher, when introduced upon inoculation of the organisms, resulted in total inhibition throughout the experiment (Fig. 4). Lower concentrations of 1, 5, and 10 ppb Hg did not cause complete inhibition as did the higher concentrations, but did significantly retard growth, the severity of which was proportional to the mercurial concentration (1 ppb retarded growth by 50%, 5 ppb by 60%, and 10 ppb by 73%). Similar results were seen when the same levels of mercury were added to cultures during their log phase and then maintained at chronic levels. Cell numbers were reduced rapidly when mercury was added at concentrations of 100 ppb (Fig. 5). The lesser concentrations reduced growth in varying degrees depending on the mercury concentration much as did the lower concentrations in the limited-exposure experiments.

Mercury was lost from cultures due to volatilization (Table 1). To measure mercury which was volatilized in the chronic-exposure experiments, 3000 ml Erlenmeyer flasks were used as culture vessels

Fig. 4. Growth of cultures of *Amphidinium carterae* exposed to chronic concentrations of mercury beginning at the time of cell inoculation.

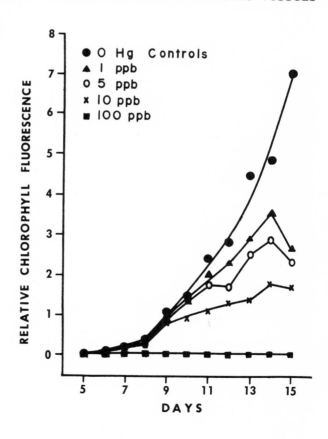

and were stoppered and connected in series (Fig. 6). Air was pumped through the flasks; the culture with the least amount of mercury was first in the series and the culture with the greatest amount was last so that the air containing a greater amount of mercury would not be pumped into flasks containing a lesser amount. The air, after leaving the last culture flask, was bubbled through a 35% nitric acid solution to trap any volatile mercury (Fig. 6). The loss of mercury from the cultures exposed to chronic levels was found to be substantial. In a 15-day experiment, when mercury was added at the time of inoculation, 23% of the total amount of mercury

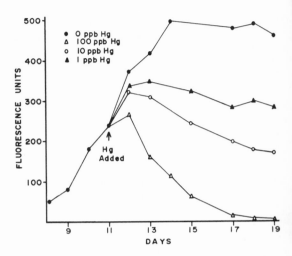

Fig. 5. Growth of *Amphidinium carterae* cultures exposed to chronic concentrations of $HgCl_2$ beginning during their log phase.

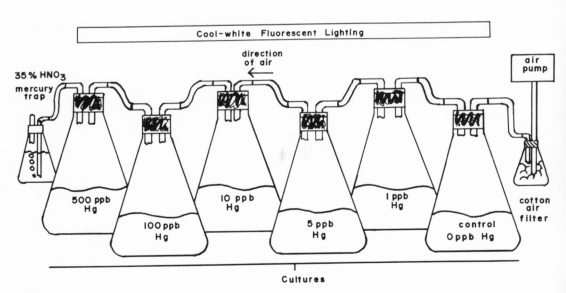

Fig. 6. Experimental apparatus for determining the distribution of mercury in cultures of phytoplankton exposed to chronic levels of mercury.

added to the cultures was recovered in the acid trap. In a 21-day experiment when mercury was added during the log phase of cell growth, 10% of the total mercury was recovered in the trap indicating a considerable loss from the cultures due to volatilization (Table 2). These values are a conservative estimate, possibly due to the inefficiency of the acid trap and loss of volatilized mercury when the flasks were unstoppered to take samples for cell counts and mercury analyses.

DISCUSSION

It is apparent from the literature and our results that mercury inhibits cellular processes of phytoplankton, even at low concentrations. Our data show that the degree of inhibition varied as a function of the concentration of mercury, the length of exposure, and the phase of the growth cycle. These data clearly indicate that the toxic effects of mercury increased as the concentration of the mercury increased. The higher concentrations of mercury (above 50 ppb) had the most adverse effects on primary production and growth of phytoplankton. Other workers have shown similar increased toxic effects on algae with increased concentration of mercury (Harriss, White, and MacFarlane, 1970; Harriss, 1971; Kamp-Nielsen, 1971; Nuzzi, 1972). The experimental use of concentrations of mercury of 100, 500 ppb or higher is probably not reflective of widespread natural conditions, although mercury concentrations of as high as 400 ppb have been measured in some British coastal waters (John S. Gray, personal communication). Such studies, however, are important, as they give short-term indices of the toxic properties of the metal.

The data also show that the toxic effects of a given concentration of mercury increased as the time of exposure to mercury was extended (Fig. 2). Kamp-Nielsen (1971) found similar results in the effects of mercury on the photosynthesis of *Chlamydomonas reinhardi* (Chlorophyta). A 4-hr exposure of 300 to 600 ppb Hg^{++} (as $HgCl_2$) caused a 25% reduction in photosynthesis in *C. reinhardi*, whereas a 21-hr exposure at these concentrations resulted in a 75% repression. Boney, Corner, and Sparrow (1959) reported a 12% mortality of sporelings of *Plumaria elegans* (Rhodophyta) when exposed to 1 ppm Hg^{++} (as $HgCl_2$) for 0.5 hrs. When the exposure time was extended to 18 hrs, 100% mortality resulted.

An important observation was the loss of mercury from our culture flasks during the growth period. A similar observation was also seen by Ben-Bassat et al. (1972). The loss was probably due to a change in the form of mercury from the nonvolatile Hg^{++} to a volatile form. The change in form would have been due to biological or nonbiological processes. Jensen and Jernelov (1969) and Jernelov (1970) reported the biological conversion of Hg^{++} to methyl mercury, probably by microorganisms in sediments. Others have demonstrated

TABLE 2
Results of two experiments in which mercury was added regularly to *Amphidinium carterae* cultures to maintain a given chronic concentration of mercury

Duration of Experiment	Total Hg added to cultures	Hg remaining in media at termination		Hg recovered on filters		Hg removed by sampling		Hg recovered in trap		Hg unaccounted for[1]	
		μg	%	μg	%	μg	%	μg	%	μg	%
15 days[2]	5381 μg	331	6.2	240	4.5	90	1.7	1229	23	3491	65
21 days[3]	7205 μg	870	12	2913	40	72	1.0	720	10	2630	37

[1] Loss probably due to adsorbtion on glass, escape of volatile Hg when sampling and inefficiency of trap.

[2] Mercury added at time of innoculation of cells.

[3] Mercury added during log phase of cell growth.

this conversion in cultures of bacteria (Yamada and Tonomura, 1972; Vonk and Sijpesteijn, 1973) and fungi (Landner, 1971; Vonk and Sijpesteijn, 1973). Also, some mercury-resistant strains of the bacterium *Pseudomonas* have been shown to reduce organo-mercury compounds to free metallic mercury (Hg) (Furukawa, Suzuki, and Tonomura, 1969; Tonomura, Furukawa, and Yamada, 1972). Only the study of Ben-Bassat et al. (1972), with axenic cultures of *C. reinhardi*, has indicated the possible phycological volatilization of mercury. The change in the form of mercury was more likely to have occurred by nonbiological processes, especially at mercury concentrations above 25 ppb, where, under our culture conditions, the most stable species of mercury would have been the reduced and volatile form (metallic mercury) (Hem, 1970). This is clearly indicated in our data, which shows a greater percentage of mercury lost from cultures receiving the higher doses of mercury (Table 1).

A result of the volatilization of mercury was the ability of *A. carterae* to recover from the toxic effects induced by an initial concentration of 100 ppb mercury (Fig. 3). Ben-Bassat et al. (1972) noted similar results in *C. reinhardi*. Hannan and Patouillet (1972) reported in their study that the toxic effects of mercury were irreversible. The length of the lag phase of growth was probably prolonged in these cultures until sufficient mercury had evaporated to allow for normal cell metabolism. Also, the normal lag phase (6 to 7 days) of *A. carterae* may have been adequate time to allow sufficient mercury to escape and thus account for why there was no apparent growth inhibition at 1 to 10 ppb mercury in cultures that received mercury at the time of inoculation (Fig. 3).

We were able to compensate for the inevitable disappearance of mercury by growing *A. carterae* exposed to a chronic dosage of mercury. We observed a difference in the degree of mercury toxicity, depending at which phase of the growth cycle we added the mercury. Mercury was less toxic when added during the log phase of growth than when added concurrently with the inoculation of the cells (Figs. 4 and 5). The differences in toxicity, however, might also be explained by the difference between the concentration of cells at the time of inoculation and at the log phase, as during the log phase of growth there would be less mercury in solution per cell. Others have reported an inverse relationship between the degree of mercury toxicity and phytoplankton population densities (Harriss, White, and MacFarlane, 1970; Kamp-Nielsen, 1971; Shieh and Barber, 1973). Their data were based, however, on mercury added only once during their experiments.

Since our primary productivity experiments in the laboratory correlated well with similar experiments in the field, we think it is reasonable to expect our laboratory growth data to be applicable under field conditions. Our laboratory data indicated that a chronic dosage of as little as 1 ppb mercury could inhibit phytoplankton growth by 50% (Fig. 4). We recognize that concentrations of 1 ppb mercury or higher are not widely found in natural waters, although such have been measured near industrial outfalls (Harriss,

1971). Should our laboratory data hold true in nature, we could expect a corresponding decrease in productivity at other levels of the aquatic and marine food webs.

Finally, as a warning in light of the above experiments, we suggest that proper precautionary measures be taken by investigator during future laboratory experiments with mercury to prevent unnecessary toxic contamination of the air by volatile mercury.

ACKNOWLEDGMENTS

This work was supported in part by Contract AT(38-1)-77, awarded to Richard G. Zingmark by the U. S. Atomic Energy Commissio The authors acknowledge the support of the Oceanographic Program of Duke University Marine Laboratory for use of the R/V EASTWARD on Cruise E-31G-71/72. The Oceanographic Program is supported by National Science Foundation Grants GA-27725 and GD-283

LITERATURE CITED

Anonymous. 1970. Mercury in the environment. Environmental Scien and Technology 4:890-92.

Ben-Bassat, D., G. Shelef, N. Gruner, and H. I. Shuval. 1972. Growth of *Chlamydomonas* in a medium containing mercury. Natur 240:43-44.

Boney, A. D. 1971. Sub-lethal effects of mercury on marine algae. Marine Pollution Bulletin 2:69-71.

_____ and E. D. S. Corner. 1959. Application of toxic agents in the study of the ecological resistence of intertidal red algae J. Mar. Biol. Ass. U. K. 38:267-75.

_____, _____, and B. W. P. Sparrow. 1959. The effects of various poisons on the growth and viability of sporelings of the red alga *Plumaria elegans*. Biochem. Pharmac. 2:37-49.

Furukawa, K., T. Suzuki, and K. Tonomura. 1969. Decomposition of organic mercurial compounds by mercury-resistant bacteria. Agr. Biol. Chem. 33:128-30.

Glooschenko, W. A. 1969. Accumulation of ^{203}Hg by the marine diat *Chaetoceros costatum*. J. Phyco. 5:224-26.

Guillard, R. R. L. and J. H. Ryther. 1962. Studies on the marine planktonic diatoms. I. *Cyclotella nana* Hustedt and *Detonula confervacea* (Cleve) Gran. Can. J. Microbiol. 8:229-39.

Hannan, P. J. and C. Patouillet. 1972. Effect of mercury on algal growth rates. Biotech. Bioengineering 14:93-101.

Harriss, R. C. 1971. Ecological implications of mercury pollution in aquatic systems. Biological Conservation 3:279-83.

_____, D. B. White, and R. B. MacFarlane. 1970. Mercury compounds reduce photosynthesis by plankton. Science 170:736-37.

Hem, J. D. 1970. Chemical behavior of mercury in aqueous media, pp. 19-24. In: *Mercury in the Environment*. Geological Survey Professional Paper No. 713.

Jensen, S. and A. Jernelov. 1969. Biological methylation of mercury in aquatic organisms. *Nature* 223:753-54.

Jernelov, A. 1970. Release of methyl mercury from sediments with layers containing inorganic mercury at different depths. *Limnol. Oceanogr.* 15:958-60.

Kamp-Nielsen, L. 1971. The effect of deleterious concentrations of mercury on the photosynthesis and growth of *Chlorella pyrenoidosa*. *Physiologia Pl*. 24:556-61.

Kiefer, D. A. 1973. The *in vivo* measurement of chlorophyll by fluorometry. In: *Estuarine Microbial Ecology*, Vol. 1 (L. H. Stevenson and R. R. Colwell, eds.). Columbia, S. C.: University of South Carolina Press.

Klein, D. H. and E. D. Goldberg. 1970. Mercury in the marine environment. *Environmental Science and Technology* 4:765-68.

Landner, L. 1971. Biochemical model for the biological methylation of mercury suggested from methylation studies *in vivo* with *Neurospora crassa*. *Nature* 230:452-54.

Lorenzen, C. D. J. 1966. A method for the continuous measurement of *in vivo* chlorophyll concentration. *Deep-Sea Res*. 12:223-27.

Nuzzi, R. 1972. Toxicity of mercury to phytoplankton. *Nature* 237:38-40.

Peakall, D. B. and R. J. Lovett. 1972. Mercury: Its occurrence and effects in the ecosystem. *BioScience* 22:20-25.

Pugh, P. R. 1970. Liquid scintillation counting of ^{14}C-diatom material on filter papers for use in productivity studies. *Limnol. Oceanogr*. 15:652-55.

Shieh, Y. J. and J. Barber. 1973. Uptake of mercury by *Chlorella* and its effect on potassium regulation. *Planta* 109:49-60.

Strickland, J. D. H. and T. R. Parsons. 1968. *A Practical Handbook of Seawater Analysis* (J. C. Stevenson, ed.), Bull. 167. Ontario: Fisheries Research Board Canada.

Tonomura, K., K. Furukawa, and M. Yamada. 1972. Microbial conversion of mercury compounds. In: *Environmental Toxicology of Pesticides* (F. Matsumura, G. M. Boush, and T. Misato, eds.), pp. 115-33. New York: Academic Press.

Ukeles, R. 1962. Growth of pure cultures of marine phytoplankton in the presence of toxicants. *Appl.Microbiol*. 10:532-37.

Vonk, J. W. and A. K. Sijpesteijn. 1973. Studies on the methylation of mercuric chloride by pure cultures of bacteria and fungi. *Antonie van Leeuwenhoek* 39:505-13.

Windom, H. L. 1973. Mercury distribution in estuarine-nearshore environment. *J. Waterw. Harb. Coast. Engin. Div*., ASCE 99: 257-64.

Yamada, M. and K. Tonomura. 1972. Formation of methylmercury compounds from inorganic mercury by *Clostridium cochlearium*. *J. Ferment. Technol*. 50:159-66.

Low temperature resistance adaptations in the killifish, *Fundulus heteroclitus*

B. L. Umminger

It is well known that when a fish becomes acclimated to low temperatures, there are many biochemical and physiological systems involved. Probably best known is the fact that whole body metabolism or oxygen consumption declines in the cold (Prosser, 1967). Likewise, osmoregulatory ability is reduced at low temperatures (Umminger, 1969a, 1970a, 1971a, b, c) and can be so severely impaired as to cause cold death due to osmotic imbalance (Doudoroff, 1945; Brett, 1952; Meyer, Westfall, and Platner, 1956; Woodhead, 1964; Umminger, 1971d). The central nervous system becomes sluggish with spinal reflexes often being blocked in the cold (Prosser, 1967). Swimming speed is reduced at low temperatures of acclimation (Davis et al., 1963) and many fishes become torpid (Nikolsky, 1963) or cease eating (Keast, 1968). Furthermore, responses to exogenous hormone administration, although not qualitatively altered, occur more slowly at lower temperatures (Mazeaud, 1964). Recently, interest has centered on cold-induced changes in specific pathways of intermediary metabolism (Prosser, 1967; Hochachka, 1967) and on alterations of enzyme kinetics (Hochachka and Somero, 1968; Baldwin and Hochachka, 1970; Freed, 1971; Hazel, 1972). My own work with resistance adaptations in killifish points out the adaptive changes in blood and tissue chemistry induced by exposure to subzero temperatures.

Changes in Blood Serum Chemistry

Initial studies (Umminger, 1969a, b, 1970a, b, c) dealt with the physiological problems faced by the killifish in surviving subzero temperatures in salt water in a supercooled state. To gain insight into the mechanism permitting *Fundulus heteroclitus* to survive in a supercooled state, serum physiochemical properties (osmolality and pH), serum inorganic ions (sodium, potassium, calcium, magnesium, chloride, phosphate, and bicarbonate), and serum organic constituents (glucose, nonglucose free carbohydrates, total cholesterol, total amino acids, urea, total nonprotein nitrogen, total protein, and electrophoretic fractions of proteins) were studied in parallel groups of adult male killifish acclimated to various temperatures in salt water under otherwise constant laboratory conditions.

The results of these investigations pointed out that the one component of the serum that is most strikingly altered in the subzero cold is glucose. It increases to a far greater degree than any other serum constituent in the cold and the major part of this increase occurs only when subzero temperatures are attained (Umminger, 1969a, b). For example, in saltwater-adapted killifish, serum glucose concentrations at 20°C and 4°C are practically identical, with a slight increase occurring at 2°C. Only when subzero temperatures are encountered is the dramatic sixfold increase in serum glucose concentration elicited (Umminger, 1969b). Glucose is therefore implicated as a serum constituent necessary in large quantities for survival at subzero temperatures and may be the compound permitting permanent supercooling of the blood of this fish.

Changes in Tissue Chemistry

Subsequent research involving tissue chemistry showed that the hyperglycemic response to subzero temperatures is produced by a progressive depletion of hepatic, but not muscle, glycogen. At warmer temperatures liver glycogen is continually being replenished by breakdown products of ingested food. However, at subzero temperatures the killifish refuses to eat so that liver glycogen reserves are eventually exhausted. When liver glycogen is totally used up, serum glucose can no longer be maintained at high concentrations. Eventually serum glucose levels fall to normal, causing the fish to die. It is characteristic that all fish living in the subzero cold show a marked hyperglycemia and livers rich with glycogen, whereas those fish dying under the same conditions have low concentrations of serum glucose and totally depleted livers. However, if glucose is added daily to the water in which the killifish live, this sugar is taken into their bodies, elevating blood glucose concentrations. Such treatment with exogenous glucose prolongs the survival of killifish at −1.5°C since it prevents the decline of serum glucose levels (Umminger, 1970b).

It is significant that winter fish exposed to the subzero cold have almost twice as much glucose in their blood as summer fish under the same conditions. Moreover, when kept at -1.5C, a winter fish can be expected to survive for an average of 63 days whereas a summer fish can be expected to survive for an average of only 28 days. There is a seasonal cycle in the ability of the killifish to tolerate subzero temperatures which is related to the amount of blood sugar accumulated (Umminger, 1970b).

Enzymatic Control During Cold Acclimation

Experiments elucidating the enzymatic control of the cold-induced hyperglycemia and glycogenolysis in killifish have been conducted in conjunction with my graduate student, David P. Benziger. We have found that in the saltwater-adapted killifish, the cold-induced hyperglycemia and breakdown of liver glycogen are accompanied by an increase in the specific activity of hepatic glycogen phosphorylase (which catalyzes the breakdown of glycogen to glucose-1-phosphate) whereas the specific activity of hepatic glucose-6-phosphatase (which converts glucose-6-phosphate to the glucose which is subsequently released into the blood) is unchanged. It is therefore felt that glycogen phosphorylase, and not glucose-6-phosphatase, is the key enzyme regulating carbohydrate reserves in the killifish at subzero temperatures (Benziger and Umminger, 1973).

Hormonal Control During Cold Acclimation

The hormonal regulation (Fig. 1) of the cold-induced hyper-glycemia in killifish is not under pituitary control (Umminger, 1971e, 1972), but results from a hypertrophy and degranulation of glucagon-containing alpha cells and an atrophy of insulin-producing beta cells in the pancreatic islets (Umminger and Bair, 1973). Glucagon is known to elicit hyperglycemia and hepatic glycogenolysis in teleosts just as in other vertebrates (Houssay, 1959; Falkmer, 1961; Epple, 1969; Larsson and Lewander, 1972; Plisetskaya, 1972). This further suggests that the increased activity of glycogen phosphorylase observed in cold-acclimated killifish might also be mediated by an increase in circulating levels of glucagon. Glucagon is known not only to activate hepatic glycogen phosphorylase in mammals (Sutherland and Cori, 1948), but also to inhibit glycogen synthetase nearly completely (DeWulf and Hers, 1968). Similar effects of glucagon on glycogen phosphorylase activity have been demonstrated in fish (Umminger and Benziger, 1973).

These cold-induced changes in carbohydrate metabolism observed in the killifish, as well as their enzymatic and hormonal control mechanisms, should not be interpreted as generalizations to be applied to all fish. For example, in the goldfish, *Carassius auratus*, the cold-induced hyperglycemia is not very pronounced and is accompanied by a synthesis, not a breakdown, of hepatic glycogen (Benziger, 1974). It is felt that the slight increase in serum

glucose in the cold in goldfish is due to gluconeogenesis and not
to glycogenolysis (Fry and Hochachka, 1970). As would be expected,
in this fish the specific activity of hepatic glycogen phosphorylase
declines in the cold (Fig. 2). Benziger has attempted to characterize the kinetic properties of hepatic glycogen phosphorylase in
goldfish as affected by low temperature. In short, he finds that
there are no adaptive changes in K_m or activation energy of this
enzyme in fish acclimated to temperatures from 1°C to 30°C, but that

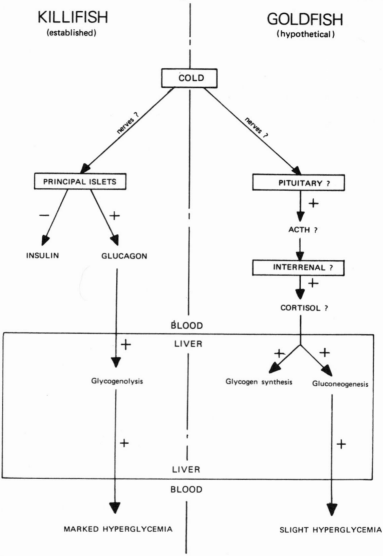

Fig. 1. Hormonal regulation of carbohydrate metabolism in the
killifish and goldfish at temperatures near freezing. The
scheme for the killifish is established, but the scheme
for the goldfish is largely hypothetical. A + represents
activation or stimulation; a – indicates inhibition;
mechanisms marked with ? are hypothetical.

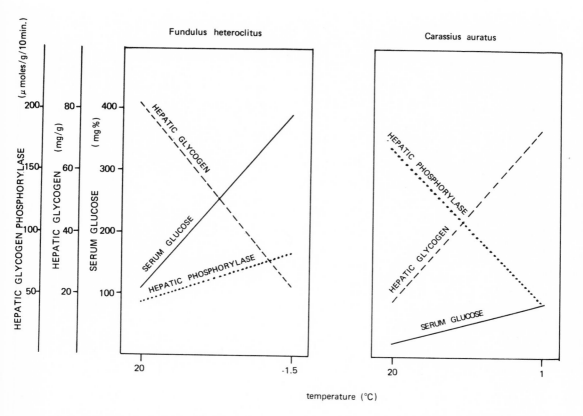

Fig. 2. Effects of temperature on serum glucose concentrations (mg%), hepatic glycogen concentrations (mg/g wet weight), and the specific activity of hepatic glycogen phosphorylase (μmoles of product/g tissue/10 min) in *F. heteroclitus* and *C. auratus*.

there are significant changes in the total amount of enzyme and in the proportion of the enzyme existing in the active (phosphorylase a) and inactive (phosphorylase b) forms. For example, in the goldfish, one finds that the decline in specific activity of the enzyme in the cold is accompanied by an increase in the amount of the enzyme in the inactive form (Benziger, 1974).

Clearly, the responses of carbohydrate metabolism to low temperature are different in killifish and goldfish (Fig. 2). Likewise, Dean and Goodnight (1964) found that low temperature did not affect serum glucose and hepatic glycogen in the same way in four species of freshwater fish. Moreover, as pointed out previously (Umminger, 1969b), although a cold-induced hyperglycemia is present in most teleosts, its degree is either slight or marked depending on species, and some fish lack the response altogether. The manner in which carbohydrate metabolism is altered by low temperature in fish is perhaps determined to a large degree by their different ecologies and physiological demands. For example, the saltwater-adapted killifish needs large quantities of serum glucose in the cold to serve as a supercooling stabilizer (see discussion below) whereas goldfish do not. The killifish breaks down its liver glycogen reserves rapidly due to an increase in the specific activity of glycogen phosphorylase, but the goldfish accumulates hepatic glycogen, possibly as a food reserve, by decreasing the specific activity of glycogen phosphorylase (Benziger and Umminger, 1971, 1972, 1973; Benziger, 1974).

Since low temperature acclimation alters carbohydrate stores in different ways in different species, then the mechanisms of their hormonal regulation are probably different as well. For example, the pituitary-independence and islet-dependence of the cold-induced hyperglycemia in killifish probably do not apply to goldfish (Fig. 1). First of all, low temperature acclimation in the goldfish is accompanied by only a slight hyperglycemia (probably due to gluconeogenesis). Second, there is a synthesis of hepatic glycogen. These sorts of responses can be induced in fish by cortisol administration (Chester Jones et al., 1969) and are inconsistent with responses elicited by glucagon as discussed above. Therefore, it is highly plausible that the changes in carbohydrate stores in goldfish are mediated by cortisol and are dependent upon the pituitary for ACTH release at low temperatures. Clearly, this possibility needs further investigation.

Possible Neural Involvement

A final point concerning the control of carbohydrate metabolism in the killifish at subzero temperatures involves the changes induced in the pancreatic islets. What causes glucagon release and insulin retention at subzero temperatures? Recent investigations by Klein (1971) have shown that the pancreatic islets of *Xiphophorus helleri* are innervated, with definite neuroinsular junctions, by at least two types of neuron (possibly sympathetic and parasympathetic) If the islets of fish are indeed under nervous control as these studies suggest, then perhaps low environmental temperatures are sensed by peripheral thermoreceptors and this information processed by the central nervous system which subsequently alters the pattern of nervous stimulation to the pancreatic islets. Of course, control of islet tissue function at low temperatures by other hormones (such as catecholamines) or by a direct effect of temperature on the islets cannot be ruled out altogether.

Adaptive Significance

The adaptive significance of the marked hyperglycemia in killifish at subzero temperatures appears to be that glucose acts as a supercooling stabilizer. When killifish are acclimated to -1.5°C, their serum osmolarity is such that the blood should freeze at -0.8°C and yet the fish remain unfrozen. In other words, the fish are supercooled by 0.7°C. Since supercooling is a physically unstable state, it has been suggested that supercooled fish release into their blood some substance or substances which stabilize the supercooled blood, thus reducing the probability of freezing (Gordon, Amdur, and Scholander, 1962; Umminger, 1969b; DeVries and Wohlschlag, 1969; Smith, 1970). The above studies on killifish suggest that glucose is the compound responsible. In fact, physio-chemical studies of supercooled aqueous solutions have demonstrated that the presence of glucose, in concentrations similar to those found in the blood of supercooled killifish, slows down the rate of ice crystal growth when the solutions are seeded with a crystal of ice (Tamman and Büchner, 1935; Lusena, 1955). Thus, it is reason-able to suspect that glucose may stabilize the supercooled blood of killifish by preventing spontaneous nucleation and subsequent freezing.

Similar studies have implicated glucose as the supercooling stabilizer in the blood of cod *(Gadus morhua)*.[*] Likewise, studies on the blood chemistry of the supercooled antarctic fish *Notothenia neglecta* have demonstrated that concentrations of reducing sugars (predominantly glucose) are ten times higher than those in fish living in the temperate zone (Smith, 1970). It is tempting to think that glucose may have some role in stabilizing the supercooled blood of this antarctic fish as well (Smith, 1972). However, as plausible as the above suggestion sounds, recent studies by DeVries (1971) and Hargens (1972) indicate that the supercooling stabilizer of the majority of both antarctic and arctic fish, respectively, is a glycoprotein, present in low concentrations, which appears to poison ice crystal growth surfaces in the supercooled blood (Scholander and Maggert, 1971).

Other organisms which live at very low temperatures that have been investigated chemically are the insects and plants. Terrestrial insects make extensive use of supercooling as a mechanism to avoid freezing during extremely cold winters (Asahina, 1966). Chemical analyses of the tissues and hemolymph of supercooled insects often reveal high concentrations of either glycerol, sorbitol, trehalose (insect blood sugar), fructose, glucose, or possibly mannitol (Sømme, 1966). In addition, practically all frost-tolerant plants exhibit increases in cellular sugar concentrations in winter. The protective sugar in plants is usually sucrose, but raffinose, stachyose, glucose, fructose, mannitol, sorbitol, and glycerol are also found to increase in many cold-hardened plants (Levitt, 1966). Therefore, in all organisms adequately studied, the ability to survive subzero temperatures in a supercooled state is accompanied by the accumulation of sugars, polyhydric alcohol derivatives of sugars, or glycoproteins in the body fluids and tissues. An exami-nation of the chemical structure of all solutes known to promote supercooling in organisms reveals that all are hydroxyl-rich com-pounds which readily form hydrogen bonds and are hydrophilic. Since hydrogen bonding is needed for water molecules to arrange themselves in the lattice structure requisite for the formation of ice crystals, it has been suggested that the numerous hydroxyl groups on these supercooling stabilizers might somehow interfere

[*]Harden Jones and Scholes, personal communication.

with water-to-water hydrogen bonding to inhibit the formation of ice nuclei or promote supercooling by poisoning ice crystal growth surfaces (Umminger, 1969b; Smith, 1970; DeVries, 1971).

It should be pointed out that even though the hydroxyl groups of glucose can stabilize the supercooled blood of the killifish by preventing spontaneous nucleation in the absence of ice crystals, glucose is totally ineffective in promoting supercooling when ice crystals are present in the water in which the fish lives. The accumulation of large amounts of serum glucose is an adaptation only for the avoidance of spontaneous nucleation of the blood; it is of no use if ice is encountered externally. However, for antarctic fish, such as *Trematomus borchgrevinki*, the glycoproteins present in the blood are much more effective. Whereas the hydroxyl groups of glucose can only slow down the rate of ice crystal growth, the active glycoprotein has hydroxyl groups which are highly structured along the protein backbone of this molecule and can stop ice crystal growth completely (DeVries, 1971). For this reason, supercooled antarctic *Trematomus* can actually tolerate contact with ice. When killifish come into contact with ice, they quickly and vigorously swim away to avoid seeding their supercooled blood.

That temperate zone and polar fish should have evolved different compounds to serve as supercooling stabilizers is not surprising. For example, the polar species need a hydroxyl-rich compound that is relatively stable and long-lived (such as a glycoprotein), since these fish encounter subzero temperatures for most of the year. The killifish, *Fundulus heteroclitus*, on the other hand, is a temperate zone species that only occasionally encounters subzero temperatures during the course of the winter. In New Haven, Connecticut, where my original studies were performed, winter water temperatures rarely stay below 0C for more than a week at a time. Therefore, the killifish needs a rather labile supercooling stabilizer that can be quickly mobilized into the bloodstream when cold weather begins and that can be quickly metabolized when the water warms up once more. Glucose is a hydroxyl-rich substance which satisfies these criteria. In addition, *Fundulus heteroclitus* is descended from a tropical line of fishes (Miller, 1955) so that sufficient time may not have elapsed for an active glycoprotein to evolve. This again points out the need to consider each species in terms of its own ecology and phylogeny in interpreting its physiological response to the cold. With this in mind, it should be pointed out that, even in arctic fish, the same supercooling stabilizer may not be used in all species. For example, Gordon, Amdur, and Scholander (1962) found that the nonprotein nitrogen (NPN) fraction of the blood was elevated in supercooled arctic fjord cod *(Gadus ogac)* but not in supercooled sculpin *(Myoxocephalus scorpius)*. Since a high NPN level is characteristic of the serum of fish containing a glycoprotein supercooling stabilizer because the active glycoproteins are not precipitated by known protein-precipitating agents (DeVries and Wohlschlag, 1969), the fjord cod probably has this glycoprotein in its blood. However, the sculpin probably does not and must use some completely different substance.

Conclusion

The above summary of my research on freezing resistance in killifish is intended to make the following points: (1) observed changes in the chemical composition of fish during cold acclimation can and should be interpreted in terms of enzymatic, hormonal, and neural regulatory mechanisms in order to appreciate fully the physiological reorganization of the fish which occurs at extremely low temperatures; (2) the observed changes should be interpreted in terms of their adaptive value to the fish; and (3) the chemical changes and control mechanisms used by one species of fish in coping with an extremely low temperature are not necessarily those employed by all species of fish, undoubtedly because the phylogenetic histories, ecologies, and physiological requirements of all species are not the same.

Acknowledgment

This work was supported by Grant GB 26321 from the National Science Foundation.

LITERATURE CITED

Asahina, E. 1966. Freezing and frost resistance in insects. In: Cryobiology (H. T. Meryman, ed.), pp. 451-86. New York: Academic Press.

Baldwin, J. and P. W. Hochachka. 1970. Functional significance of isoenzymes in thermal acclimatization. Acetylcholinesterase from the trout brain. Biochem. J. 116:883-87.

Benziger, D. 1974. Effects of thermal acclimation on glycogenolytic enzymes and carbohydrate reserves in teleost fishes. Ph.D. dissertation, University of Cincinnati.

_____ and B. L. Umminger. 1971. Glycogenolytic enzymes in the livers of thermally-acclimated teleosts. Amer. Zool. 11:670.

_____ and _____. 1972. Hepatic phosphorylase activity in thermally-acclimated fish. Amer. Zool. 12:xxviii-xxix.

_____ and _____. 1973. Role of hepatic glycogenolytic enzymes in the cold-induced hyperglycemia of the killifish, Fundulus heteroclitus. Comp. Biochem. Physiol. 45A:767-72.

Brett, J. R. 1952. Temperature tolerance in young Pacific salmon, genus Oncorhynchus. J. Fish. Res. Bd. Canada 9:265-323.

Chester Jones, I., D. K. O. Chan, I. W. Henderson, and J. N. Ball. 1969. The adrenocortical steroids, adrenocorticotropin and the corpuscles of Stannius. In: Fish Physiology. Vol. 2. The Endocrine System (W. S. Hoar and D. J. Randall, eds.), pp. 321-76. New York: Academic Press.

Davis, G. E., J. Foster, C. E. Warren, and P. Doudoroff. 1963. The influence of oxygen concentration on the swimming performance of juvenile Pacific salmon at various temperatures. Trans. Amer. Fish. Soc. 92:111-24.

Dean, J. M. and C. J. Goodnight. 1964. A comparative study of carbohydrate metabolism in fish as affected by temperature and exercise. Physiol. Zool. 37:280-99.

DeVries, A. L. 1971. Glycoproteins as biological antifreeze agents in antarctic fishes. Science 172:1152-55.

_____ and D. E. Wohlschlag. 1969. Freezing resistance in some antarctic fishes. Science 163:1073-75.

DeWulf, H. and H. G. Hers. 1968. The interconversion of liver glycogen synthetase a and b in vitro. Eur. J. Biochem. 6: 552-57.

Doudoroff, P. 1945. The resistance and acclimatization of marine fishes to temperature changes. II. Experiments with Fundulus and Atherinops. Biol. Bull. 88:194-206.

Epple, A. 1969. The endocrine pancreas. In: Fish Physiology. Vol. 2. The Endocrine System (W. S. Hoar and D. J. Randall, eds.), pp. 275-319. New York: Academic Press.

Falkmer, S. 1961. Experimental diabetes research in fish. On the morphology and physiology of the endocrine pancreatic tissue of the marine teleost Cottus scorpius with special reference to the role of glutathione in the mechanism of alloxan diabetes using a modified nitroprusside method. Acta Endocrinol. 37: 1-122.

Freed, J. M. 1971. Properties of muscle phosphofructokinase of cold- and warm-acclimated Carassius auratus. Comp. Biochem. Physiol. 39B:747-64.

Fry, F. E. J. and P. W. Hochachka. 1970. Fish. In: Comparative Physiology of Thermoregulation. Vol. 1. Invertebrates and Nonmammalian Vertebrates (G. C. Whittow, ed.), pp. 79-134. New York: Academic Press.

Gordon, M. S., B. H. Amdur, and P. F. Scholander. 1962. Freezing resistance in some northern fishes. Biol. Bull. 122:52-62.

Hargens, A. R. 1972. Freezing resistance in polar fishes. Science 176:184-86.

Hazel, J. R. 1972. The effect of temperature acclimation upon succinic dehydrogenase activity from the epaxial muscle of the common goldfish (Carassius auratus L.). II. Lipid reactivation of the soluble enzyme. Comp. Biochem. Physiol. 43B:863-82.

Hochachka, P. W. 1967. Organization of metabolism during temperature compensation. In: Molecular Mechanisms of Temperature Compensation (C. L. Prosser, ed.), pp. 177-203. Washington, D. C.: American Association for the Advancement of Science, Publ. No. 84.

_____ and G. N. Somero. 1968. The adaptation of enzymes to temperature. Comp. Biochem. Physiol. 27:659-68.

Houssay, B. A. 1959. Comparative physiology of the endocrine pancreas. In: Comparative Endocrinology (A. Gorbman, ed.), pp. 639-67. New York: Wiley.

Keast, A. 1968. Feeding of some Great Lakes fishes at low temperatures. J. Fish. Res. Bd. Canada 25:1199-1218.

Klein, C. 1971. Innervation des cellules du pancréas endocrine du poisson téléostéen *Xiphophorus helleri* H. Zeit. Zellforsch. 113:564-80.

Larsson, Å. and K. Lewander. 1972. Effects of glucagon administration to eels (*Anguilla anguilla* L.). Comp. Biochem. Physiol. 43A:831-36.

Levitt, J. 1966. Winter hardiness in plants. In: Cryobiology (H. T. Meryman, ed.), pp. 495-563. New York: Academic Press.

Lusena, C. V. 1955. Ice propagation in systems of biological interest. III. Effects of solutes on nucleation and growth of ice crystals. Arch. Biochem. Biophys. 57:277-84.

Mazeaud, F. 1964. Vitesse de production de l'hyperglycémie adrénalinique en fonction de la température chez la carpe. Intensité de la résponse en fonction de la dose d'hormone. C. R. Soc. Biol. 158:36-40.

Meyer, D. K., B. A. Westfall, and W. S. Platner. 1956. Water and electrolyte balance of goldfish under conditions of anoxia, cold and inanition. Amer. J. Physiol. 184:553-56.

Miller, R. R. 1955. An annotated list of the American cyprinodont fishes of the genus *Fundulus*, with the description of *Fundulus persimilis* from Yucatan. Occasional Papers, Museum Zool., Univ. Michigan, No. 568:1-25.

Nikolsky, G. V. 1963. The Ecology of Fishes. New York: Academic Press.

Plisetskaya, E. M. 1972. The effect of glucagon on the blood sugar and liver glycogen in the scorpion fish *Scorpaena porcus*. Zh. Evol. Biokhim. Fiziol. 8:447-49.

Prosser, C. L. 1967. Metabolic and central nervous acclimation of fish to cold. In: The Cell and Environmental Temperature (A. S. Troshin, ed.), pp. 375-83. New York: Pergamon Press.

Scholander, P. F. and J. E. Maggert. 1971. Supercooling and ice propagation in blood from arctic fishes. Cryobiology 8:371-74.

Smith, R. N. 1970. The biochemistry of freezing resistance of some antarctic fish. In: Antarctic Ecology, Vol. 1 (M. W. Holdgate, ed.), pp. 329-36. New York: Academic Press.

_____. 1972. The freezing resistance of antarctic fish. I. Serum composition and its relation to freezing resistance. Br. Antarct. Surv. Bull. 28:1-10.

Sømme, L. 1966. The effect of temperature, anoxia, or injection of various substances on haemolymph composition and supercooling in larvae of *Anagasta kuehniella* (Zell.). J. Insect Physiol. 12:1069-83.

Sutherland, E. W. and C. F. Cori. 1948. Influence of insulin preparations on glycogenolysis in liver slices. J. Biol. Chem. 172:737-50.

Tamman, G. and A. Büchner. 1935. Die Unterkühlungsfähigkeit des Wassers und die lineare Kristallisationsgeschwindigkeit des Eises in wässrigen Lösungen. Zeit. Anorgan. Allgem. Chemie 222:371-81.

70

Umminger, B. L. 1969a. Physiological studies on supercooled killifish *(Fundulus heteroclitus)*. I. Serum inorganic constitutents in relation to osmotic and ionic regulation at subzero temperatures. J. Exp. Zool. 172:283–302.

_____. 1969b. Physiological studies on supercooled killifish *(Fundulus heteroclitus)*. II. Serum organic constitutents and the problem of supercooling. J. Exp. Zool. 172:409–24.

_____. 1970a. Osmoregulation by the killifish, *Fundulus heteroclitus* in fresh water at temperatures near freezing. Nature 225:294–95.

_____. 1970b. Physiological studies on supercooled killifish *(Fundulus heteroclitus)*. III. Carbohydrate metabolism and survival at subzero temperatures. J. Exp. Zool. 173:159–74.

_____. 1970c. Effects of temperature on serum protein components in the killifish, *Fundulus heteroclitus*. J. Fish. Res. Bd. Canada 27:404–9.

_____. 1971a. Osmoregulatory role of serum glucose in freshwater-adapted killifish *(Fundulus heteroclitus)* at temperatures near freezing. Comp. Biochem. Physiol. 38A:141–45.

_____. 1971b. Patterns of osmoregulation in freshwater fishes at temperatures near freezing. Physiol. Zool. 44:20–27.

_____. 1971c. Osmoregulatory overcompensation in the goldfish, *Carassius auratus*, at temperatures near freezing. Copeia, 1971, pp. 686–91.

_____. 1971d. Chemical studies of cold death in the Gulf killifish *Fundulus grandis*. Comp. Biochem. Physiol. 39A:625–32.

_____. 1971e. Lack of pituitary involvement in the cold-induced hyperglycemia of the killifish, *Fundulus heteroclitus*. Experientia 27:701–2.

_____. 1972. Physiological studies on supercooled killifish *(Fundulus heteroclitus)*. IV. Carbohydrate metabolism in hypophysectomized killifish at subzero temperatures. J. Exp. Zool. 181:217–22.

_____ and R. D. Bair. 1973. Role of islet tissue in the cold-induced hyperglycemia of the killifish, *Fundulus heteroclitus*. J. Exp. Zool. 183:65–70.

_____ and D. Benziger. 1973. In vitro stimulation of hepatic glycogen phosphorylase activity by epinephrine and glucagon in the brown bullhead, *Ictalurus nebulosus*. Amer. Zool. 13:1280.

Woodhead, P. M. J. 1964. The death of North Sea fish during the winter of 1962–63, particularly with reference to the sole, *Solea vulgaris*. Helgoland. Wiss. Meeresuntersuch. 10:283–300.

ADDENDUM

Recent studies by Parker (1972) have elucidated the physiochemi-cal mechanism whereby glucose acts as a supercooling stabilizer: hydrogen bonding between glucose and the surface of ice lattices occurs asymmetrically. When ice lattices are constructed around the glucose molecule, fit does not occur below the cryoprotectant whereas it does occur above it. Whereas one side of the glucose molecule becomes easily incorporated into the ice lattice, the other side does not. thus retarding the incorporation of additional water molecules into the ice lattice. Regarding the role of insulin in low temperature acclimation in fish, Moule and Yip (1973) have also accumulated evidence that insulin production and release may be retarded in the cold in the brown bullhead: little or no conversion of proinsulin into insulin occurred at temperatures below 12°C.

Moule, M. L. and C. C. Yip. 1973. Insulin biosynthesis in the bullhead, *Ictalurus nebulosus*, and the effect of temperature. Biochem. J. 134:753-61.

Parker, J. 1972. Spatial arrangement of some cryoprotective com-pounds in ice lattices. Cryobiology 9:247-50.

Studies on the salinity resistance of the copepod *Euterpina acutifrons* (Dana)

G. S. Moreira

Although environmental factors have been known to affect the distribution and the physiological processes of the planktonic copepods, relatively few studies have been made to determine the effect of a wide range of salinities, temperatures, and oxygen content on the survival of these animals under laboratory conditions. Experiments on the degree to which free-living planktonic copepods can endure dilution have been done by several researchers (Marshall, Nicholls, and Orr, 1935; Hopper, 1960; Lance, 1963, 1964; Tundisi and Tundisi, 1968; and Yamashita, 1972). Most of the available results are obtained from benthic harpacticoids (Ranade, 1957; Eltringham and Barnett, 1958; Battaglia, 1958; Matutani, 1962; and Jansson, 1968). The scarcity of these studies probably is due to the difficulties which are usually encountered in maintaining planktonic copepods in the laboratory.

Euterpina acutifrons is a planktonic species with a wide distribution in coastal temperate and warm waters (Lang, 1948). It is one of the dominant forms along the eastern coast of North and South America, and it is easily cultured in the laboratory. It is known from field studies that it occurs in salinities ranging from 8 o/oo (Tundisi, 1972) to 38 o/oo (El-Maghraby, 1965). In the mangrove region of the Cananeia estuary (south of Brazil), this species does not penetrate into the "marigot," with very low salinities, but is very abundant in the regions with salinity above 13 o/oo. Tundisi and Tundisi (1968) have studied the tolerance to various dilutions of the females of *E. acutifrons* and other copepods from Cananeia estuary in a short-term experiment.

The purpose of the present work has been to determine the effect of a wide range of salinities on the mortality of females of *E. acutifrons* from two different regions. Emphasis was placed on the effects of temperature acclimation on the salinity resistance of these copepods.

MATERIAL AND METHODS

Animals Studied

Euterpina acutifrons were obtained from horizontal plankton tows in two different places: (1) São Sebastião Channel, in front of the laboratory of the Marine Biology Institute (lat. 23°49'6" S and long. 45°25'3" W); and (2) Santos Bay (lat. 23°59'0" S and long. 45 21'0" W). Temperatures were essentially the same in both areas (around 29°C, but salinities differed greatly. During the collection of the samples at São Sebastião Channel, the salinity was 35 o/oo. This value remains more or less constant there throughout the year. Santos is an estuarine region, with an average salinity of around 23 o/oo. This salinity varies with the tide and the seasons, sometimes going up to 28 o/oo or falling down to 17 o/oo. Thus, copepod from São Sebastião Channel and those from Santos Bay differ greatly in their life histories. No distinctive morphological differences were noted between these two populations.

The plankton samples were collected during the summer in both areas and were brought in large bottles inside an isopor box to the laboratory of the Physiology Department of the Biosciences Institute in São Paulo. Soon after the arrival, the copepods were picked up from the samples and placed in culture dishes (5.0 cm bottom diameter, 8.5 cm top diameter, and 5.7 cm high), filled with unfiltered seawater and covered with a 10 cm Petri dish top. Each dish received about 50 animals. *Platymona* sp. (cultured in Guillard medium) was used as food for the copepods in a concentration of about 87,500 cells/ml. About one half of the sampled *Euterpina* of each place (Santos and Sãn Sebastião) was placed in a constant temperature box (15°C temperature, cold acclimation) and the others in a constant temperature room (25°C temperature, warm acclimation). The animals from Santos remained at 23 o/oo salinity and those from São Sebastião at 35 o/oo salinity, until used for experiments one week after the capture.

Resistance Experiments

For these experiments, 14 different concentrations of salinity were prepared, varying from approximately 0 o/oo (distilled water) to 65 o/oo, with 5 o/oo intervals. The dilutions were obtained adding distilled water to seawater. Greater concentrations were obtained by freezing the seawater and collecting the liquid phase. Salinities were determined by titrations against silver nitrate

(Harvey, 1955). For each salinity, 100 ml of the water were placed
in culture dishes (already described) and about 10 adult female
E. acutifrons were added. At the beginning of the experiments,
results were recorded in duplicate, two identical ranges of salinity
being used for the same conditions. As variation in survival was
very small, it was assumed that a reasonably accurate estimation of
the salinity resistance of this species under laboratory conditions
could be obtained from a single result. A total of 56 experiments
were made, 28 with animals from São Sebastião Channel (14 with the
cold-acclimated animals and 14 with the warm-acclimated animals) and
28 with the animals from Santos Bay (14 with the cold-acclimated and
14 with the warm-acclimated animals). The food concentration was
maintained at the same level in all the dishes. Dead animals were
counted and removed at one-day intervals in the salinities close to
the acclimation salinity and progressively smaller near the extreme
of the range. Experiments lasted from a few minutes to 60 days.
Copepods were considered dead when no movement of any external or
internal part of the body could be detected upon close examination
under microscope. "Dead" animals did not recover when they were
returned to the appropriate acclimation conditions. Thorough inspection of each animal was necessary before death could be determined
in low or high salinity experiments since copepods that appear to be
dead from the effects of salinity can recover completely within 1 to
2 hrs, depending on the concentration of the previous solution and
the duration of immersion in it.

RESULTS AND DISCUSSION

The results of the salinity resistance tests from both populations (Santos Bay and São Sebastião Channel) are shown in Table 1
and Figure 1, as well as the effect of the temperature acclimation
on the resistance. Female *Euterpina acutifrons* from both areas
studied survived better in normal and high salinities in high temperature (25°C) and survived better in lower salinities in low temperature (15°C). When these copepods were placed in the extreme
salinities, they fell into a state of apparent death, they ceased
their activity and sank to the bottom of the dish and lay motionless.
In the lower salinities they swelled in a few minutes, but after
some hours they regained their original body volume, except in distilled water, where they burst. Inversely, in the higher salinities
they shrank at the beginning and except for the animals from Santos
Bay in salinities higher than 50 o/oo they also regained their original body volume. The animals from Santos Bay in salinities higher
than 50 o/oo died in a few hours. Experiments on volume regulation
of these copepods were done recently by Moreira and Yamashita (1973).
Usually in lower concentrations the animals secreted a few small
strands of mucus from the hind and ventral part of the abdomen,
where *Platymona* cells became attached. The secretion of mucus in

lower salinities was noticed also by Jansson (1968) in his study of
mesopsammic copepods. It is well known that many animals secrete
mucus under salinity stress (Kinne, 1964) and that this mucus great
reduces the permeability of the integument (Potts and Parry, 1964).
   Nauplii were born in almost all the experimental culture dishe
kept in warm temperature, but they developed only in the range from
15 to 45 o/oo. In the dishes kept in low temperature (15°C) the
nauplii did not leave the egg or died before the first moult, with
a few exceptions. These nauplii were picked out from the experimental culture dishes in order to maintain the same conditions in
the tested concentrations. They were transferred to other dishes.
Haq (1972), studying *Euterpina acutifrons* from Menai Straits,
Anglesey, England, found the better survival of the larval phase in

TABLE 1
Salinity resistance of *Euterpina acutifrons* at different salinities

Locale of Capture	Test salinities o/oo	$LD_{50}(15°C)$[*]	$LD_{50}(25°C)$
	0	10 min	8 min
	5	1 d 2 hr 17 min	23 hr
	10	6 d	5 d 12 hr
	15	14 d	11 d
	20	21 d	15 d
	25	29 d	23 d
	30	30 d	32 d
Santos Bay	35	17 d	28 d
(23 o/oo)	40	14 d	18 d
	45	10 d	15 d
	50	1 d 8 hr 35 min	2 d
	55	4 hr 50 min	22 hr 5 min
	60	3 hr	3 hr 30 min
	65	16 min	17 min
	0	9 min	8 min
	5	1 d	23 hr
	10	2 d 6 hr 25 min	1 d
	15	8 d	7 d
	20	10 d	9 d
	25	9 d	7 d
	30	10 d	12 d
São Sebastião	35	12 d	13 d
Channel(35 o/oo)	40	9 d	10 d
	45	4 d	7 d
	50	5 d	8 d
	55	6 d	7 d
	60	5 d	6 d
	65	1 d	1 d 4 hr

[*]LD_{50} indicates time required for 50% mortality.

temperatures from 14 to 20C. This difference in relation to the
present results probably is due to the different life histories of
both populations.

The graphs of the animals from Santos show a continuous curve
with the peak in the salinity of 30 o/oo. The salinities in the
Santos Bay are usually under 28 o/oo and these animals were collected
and acclimated in the laboratory to a salinity of 23 o/oo. This
shows that although these animals are very abundant during all the
year in the Santos Bay, the preference of the female adults is for
slightly higher salinities. In São Sebastião, the peak is in 35 o/oo,
for both cold- and warm-acclimated animals, i.e., the same salinity
of the biotope from which the population came. The curve of the
resistance to different salinities for the animals from São Sebastião
is not continuous, presenting a lower survival for 25 and 45 o/oo
when compared with the adjacent points, showing that the effect of
increasing or decreasing concentrations of salinity may be more
critical in some points. The tests in these critical concentrations
were repeated several times, with other animals and a newly prepared
medium, and the results were always the same. Maybe this difference
between the curves of Santos and São Sebastião occurs becuase the
animals of Santos live in a very changeable biotope and are less
sensitive to small variations of the environment than the animals
from São Sebastião, which live in a much more constant biotope.

The most striking difference between both populations studied
is that the animals from Santos are much more tolerant to lower
salinities than animals from São Sebastião. Inversely, the animals
from São Sebastião are much more tolerant to high salinities than
animals from Santos. This fact was expected, since the population
from Santos was living in lower salinities than the population from

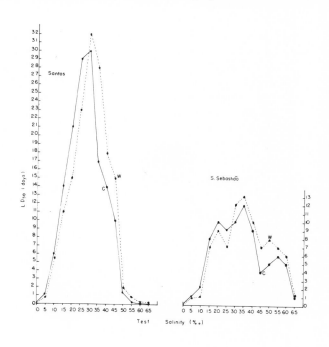

Fig. 1. Survival of
Euterpina acutifrons
females exposed to
different salinities.
LD_{50} indicates time
required for 50%
mortality. Animals
from Santos Bay and
São Sebastião Channel.
—— cold acclimated
----- warm acclimated

São Sebastião. Another difference is in the time to death. Animal
from Santos lived longer than animals from São Sebastião in all
salinities lower than 50 o/oo. Some females from Santos lived in
30 o/oo salinity for 60 days, whereas the greater survival for
females from São Sebastião in the laboratory was 37 days. Yamashit
(1972) demonstrated that the life cycle (i.e., time since the birth
till the first posture) is longer in lower salinities than in norma
and high salinities. She also studied the salinity resistance of
female *E. acutifrons* collected in the São Sebastião Channel during
the winter and found the same type of curve, but a greater mortalit
in all salinities. This suggests a seasonal variation in the lengt
of the lifetime. The short-term experiment conducted by Tundisi an
Tundisi (1968) in Cananeia also showed a greater mortality when com
pared with the present data. After 6 hrs, they found more than 50%
mortality for all dilutions lower than 8 o/oo, while in our experiments after 6 hrs only the animals in distilled water and those fro
Santos in salinities higher than 55 o/oo were dead.

Lance (1964) studied the effect of temperature on the resistan
of planktonic copepods to different salinities. She showed that
Acartia tonsa, *A. discaudata*, and *A. bifilosa* are most tolerant whe
the experimental and the environmental temperatures were close, and
least tolerant when the experimental and field temperatures differe
markedly. She worked only with salinities of 36 o/oo and dilutions
of this water (Plymouth water). Their results for *Acartia* differed
from mine for *E. acutifrons* where cold acclimation (temperature
markedly different from the environmental) increased the resistance
to low salinities. According to Kinne (1970), several aquatic
invertebrates which live in habitats with greatly fluctuating temperature and salinity conditions can tolerate subnormal temperature
better at the lower end of their salinity range, and supranormal
temperatures better at the upper end of their salinity range. Thes
low/low and high/high combinations have been found also in the
benthonic copepod *Tigriopus fulvus* (Ranade, 1957). It seems that
beneficial effects of such combinations are evident, since in shall
coastal water and in rockpools an increase in temperature is usuall
accompanied by an increase in salinity.

LITERATURE CITED

Battaglia, B. 1958. Modificazione della vitalità in alcune forme
    de *Tisbe reticulata* (Copepoda, Harpacticoida) in rapporto con
    cambiamenti di salinità e temperatura. Arch. Oceanogr. Limnol
    11:251-63.
El-Maghraby, A. M. 1965. The seasonal variation in length of some
    marine planktonic copepods from Eastern Mediterranean at
    Alexandria. Crustaceana 8:37-47.
Eltringham, S. K. and P. R. O. Barnett. 1958. Survival at reduced
    salinities and osmoregulation in *Limnoria* and *Platychelipus*
    (Harpacticoida). Ann. Rep. Challenger Soc. 3, No. 10.

Haq, S. M. 1972. Breeding of *Euterpina acutifrons*, a harpacticid copepod, with special reference to dimorphic males. Mar. Biol. 15:221-35.

Harvey, H. W. 1955. The Chemistry and Fertility of Seawaters. New York: Cambridge University Press.

Hopper, A. F. 1960. The resistance of marine zooplankton of the Caribbean and South Atlantic to changes in salinity. Limnol. Oceanogr. 5:43-56.

Jansson, B. O. 1968. Quantitative and experimental studies of the interstitial fauna in four Swedish sandy beaches. Ophelia 5: 1-71.

Kinne, O. 1964. The effects of temperature and salinity on marine and brackish water animals. II. Salinity and temperature salinity combinations. Oceanogr. Mar. Biol. Ann. Rev. 2:281-339.

_____. 1970. Temperature. Invertebrates. In: Marine Ecology, Vol. 1 (Otto Kinne, ed.), pp. 407-514. London: Wiley-Interscience.

Lance, J. 1963. The salinity tolerance of some estuarine planktonic copepods. Limnol. Oceanogr. 8:440-49.

_____. 1964. The salinity tolerances of some estuarine planktonic crustaceans. Biol. Bull., Woods Hole 127:108-18.

Lang, K. 1948. Monographie der Harpacticoiden. 2 vols. Lund: Hakan Ohlsson.

Marshall, S. M., A. G. Nicholls, and A. P. Orr. 1935. On the biology of *Calanus finmarchicus*. VI. Oxygen consumption in relation to environmental conditions. J. Mar. Biol. Ass. U. K. 20:1-27.

Matutani, K. 1962. Studies on the temperature and salinity resistance of *Tigriopus japonicus*. IV. Heat resistance in relation to salinity of *Tigriopus* acclimated to dilute and concentrated sea water. Physiol. Ecol. Kyoto 10:63-67.

Moreira, G. S. and C. Yamashita. 1973. Experimental studies on physiological and behavioural response mechanisms of the planktonic copepod *Euterpina acutifrons* (Dana) to various salinities. Bolm Zool. Biol. Mar. In press.

Potts, W. T. W. and G. Parry. 1964. Osmotic Regulation in Animals. London: Pergamon Press Ltd.

Ranade, M. R. 1957. Observations on the resistance of *Tigriopus fulvus* (Fischer) to changes in temperature and salinity. J. Mar. Biol. Ass. U. K. 36:372-75.

Tundisi, J. and T. M. Tundisi. 1968. Plankton studies in a mangrove environment. V. Salinity tolerances of some planktonic crustaceans. Bolm Inst. Oceanogr. S Paulo 17:57-65.

Tundisi, T. M. 1972. Aspectos ecológicos do zooplâncton da região lagunar de Cananéia com especial referência aos Copepoda (Crustacea). Ph.D. dissertation, São Paulo University.

Yamashita, C. 1972. Fisioecologia e Fisioetologia de *Euterpina acutifrons* (Dana) (Crustacea, Copepoda). Influência da salinidade. Master's thesis, São Paulo University.

Nitrogen turnover and food relationships of the pinfish
Lagodon rhomboides in a North Carolina estuary

R. M. Darnell and **T. E. Wissing**

Estuarine systems (including marshes, flats, and open waters) are widely recognized as areas of high fertility (Ketchum, 1967) and productivity (Teal, 1962), and as feeding and nursery areas for a variety of coastal fishes and invertebrates (Darnell, 1958). It has been suggested that most of the nutrients passing to the consumer species do so through the medium of detrital food chains (Darnell, 1961; Odum and de la Cruz, 1967). Yet, with a few notable exceptions, our knowledge of the quantitative dynamics of nutrient turnover by individual detritus-feeding species is scanty. To some extent we have ignored the heart of this estuarine problem.

Nutrient transfer may be expressed in terms of energy, of any of the chemical elements making up protoplasm, or of a combination of these. The element nitrogen has been held by Gerking (1954) to be of special value in expressing nutrient relationships of fishes since most of the body nitrogen is chemically bound as protein and since protein accumulation is one of the best measures of growth. Pandian (1967c), however, has suggested that studies of energy alone or of energy and nitrogen together may permit a more complete interpretation of secondary production since proteins, through deamination, may be converted to fats and carbohydrates within the fish body. While recognizing the validity of such arguments, there still exists the need for transfer studies of individual elements, each of which contributes to the overall picture of animal production. The element nitrogen provides a convenient handle for examining the interface of physiology and ecology for the following reasons:

(1) it is relatively simple to measure; (2) it may readily be expressed in terms of its protein equivalent; and (3) together with phosphorus it is often considered a factor which limits aquatic production.

To date, nitrogen turnover has been investigated in three species of freshwater fishes—the mirror carp (Karsinkin, 1935), the bluegill (Gerking, 1952, 1954, 1955, 1962), and the Indian snakehead (Pandian, 1967a, b, c)—as well as in three species of marine fishes—the angelfish and the Nassau grouper of the Bermuda reefs (Menzel, 1959, 1960) and the Indian tarpon (Pandian, 1967b, c). The general techniques for these studies are more or less the same. Growth rates (expressed as increase in body nitrogen or protein) are determined for fishes maintained on various feeding regimes in the laboratory, and the laboratory growth rates are compared with estimated rates of individual growth in field populations. From such comparisons one may estimate rates of nitrogen, protein, and total food intake in the field. Various growth and conversion efficiencies may be computed, and food intake may be related to population size, food availability, and other field parameters. It is important to note that five of the above species are primarily carnivores which feed upon arthropods and that of them only the young tarpon is normally a resident of brackish waters. At the present time there exists no literature on nitrogen turnover by broadly multivorous fishes in which organic detritus forms a significant percentage of the diet or by fishes which spend most of their lives in the brackish coastal waters.

The present study was designed to provide information on the nitrogen turnover of one of the most abundant estuarine multivores of the south Atlantic and northern Gulf coasts, the pinfish *Lagodon rhomboides*. This fish has been referred to as an important regulator species in the estuarine communities (Darnell, 1970), and quantitative information concerning its food relations provides considerable insight into the dynamics of the estuarine ecosystem. Food habits and life history information relating to the pinfish have been summarized by Gunter (1945), Caldwell (1957), Darnell (1958, 1964, 1970), and Carr and Adams (1973). In the present study considerable effort was made to assess variability of fish growth in the field population and to ascertain the factors underlying this variability.

GENERAL ECOLOGICAL BACKGROUND

The investigation was carried out on the North Carolina coast during the summers (June through August) of 1960 through 1963. The study area was a small protected bed of marine grasses near Piver's Island in the estuary of the Newport River, 2 miles from the Atlantic Ocean. This grass bed (180 m × 90 m) was bordered on the west by a sand shoal and on the east by a channel 20 m in width which separated the bed from Piver's Island. Near the center of the bed (and

directly south of the main study area) was a dredge hole about 50 m
in diameter and 4 m deep. Main channels of the estuary flowed
north, west, and south of the protected area, but considerable tidal
exchange took place across the grass bed (Fig. 1).

Fig. 1. Maps showing the
study area and its
immediate environment.
A. Piver's Island, study
area, and surrounding
open water and marsh
areas. B. Study area
showing distribution and
density of sea grasses
(primarily *Zostera
marina*).

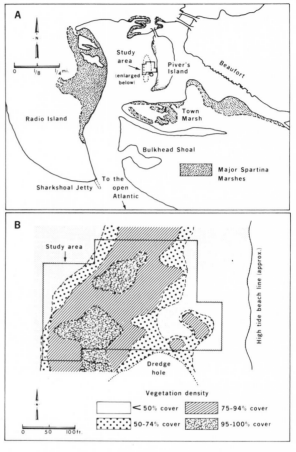

The study area was subject to two tidal cycles per day whose
amplitude varied from 0.4 to 1.1 m (average, 0.8 m). At normal high
tide the area was under about 1 m of water, and at normal low tide
at least a portion of the bed was exposed. Tidal exchanges periodically subjected the area to strong water currents. Water temperature during the study period varied between 24 and 27C, and salinity
ranged from 20 o/oo to 30 o/oo. The water contained much particulate
matter and was especially turbid during periods of strong tidal
current flow. When the water was still, the bottom was generally
visible at 1 to 2 m depths.

Occasional extreme hydrographic conditions occurred, reflecting
wind direction and velocity as well as rainfall. Thus, a strong
northerly wind coincident with a very low tide would sometimes expose all of the study area to the air. Exceptionally high water
associated with coastal or oceanic storms would cover the area with
2 or 3 m of water. Heavy rains in the watershed or on the study

area or at low tide would considerably lower the salinity. Short-term extreme environmental conditions were characteristic of this estuarine environment, and although the community generally could withstand such changes, significant biological effects were noted (see discussion section).

The substratum of the grass bed was composed of silty sand mixed with varying quantities of decaying organic matter. Molluscs, ostracods, and tube-dwelling polychaetes were found in association with this substratum. The eelgrass *(Zostera marina)* was the pre-dominant vascular plant, constituting over 95% of the larger vegeta-tion. *Ruppia maritima* and *Diplanthera* sp. were occasionally seen as were the larger algae *Ulva lactuca*, *Gelidium* sp., and *Padina* sp. Extensive *Spartina* marshes located both upstream and downstream in the estuary undoubtedly contributed much of the organic matter brought in by water currents, but no live *Spartina* was present in the study area. Individual blades of eelgrass were generally clothed with filamentous algae of which the most abundant were species of *Cladophora*, *Chaetomorpha*, and *Ectocarpus*. Associated with the filamentous algae were quantities of decomposing organic material, diatom colonies, small polychaetes, bryozoans, and a variety of microcrustaceans (many unidentified larval forms as well as copepods, gammarid and caprellid amphipods, small shrimp, and crabs). Larger invertebrates inhabiting the grasses included scallops, palaemonid and peneid shrimp, and blue crabs.

During the course of the field studies at least thirty-nine species of fishes from the grass bed were recorded (Table 1). This list includes both the common inhabitants of the bed and stray indi-viduals from other estuarine and marine habitats. Our own records and those of Tagatz and Dudley (1961) reveal that many additional species frequent the nearby *Spartina* marshes, mud flats, channels, and open Atlantic beach habitats and might be expected on the beds occasionally. Sharks and other predatory forms known to frequent the nearby channels, especially at night, undoubtedly were able to evade the slow-moving collecting gear. Thus, with effort the number of fish species could have been greatly extended.

The most abundant fish species encountered were the pinfish, planehead filefish, brown pipefish, and spot, with striped killifish appearing in abundance at low tide and silversides at high tide. To these was added a number of common and rare species, most of which were carnivorous. All the fishes were derived from marine or brack-ish water habitats; none were of freshwater derivation. Most were the young of marine species which utilize the estuaries as nursery grounds. The most important predators were the young barracudas (which appeared at the end of the third summer and were present thereafter) and the bluefish, toadfish, and flounders (which moved in from the channels at night). Among the shore birds only the laughing gull *(Larus articilla)* was commonly seen on the area.

In summary, the eelgrass bed which was subject to tidal flushin twice daily was in a position to remove quantities of suspended matter from the tidal current. To this extent it was a nutrient tra

TABLE 1
Life history stages (ad, adult; juv, juvenile) of fish species collected
on eelgrass bed and relative abundance (a, abundant; c, common; r, rare)

Species	Life history stage	Relative abundance
Dasyatidae		
Dasyatis sayi, bluntnose stingray	juv	r
Clupeidae		
Brevoortia tyrannus, Atlantic menhaden	juv	r
Engraulidae		
Anchoa hepsetus, striped anchovy	ad	c
Anchoa mitchilli, bay anchovy	ad	c
Synodontidae		
Synodus foetens, inshore lizardfish	juv	c
Belonidae		
Strongylura marina, Atlantic needlefish	juv	r
Hemirhamphidae		
Hyporhamphus unifasciatus, halfbeak	juv	r
Cyrinodontidae		
Fundulus heteroclitus, mummichog	juv, ad	r
Fundulus majalis, striped killifish	juv, ad	a
Fistularidae		
Fistularia tabacaria, cornetfish	juv	r
Syngnathidae		
Hippocampus sp., seahorse	juv, ad	r
Syngnathus fuscus, northern pipefish	juv, ad	a
Syngnathus scovelli, Gulf pipefish	juv, ad	r
Serranidae		
Centropristes striatus, black sea bass	juv	c
Epinephelus sp., grouper	juv	r
Pomatomidae		
Pomatomus saltatrix, bluefish	juv	r
Carangidae		
Trachinotus carolinus, pompano	juv	r
Gerridae		
Eucinostomus gula, silver jenny	juv	r
Pomadasyidae		
Orthopristis chrysopterus, pigfish	juv	c
Sciaenidae		
Bairdiella chrysura, silver perch	juv	r
Leiostomus xanthurus, spot	juv	a
Sparidae		
Diplodus holbrooki, spottail pinfish	juv	r
Lagodon rhomboides, pinfish	juv, ad	a
Gobiidae		
Unidentified gobies	juv, ad	r

Table 1 (cont.)

Species	Life history stage	Relative abundance
Triglidae		
Prionotus spp., searobins	juv	r
Sphyraenidae		
Sphyraena barracuda, great barracuda	juv	r-c
Mugilidae		
Mugil cephalus, striped mullet	juv, ad	c
Mugil curema, white mullet	juv, ad	c
Atherinidae		
Membras martinica, rough silverside	juv, ad	c
Menidia menidia, Atlantic silverside	juv, ad	c
Bothidae		
Paralichthys spp., flounders	juv, ad	r
Cynoglossidae		
Symphurus plagiusa, blackcheek tonguefish	ad	c
Balistidae		
Alutera schoepfi, orange filefish	juv	r
Monacanthus hispidus, planehead filefish	juv	a
Tetraodontidae		
Chilomycterus schoepfi, striped burrfish	juv	r
Diodontidae		
Sphaeroides maculatus, northern puffer	juv	r
Batrachoididae		
Opsanus tau, oyster toadfish	juv, ad	c

In contrast to the paucity of vascular plant species, there was high diversity in algal, invertebrate, and fish inhabitants. Many of the fishes were juveniles which complete their life histories in more saline waters. The area was also utilized by a number of transient predatory fish species.

LABORATORY STUDIES OF NITROGEN UTILIZATION

Materials and Methods

Feeding experiments. Young pinfish, 45.0 mm in standard body length were selected from the field population and maintained in the laboratory in individual transparent acrylic plastic chambers (13.0 cm long × 8.0 cm wide × 8.0 cm high). Each chamber contained several gauze-covered windows to facilitate water exchange with the larger acrylic plastic aquarium in which it was suspended. Each aquarium, containing three or four chambers, was aerated continuously, and the water was replaced every three days with fresh filtered water brought in from the eelgrass habitat. The temperature in the aquaria varied between 23.0 and 26.0°C, and averaged 24.3°C for the total experimental period.

After collection, all fish were measured, weighed, and acclimated to the experimental chambers for one day prior to the beginning of the feeding experiments. All fish were weighed on a triple-beam balance in a preweighed beaker of water, and excess moisture was blotted from each fish prior to weighing.

The experimental fish were placed on three predetermined feeding regimes as follows: (1) Group I, (3 fish) fed ca. 5% of body wt./day; (2) Group II, (4 fish) fed ca. 7.5% of body wt./day; and (3) Group III, (4 fish) fed ad libitum (which turned out to be about 12.6% of body wt./day). The experimental fish were reweighed at 10-day intervals, and feeding rates were adjusted to maintain the daily ration at a constant proportion of body weight. The ad libitum feeding rates, of course, exhibited considerable daily variation.

The fish were fed daily at 1,000 hrs on small whole redworms (*Lumbricus rubellus*), and any worms or fragments thereof remaining after 6 hrs were collected, blotted, weighed, and subtracted from the original amount fed. Except in the ad libitum chambers the worms were generally consumed entirely, and even in these chambers the excess food was normally recovered intact, i.e., a given worm was either eaten or left alone. Small redworms were selected as the food source because the package of worm tissue and plant detritus (in the gut of the worm) tended to approximate the natural mixed diet of the fish and because the pinfish normally consume quantities of annelids. In practice it was found that the small worms were ideal for individual weighing and feeding, and they remained intact in the water.

Fecal material was collected daily with a bulb pipette and stored under refrigeration. Ten-day fecal accumulations for each fish were combined on weighed filter paper, oven-dried to constant weight (at 95.0°C) and reweighed on a Roller-Smith balance. The dried samples were then refrigerated in tightly stoppered vials for later nitrogen determination.

At the termination of the feeding experiments the fish were sacrificed, weighed, measured, and oven-dried to constant weight at 95.0°C. The desiccated fish were then refrigerated in tightly stoppered vials for later nitrogen determination.

Nitrogen analysis. Two procedures were employed for nitrogen determination, the Kjeldahl technique (as modified by Hawk, Oser, and Summerson, 1954) and the Nessler procedure (as modified by Lang, 1958). It was found that values obtained by the Nessler procedure averaged 4.38% lower than those given by the Kjeldahl technique. All Nessler values have, therefore, been corrected to their corresponding Kjeldahl values so that they may be compared directly with the Kjeldahl nitrogen values reported in the literature.

Dried earthworms and 10-day fecal samples were small enough to be digested without prior handling. However, it was found necessary

to pulverize the dried fish with an agate mortar and pestle prior to
digestion. Low variability between fish subsamples indicated that
pulverization was adequate and that 10% of a given fish was a sufficiently large sample to represent the entire fish. All results
are expressed as Kjeldahl nitrogen values, and these are assumed to
represent total organic nitrogen.

Feeding Experiments

Determination of conversion factors. To facilitate calculations
associated with the laboratory-feeding experiments a number of
weight conversion curves were prepared. These curves, relating wet
weight, dry weight, and body nitrogen for both the pinfish and redworms, were computed by the method of least squares from large numbers of individual measurements. Within the ranges of values
represented, the linear regression lines appear to depict adequately
the relationships under study, although tests for linearity were not
carried out.

Figure 2 shows the relationship of dry weight and organic
nitrogen to wet body weight in pinfish of approximately 45 mm body
length. Data for these two curves were obtained by terminal analysis
of the same fish used in the laboratory-feeding experiments. Figure
3 and 4 give the relationship between dry weight, organic nitrogen
weight, and wet body weight of selected redworms of the sizes
employed in the feeding experiments.

Growth efficiency studies. Data from the laboratory-feeding experiments are presented in Table 2. All fish included in the table were
alive at the end of the 40-day feeding period. All except three of
the fish were maintained on the feeding regimes for the entire period

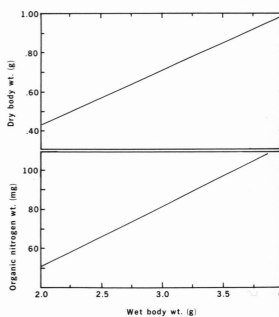

Fig. 2. Relationship of dry
weight and organic nitrogen
weight to wet body weight
in pinfish of approximately
45 mm body length.

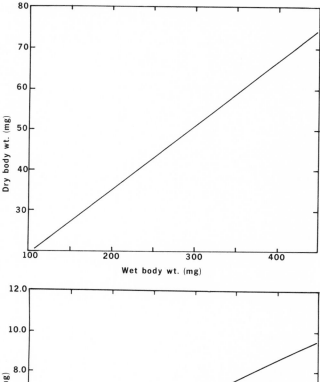

Fig. 3. Relationship
between dry and wet
body weight in red-
worms of the sizes
employed in the
feeding experiments.

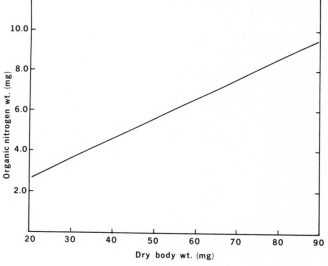

Fig. 4. Relationship
between organic
nitrogen weight and
dry body weight in
redworms of the sizes
employed in feeding
experiments.

Due to fouling of the laboratory water line, three fish in the
ad libitum group died, and this portion of the experiment had to be
started a second time. Since the experiment had to be terminated
at the end of the summer, these fish were fed for shorter time
periods (two for 17 days, one for 10 days). However, data from
these three fish were projected for the entire 40-day period.

As revealed in Table 2, within each of the three feeding groups
considerable variability was exhibited in the increases in body
length, weight, and nitrogen content. Despite this variability the
averages show clear trends in the various growth parameters. The
lowest feeding rate (5.05% of body weight per day) is clearly not
a subsistence diet. Although these fish survived the 40-day period
and gained slightly in body length, they showed a net loss in both
body weight and body nitrogen. In physical appearance they exhibited
evidence of starvation.

TABLE 2
Results of 40-day laboratory-feeding experiments with 45.0 mm pinfish.

Group no.	Fish no.	Feeding rate (% body wt/day)	Body length (mm) Initial	Gain	Body weight (g) Initial	Gain	Body organic nitrogen (mg) Initial*	Gain
I	A	4.84	45.0	1.5	2.70	-0.11	72.40	-6.39
	B	4.97	45.0	1.0	2.29	0.28	58.90	8.44
	C	5.34	45.0	0.0	2.44	-0.34	63.80	-16.60
Averages		5.05	45.0	0.8	2.48	-0.06	65.03	-4.85
II	D	7.32	45.0	1.0	2.46	0.31	64.30	8.98
	E	7.40	45.0	2.0	2.48	0.76	65.10	29.24
	F	7.76	45.0	2.0	2.39	0.26	62.20	15.74
	G	7.78	45.0	3.0	2.41	0.78	62.70	31.26
Averages		7.56	45.0	2.0	2.43	0.53	63.57	21.31
III	H**	11.45	45.0	6.0	2.45	0.69	64.00	17.79
	I†	10.48	45.0	4.0	2.43	1.16	63.50	37.59
	J†	13.56	45.0	4.0	2.36	1.32	61.00	42.16
	K†	15.08	45.0	4.0	2.34	1.48	60.50	39.55
Averages		12.64	45.0	4.5	2.39	1.17	62.25	34.27

*Computed from curve in Fig. 2B.
**Forty-day figures projected from actual feeding period of only 17 days.
†Forty-day figures projected from actual feeding period of only 10 days.

Growth proceeded well at the two higher feeding rates (7.56 and 12.64%) with the fish showing good gains in body length, weight, and nitrogen content. At the 7.56% feeding rate the respective gains were 2.0 mm, 0.53 g, and 21.31 mg; and at the higher feeding rate the gains were 4.5 mm, 1.17 g, and 34.27 mg. It is noted that increasing the feeding rate from 5.05 to 7.56% results in a proportionately greater change in growth than does increasing the feeding rate from 7.56 to 12.64% per day.

In order to determine the amount of ingested food that was actually absorbed, the egested fecal material was subtracted from the amount of food ingested. Surprisingly, the starving fish produced large quantities of nitrogen-rich fecal material, which may reflect a lowered metabolic rate or some interference with the absorptive mechanism. As revealed in Table 3, absorption efficiences were uniformly high in the two higher feeding rate groups, and less than 8% of the ingested nitrogen was lost with the feces.

The final column of Table 3 provides data concerning the utilization of nitrogen for growth (100 × the ratio of growth nitrogen to absorbed nitrogen). These values most clearly express the fish's ability to make use of the food nitrogen. Individual variability is again encountered within each feeding group. On the average, fish maintained on the lowest ration showed negative nitrogen utilization (−4.81%). These fish actually lost nitrogen during the study. Those on the intermediate ration exhibited high utilization (11.82%), and they nearly achieved the rate of utilization shown by high-ration group (12.71%). Although fish grew faster at the higher feeding rate, the nitrogen utilization efficiency increased only slightly.

The average nitrogen absorption and utilization efficiencies are plotted against feeding rate in Figure 5. From these data it is possible to estimate the subsistence diet in terms of nitrogen to be a feeding rate of 5.75% of the body weight per day. At this break-even point, the efficiency of nitrogen absorption is estimated to be 91%. However, it must be emphasized that we are dealing with averages and that the data clearly show that some individual fish may survive and grow at a lower rate of food intake.

Food conversion efficiencies (100 × the ratio of gain in weight to food consumed) for wet weight, dry weight, and organic nitrogen are given in Table 4. Unlike the previous data, these do not account for fecal loss. The general trends are quite similar to those presented above for nitrogen utilization efficiency. Whether one is dealing with wet, dry, or nitrogen weights, negative growth is indicated for the lowest diet, and the greatest gain in efficiency occurs between the lowest and the middle feeding rate.

The average food conversion efficiency values are plotted in Figure 6. Subsistence diets based on these data are in close agreement with the previously calculated value of 5.75% of body weight per day. It is interesting to note that at the two higher feeding rates the conversion efficiencies for dry weight and organic nitrogen are very close (i.e., within the range of between 11.92 and 14.41%) but that increasing the rate (from 7.56 to 12.64% of body rate per day) does slightly increase the conversion efficiencies. However, the wet weight data give lower efficiency values (6.54 and 8.73%), and the differences between the wet weight efficiencies at the two higher feeding rates are relatively greater than would have been expected from the dry weight and nitrogen data. Comparison of Figures 2 and 3 reveals that redworms have a higher water content than pinfish which accounts, in part, for the lower efficiencies when dealing with wet weight data.

TABLE 3

Forty-day food consumption, fecal production, nitrogen absorption efficiency, and nitrogen growth utilization for 45.0 mm pinfish in laboratory-feeding experiments*

Group no.	Fish no.	Feeding rate (% body wt/day)	Worm Ingested (g) Wet wt (g)	Dry wt (g)	Organic nitrogen content (mg)	Fecal Production Dry wt (g)	Organic nitrogen content (mg)	Efficiency of organic nitrogen absorption (%)	Utilization of organic nitrogen for growth (%)
I	A	4.84	4.81	0.88	115.69	0.37	15.35	86.73	- 6.37
	B	4.97	4.83	0.88	116.20	0.20	9.55	91.78	7.91
	C	5.34	4.78	0.88	115.79	0.22	11.79	89.82	-15.96
Averages		5.05	4.81	0.88	115.89	0.26	12.23	89.44	- 4.81
II	D	7.32	7.71	1.27	160.81	0.24	9.58	94.04	5.94
	E	7.40	8.23	1.44	169.28	0.16	4.54	97.32	11.68
	F	7.76	7.82	1.37	163.37	0.20	7.57	95.31	10.10
	G	7.78	8.15	1.43	169.60	0.22	9.78	94.23	19.56
Averages		7.56	7.98	1.38	165.76	0.20	7.87	95.22	11.82
III	H	11.45	15.61	1.98	222.92	0.33	12.85	94.24	8.47
	I	10.48	11.95	2.03	227.31	0.25	9.05	96.02	17.22
	J	13.56	12.80	2.15	238.24	0.33	17.64	92.60	19.11
	K	15.08	14.13	2.38	263.24	0.24	16.92	93.57	16.06
Averages		12.64	13.62	2.14	237.93	0.29	14.11	94.11	12.71

*See footnotes for Table 2.

Fig. 5. Average nitrogen absorption and growth efficiencies as a function of feeding rate in the experimental pinfish. Subsistence diet is graphically estimated to be 5.75% of the body weight/day, and the nitrogen absorption efficiency at this feeding rate is estimated to be 90.8%.

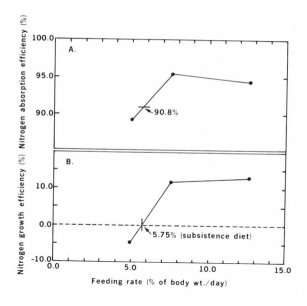

Fig. 6. Food conversion efficiencies (in terms of nitrogen weight, dry weight, and wet weight) as a function of feeding rate in the experimental pinfish. The level of the subsistence diet is indicated on each curve.

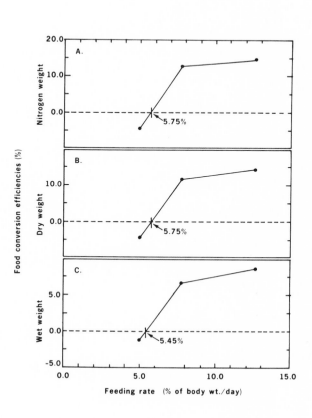

TABLE 4

Food conversion efficiencies for 45.00 mm pinfish in laboratory-feeding experiments*

Group no.	Fish no.	Feeding rate (% body wt/day)	Wet weight data (g)			Dry weight data (g)			Organic nitrogen data (mg)		
			Food consumed (A)	Gain (B)	Conversion efficiency (B/A)x 100	Food consumed (C)	Gain (D)	Conversion efficiency (D/C)x 100	Organic nitrogen consumed (E)	Gain (F)	Conversion efficiency (F/E)x 100
I	A	4.84	4.81	-0.11	- 2.29	0.88	-0.07	- 8.30	115.69	- 6.39	- 5.52
	B	4.97	4.83	0.28	5.80	0.88	0.06	7.32	116.20	8.44	7.26
	C	5.34	4.78	-0.34	- 7.11	0.88	-0.11	-12.41	115.79	-16.60	-14.34
Averages		5.05	4.80	-0.06	-1.20	0.88	-0.04	- 4.46	115.89	- 4.85	- 4.20
II	D	7.32	7.71	0.31	4.02	1.27	0.06	4.78	160.81	8.98	5.58
	E	7.40	8.23	0.76	9.24	1.44	0.22	15.52	169.28	29.24	17.27
	F	7.76	7.82	0.26	3.32	1.37	0.12	8.52	163.37	15.74	9.63
	G	7.78	8.15	0.78	9.57	1.43	0.27	18.85	169.60	31.26	18.43
Averages		7.56	7.98	0.53	6.54	1.38	0.17	11.92	165.76	21.31	12.73
III	H	11.45	15.61	0.69	4.42	1.98	0.17	8.57	222.92	17.79	7.98
	I	10.48	11.95	1.16	9.71	2.03	0.37	18.13	227.31	37.59	16.54
	J	13.56	12.80	1.32	10.31	2.15	0.34	15.74	238.24	42.16	17.70
	K	15.08	14.13	1.48	10.47	2.38	0.36	15.21	263.24	39.55	15.02
Averages		12.64	13.62	1.17	8.73	2.14	0.31	14.41	237.93	34.27	14.31

*See footnotes for Table 2.

NITROGEN TURNOVER BY THE FIELD POPULATION

Growth in the Field Population

In order to relate the laboratory growth data to the field population it was necessary to estimate growth rates under field conditions. Three methods were attempted: (1) growth of marked individuals released into the population and subsequently recaptured, (2) growth in live cages, and (3) measuring the increase in mean body size of the field population based on weekly samples.

Marking studies. Various types of marking techniques were investigated in the laboratory. Anal fin clipping, although successful, was discarded because of its possible effects on the growth rate. Latex injection was discarded because of the high mortality resulting from infection at the injection site. Sixty fish injected with 1% solutions of trypan blue or trypan red showed no infections and no mortality during a one-month laboratory holding period. The blue marks were clearly visible for 4 wks, and the red marks for 2 wks. Therefore, dye-marking was employed for growth studies of the field population.

Of a total of 222 dye-marked fish released into the field population during the summer of 1963, only 2 recognizable individuals were recaptured after having been at liberty for at least 2 wks. Growth in these 2 fish coincided well with growth observed in the live-cage fish (Fig. 7). It is assumed that heavy mortality resulted from the handling associated with making precise initial measurements.

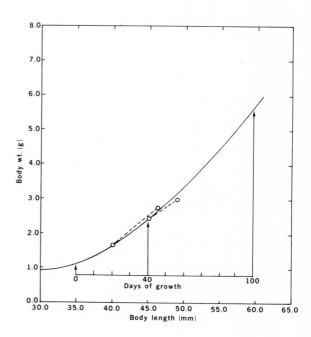

Fig. 7. Results of live-cage experiments. Body weight of pinfish in relation to body length and to days of growth (beyond an initial body length of 35.0 mm). Open circles and dashed lines represent growth of recaptured individuals dye-tagged and released into the field population. Note that one grew faster and one grew slower than the live-cage fish.

TABLE 5
Growth of pinfish in live cages.

Experiment Size	Initial size length (mm)	weight* (mg)	Final size length (mm)	weight (mg)	Daily growth rate length (mm)	weight (mg)	Daily growth as percent of average length (mm)	weight+ (mg)
40-day expt. min	35.0	1,150	41.0	1,880	0.150	18		
40-day expt. max	35.0	1,150	50.5	3,450	0.388	57		
40-day expt. avg	35.0	1,150	45.1	2,510	0.252	34	0.63	1.86
60-day expt. min	45.0	2,500	55.0	4,500	0.167	33		
60-day expt. max	45.0	2,500	66.0	7,940	0.350	91		
60-day expt. avg	45.0	2,500	59.4	5,700	0.240	53	0.46	1.30
100-day expt. avg	35.0	1,150	59.4	5,700	0.244	45.5	0.517	1.33

*Body weights are estimated from the length-weight growth curve in Fig. 7.
+Calculated from the formula 100 x (average daily growth/mean body size during growth period).

Live-cage studies. During the summer of 1963 three live cages were constructed of 0.6 cm mesh galvanized hardware cloth, each measuring 76.2 cm (length) x 61.0 cm (width) x 45.7 cm (height). A door fashioned in the top of each cage provided ready access to the contents. All three cages were wired securely together and were attached at the outside corners to metal poles driven into the bottom. The cages were placed at the edge of the deep dredged hole so that even at the lowest tide a portion of each cage would remain submerged with an ample supply of relatively cool, oxygenated water. Each cage also received a layer of substrate and eelgrass clumps covering about half the cage bottom. In spite of the fact that the corner poles were clearly marked with large labeled flags, one of the cages was severely damaged at night by a passing motorboat. Therefore, only two of the cages provided growth data.

Pinfish for the live-cage study were collected from the adjacent grass bed by means of a beach seine. One cage received six fish of 35.0 mm, and the other received five fish of 45.0 mm initial body length. At 10-day intervals the fish were measured and the substrate and eelgrass were replaced with new material.

Growth data from the live cages are summarized in Table 5. Within each cage individual fish exhibited different rates of growth. However, on the average a fish initially 35.0 mm in body length had achieved a length of 45.1 mm at the end of 40 days. The average daily growth of these fish was 0.252 mm in body length and 34 mg in

wet body weight. Fish initially 45.0 mm in body length were maintained for a period of 60 days at the end of which time they averaged 59.4 mm. The average daily growth for this group was 0.240 mm
in body length and 53 mg in wet body weight.

Ignoring the 0.1 mm overlap between the final size of the first
group and the initial size of the second (a factor which represents
less than one-half day's growth), a 100-day growth curve was constructed by placing the two curves end to end. Thus, at the end of
100 days a 35.0 mm fish would have increased to 59.4 mm for an
average daily growth rate of 0.244 mm and 45.5 mg.

Throughout the entire 100-day period the daily rate of increase
in body length was almost constant, being slightly faster during
the first 40 days than during the last 60 days. Body weight, of
course, increased much faster in the larger sized fish. Average
percent daily growth (expressed as a percentage of the average body
size during the growth period) was far greater in the smaller sized
fish. These results are expressed graphically in Figure 7, which
also includes the results of the tagging experiment.

Weekly sampling studies. Change in mean body size of young pinfish
on the eelgrass bed may reflect growth, differential mortality,
differential population movement (recruitment or emigration), or a
combination of these factors. However, coupled with other information regarding growth it could be expected to provide information
concerning the dynamics of the resident population. Therefore, an
effort was made to obtain weekly samples with standard collecting
gear (a 40-ft beach seine with 1/4-in mesh and containing a bag
with 3/8-in mesh and a heavily weighted lead line). At moderately
low tide this gear was exceedingly effective, and seldom was it
necessary to make more than one haul to obtain the necessary number
of fish.

Although not every week was represented, during three successive summers (1961 through 1963) sufficient samples were taken to
provide a clear picture of the progressive size increases during the
first 8 wks (i.e., during June and July). During 1961 the samples
also continued through the month of August, providing a 12-wk record.
All initial collections were made between June 4 and 8. Samples
included from 46 to 211 fish; most contained between 60 and 125
individuals.

Regression lines depicting the moving sample means for the
three summers are given in Figure 8. Increase in mean body length
appeared to be essentially linear during this early growth period,
but tests for linearity were not made. However, beginning about
the eighth week, a marked reduction in the trend became obvious,
and this was due largely to the absence of the expected larger sized
fishes (i.e., from about 65 mm on up). Therefore, a separate
regression line was calculated for the 8- to 12-wk period of the
1961 samples. It had originally been planned to follow modal sizes
in the samples. These modes were quite pronounced in the earlier
weeks, but they became totally indistinguishable during the second

4-wk period. Therefore, recourse was taken to the use of sample means.

Fig. 8. Estimated growth in the field population during three consecutive summers based upon mean size in weekly samples. Growth in the live cages and in the lab-feeding experiments are included for comparison. Relevant data are included below.

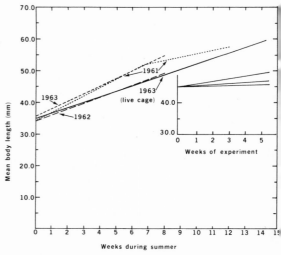

1961 (< 52 mm)	y = 34.52 + 2.61x	(0.374 mm/day)
1961 (> 52 mm)	y = 46.04 + 0.95x	(0.137 mm/day)
1962	y = 34.08 + 1.87x	(0.267 mm/day)
1963	y = 35.74 + 2.38x	(0.340 mm/day)
Live cage		(0.244 mm/day)
Lab (12.64% feeding rate)		(0.113 mm/day)
Lab (7.56% feeding rate)		(0.05 mm/day)
Lab (5.05% feeding rate)		(0.02 mm/day)

Examination of Figure 8 reveals that during the first week of June the mean size of the pinfish was about the same during all three years (i.e., in the range of 34.0 to 36.0 mm). During 1961 and 1963 the size increase rates were rapid and fairly similar (0.374 mm and 0.340 mm per day, respectively). During 1962 the increase rate was considerably less (0.267 mm per day). However, the latter figure still slightly exceeded the average rate of growth of the live-cage fish (0.244 mm per day).

Interpretation of the weekly increase data is somewhat difficult, but there is no reason to suspect that they do not represent true growth. Admitting the possibility of differential mortality of the smaller sized individuals by predators known to be present, it has also been clearly established that growth rates in the pinfish can be dramatically affected by the food supply, and this would appear to be the overriding consideration. This factor could account for annual differences in growth rate as well as differences between live-cage and general field growth. Field notes made during June of 1962 (the slow growth summer) lend additional support to this supposition. During that summer small pinfish were far easier to catch in large numbers, and it appeared that the density of the resident population was unusually high. Gulf pipefish, sea bass, and pigfish were also present in greater quantity, and a greater diversity of marine forms was indicated by the presence of half-beaks, cornetfish, northern pipefish, groupers, bluefish, pompano, sea robins, flounders, burrfish, and brown shrimp. The planehead filefish, always abundant, also appeared to be stunted that summer. Taken together, the available evidence points to unusually keen competition for the limited food supply.

We conclude that the weekly increase in mean body length shown in Figure 8 provides the best available representation of actual growth by the field population. Young pinfish in the 35 to 36 mm group apparently grew at different rates (i.e., between 0.267 and 0.374 mm/day) depending upon the available food supply in relation to population size and food demand during the particular summer growing season. Thus, growth in the live cages (0.244 mm/day) probably represents the near-minimum field growth rate which was inadvertantly restricted by food availability within the cages.

Surrounded by channels, the grass bed is apparently subject to little pinfish recruitment after the beginning of June. However, depression of the curve in the 8- to 12-wk samples (of 1961) clearly reflects loss of the larger sized fish which leave the bed for deeper waters at a body length of 60 to 70 mm.

Population Estimates

Laboratory studies indicated that clipping either the anterior or posterior third of the anal fin was a quick, easy operation which produced little mortality. Although regeneration was rapid, the clipped fin could readily be recognized for at least 3 wks and, thus, provided an ideal marker for field population estimates. However, preliminary field studies clearly demonstrated that fin-clipped fish moved very little and that they could not be expected to randomize themselves throughout the population. Therefore, an effort was made to redistribute the fish throughout the one-acre study area following the clipping procedure.

During the actual operation, fish were seined from the bed and held in plastic buckets in a floating skiff centered on the east edge of the grass bed. After a given fish was clipped, it was held in a floating live cage (0.6 m × 2.4 m) at the side of the skiff. After all fish were clipped they were distributed back onto the beds in lines radiating out from the skiff; an effort was made to release fish at intervals along each distribution line so that all portions of the bed would receive an approximately equal represen-tation. Undoubtedly, the center section of the bed near the skiff received a somewhat greater density of marked fish than did the periphery. This was not considered to be a serious problem since retrieval of the fish was to be accomplished by seining directly across the bed (from west to east between marked stakes) at angles to the distribution lines. Since the entire bed was seined at each census period it was assumed that each fish in the population had an approximately equal opportunity of appearing in the samples.

A further check on this procedure was provided by a repeated census.

On 19 July 1963 a total of 493 fish were marked and distributed as indicated above. The next day (after a full tidal cycle) the population was sampled, the number of marked and unmarked fish was noted, and the fish were redistributed along radiating lines. Two days later the census was repeated. On 19 August 1963 (after the original fins had regrown), 326 fish were marked and distributed. The population was censused the following day.

Results of these studies are presented in Table 6. The nearly identical results of the census and the recensus for the July estimate (within 0.3%) is taken as validation of the procedures employed. Thus, a population of nearly 17,000 fish per hectare in late July had dwindled to about 8,700 per hectare by late August. These losses (amounting to 48.2%) were due in some measure to predation, but the major loss is assumed to be desertion of the grass beds by the maturing fish. If one may assume a 24.1% monthly loss due to predation alone throughout the summer, the late June population may be estimated to have been around 22,590 fish per hectare. All evidence suggests that predation was heaviest on the younger fish, and although these earlier estimates are considered minimal, they at least provide order-of-magnitude figures for the early summer.

TABLE 6
Estimation of field population size by the Petersen method[*]

Date (1963)	No. of fish marked	No. of fish captured	No. of marked fish recaptured	Population estimate fish/hectare
July 19	493			
July 20		971	70	16,898
July 22		1,079	78	16,852
Aug. 19	326			
Aug. 20		391	36	8,749

[*]Size of the sampling area was 0.40 hectare.

On the basis of field collecting data and the estimated growth rates, 1963 was considered to have been an ordinary summer fairly comparable to 1961. The year 1962, however, was marked by high population density which is estimated to have been at least 1.5 to 2.0 times that of 1961 or 1963. Thus, in 1962 the late May population may have had nearly 50,000 fish per hectare, while the late August population would have had over 17,000 fish per hectare since a lower percentage of the slower growing fish would have left the grass bed sanctuary by this part of the summer.

It will be recalled that the 1962 field growth rate only slightly exceeded that of live-cage fish. However, the density of the live-cage fish was 10.8 fish/m^2, whereas the highest density postulated for the 1962 population (late May) was 5.0/m^2. Therefore if space and associated food supply were the limiting factors in both cases, the estimate of the 1962 field population is likely to be quite conservative. A closer approximation would be 75,000 fish per hectare. A higher initial population would certainly be accompanied by a higher, but unknown, mortality rate, so for purposes of further calculations we will employ the earlier and more conservative figure.

Turnover Rates

It has been shown that in the field populations, growth (expressed as daily increase in body length) is essentially linear and that the respective growth rates for the three summers were: 1961, 0.374 mm/day; 1962, 0.267 mm/day; and 1963, 0.340 mm/day. From this information it may be determined that the times required for the three populations to gain 10.0 mm in body length were 26.75 days (1961), 37.50 days (1962), and 29.43 days (1963). Given the body weight at each 10.0 mm size interval (Fig. 9), it is possible to determine the total increase in body weight during each 10.0 mm interval and during each time period. From this information the average daily gain in body weight during the interval may be calculated. Table 7 provides estimated daily growth rates for the field populations during the three summers. Data are broken down by size class intervals, and growth rates are expressed in terms of wet, dry, and organic nitrogen weights.

Fig. 9. Relationship between wet body weight and standard body length in the pinfish.

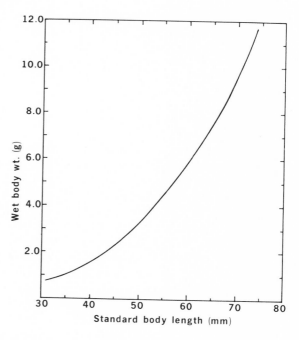

From the known growth rates and an assumed average 10% food conversion efficiency at the higher feeding levels (see Fig. 6), it is possible to estimate the daily rate of food intake (in terms of wet weight) by fishes of each size class in the field populations. The wet weight values may, in turn, be converted to dry weights and organic nitrogen weights if one assumes that the earthworms (Figs. 3 and 4) used in the laboratory are reasonably representative of the field food. The feeding rates may also be expressed as the percentage of body weight/day of food ingested. These calculations have been carried out, and the estimated daily food intake rates are given in Table 8.

TABLE 7
Estimation of daily growth rates of pinfish of the different size class intervals for the summers of 1961, 1964, and 1963

| Summer | Size class interval (mm) | Avg. daily gain in body wt within size class interval[*] | | |
		Wet wt (g)	Dry wt (g)**	Org. nitrogen wt (g)†
1961	35.0-45.0	0.0505	0.0112	0.0013
	45.0-55.0	0.0748	0.0166	0.0019
	55.0-65.0	0.1121	0.0249	0.0029
1962	35.0-45.0	0.0360	0.0080	0.0010
	45.0-55.0	0.0533	0.0118	0.0014
	55.0-65.0	0.0800	0.0178	0.0021
1963	35.0-45.0	0.0459	0.0102	0.0012
	45.0-55.0	0.0680	0.0151	0.0018
	55.0-65.0	0.1019	0.0226	0.0026

[*]Body length-body weight relationships upon which this table is based include the following: 35.0 mm, 1.15 g; 45.0 mm, 2.50 g; 55.0 mm, 4.50 g; 65.0 mm, 7.50 g.
**Values derived from Fig. 2.
†Values derived from Fig. 2.

Several sources of error inherent in the above calculations deserve mention. As indicated in Table 5 the daily increase in body length is not absolutely linear, but tends to decelerate slightly as the body size increases. The 3% error arising from this source is considered to be negligible. The second error derives from the fact that all length-weight relationships are taken from the same curve (Fig. 9). It is widely known, however, that these relationships will vary with nutrient conditions, with a well-fed fish being plumper and heavier than a semistarved fish of the same length. This phenomenon, recognized by fishery biologists as "coefficient of condition," suggests that the length-weight relationships of the pinfish vary from year to year depending upon population density in relation to food supply. It may be postulated that the fish were relatively heavier during the fast growth years when the population density was lowest, but in the absence of food availability data there is no objective basis for

such speculation. Finally, the 10% food conversion efficiency
value, derived from the laboratory-feeding experiments, may not be
absolutely valid. The earthworms employed in the laboratory may
have been more or less nutritious than the field food, and the
maximum feeding rate observed in the laboratory was considerably
less than most feeding rates postulated for the field population.
One may assume that the conversion efficiency tends to vary with
the type and quantity of food ingested, but again without objective
methods of evaluating these potential sources of error we are forced
to stick with the data as presented.

TABLE 8

Estimation of daily food intake by pinfish of different size class intervals for the summers of 1961, 1962, and 1963

Summer	Size class interval (mm)	Estimated daily food intake rate*			
		Wet wt (g)	Dry wt† (g)	Org. nitrogen† wt (g)	Wet wt (% mean body wt)
1961	35.0–45.0	0.505	0.086	0.0098	27.7
	45.0–55.0	0.748	0.127	0.0145	21.4
	55.0–65.0	1.121	0.191	0.0218	18.7
1962	35.0–45.0	0.360	0.061	0.0070	19.7
	45.0–55.0	0.533	0.091	0.0104	15.2
	55.0–65.0	0.800	0.136	0.0155	13.3
1963	35.0–45.0	0.459	0.078	0.0089	25.1
	45.0–55.0	0.680	0.116	0.0132	19.4
	55.0–65.0	1.019	0.173	0.0197	17.0

*Estimated from body weight increase data, assuming that at the higher
feeding rates the pinfish exhibits a 10% wet weight food conversion
efficiency (from Fig. 6).

†Values (calculated from Figs. 3 and 4) based on the assumption that, on the
average, the field food has the same dry weight and nitrogen concentrations
as do the earthworms used in the lab-feeding experiments.

The remarkable fact revealed in Table 8 is the exceedingly
high feeding rates estimated for the field populations. All of the
values exceed the average ad libitum feeding rate observed in the
laboratory, and some of the values are twice the ad libitum rate.
These values are considerably higher than most food intake rates
reported in the literature. The table also reveals that whereas
the quantity of food in the daily ration increases with body size,
the feeding rate expressed as percentage of body weight decreases
about 28.7% between the first two size classes and about 12.5%
between the middle and largest size classes.

When the discrepancy between the ad libitum laboratory and the
calculated field feeding rates was discovered, the last summer was
nearly over, but an experiment was set up to determine whether fish
in a group feed more freely than do fish kept in complete isolation.

Results of this experiment, together with data from the previous
ad libitum study (Table 2), are presented in Table 9. Despite the
shortness of time the results are quite informative. Vision and
behavioral interaction are clearly important factors in determining
the quantity of food the fish will ingest. In isolation the fish
does not consume enough to support itself (as pointed out earlier,
the subsistence diet is 5.75% of the body weight/day). In a freely
interacting group the fish will consume nearly a third of its body
weight/day, which is a higher feeding rate than that postulated for
any of the field populations. From these results and from our
total experience with this fish, we conclude that the calculated
field feeding rates are realistic values.

TABLE 9
Rate of food intake of laboratory fish, maintained on ad libitum feeding, as
a function of degree of isolation

Feeding group	No. of fish	Length of experiment (days)	Ad libitum feeding rate (% body wt/day)
A*	4	7–8	4.2
B**	4	40	12.6
C†	6	8	32.6

*Individual fish were completely isolated from one another; no other fish
visible.

**Fish were individually isolated; three or four other fish were always visi-
ble through transparent plastic walls (data obtained from Table 2).

†Fish were placed together in an aquarium and were free to interact.

Estimates may now be made of the standing crop, daily food
intake, and daily gain in body weight for the fish populations on
the July and August census dates in 1963. By reference to the
field growth curves, it is found that the July 19–20 fish correspond
to a mean body length of 52.5 mm (which translates to a mean body
weight of 3.87 g). Reference to Figure 10 shows that this corresponds to a daily food intake rate of 0.748 g, which represents a
daily growth rate of 0.0748 g at the 10% conversion efficiency.
Only the 1961 population size-group curve extends into late August,
but since the 1961 and 1963 curves are quite close our best estimate
of the mean size of individuals in the August 1963 population is
derived from the 1961 curve. On this basis the August 19–20 fish
are assumed to average 57.5 mm in length and to weigh 5.18 g. This
translates to a daily food intake of 0.920 g and a daily growth
rate of 0.092 g.
Expanding the above data to include the dry and organic nitrogen weights and multiplying by the appropriate census figures, one
derives information concerning standing crop, daily food intake,

and daily gain in weight in the field population. In Table 10
these results are expressed on a per hectare basis. From this table
it may be seen that in July, for a given hectare of grass bed habitat, a standing crop of 16,898 pinfish weighs 65.4 kg and daily
consumes 12.6 kg of food from which it gains 1.3 kg in body weight.
In the August population for a hectare, 8,749 fish weigh 45.3 kg,
and each day they consume 8.0 kg of food from which they gain 0.8 kg
in body weight. Corresponding figures concerning dry and organic
nitrogen weights are also given in the table. Comparing the two
groups, the August population contains only slightly more than half
the number of individuals as the July population (51.8%), but being
larger fish they constitute over two-thirds the biomass (69.3%),
and they consume over three-fifths the quantity of food (63.7%).

Fig. 10. Estimated daily
food intake rates in the
field populations and
the live-cage fishes in
relation to body length.
Data from the
experiments are included
for comparison.

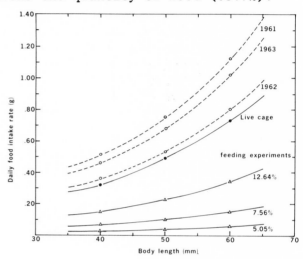

TABLE 10
Standing crop, daily food intake, and daily growth in the field population
during the mid- and late-summer of 1963[*]

Date	Measure	Standing crop g/ha	Daily food intake[†] g/ha/day	Daily gain in weight[†] g/ha/day
July	Wet wt	65,394.1	12,639.5	1,263.94
	Dry wt	16,221.8	2,162.9	280.50
	Org. N. wt	1,858.7	246.7	33.80
Aug.	Wet wt	45,320.3	8,049.2	804.92
	Dry wt	11,548.8	1,373.6	178.48
	Org. N. wt	1,338.6	156.6	21.00

[*]Figures based on the field census estimates of July 19-20 and August 19-20,
1963 (Table 6).
[†]Extrapolated from Fig. 9 and from Tables 7 and 8.

As pointed out earlier, between the July and August census
dates the population was being diminished by predation as well as

by departure of the larger fish (65 mm and over) for the deeper waters. Although the latter is considered to be the predominant factor, present data do not permit quantitative assessment of the relative effects of the two factors. Rough estimates of the population levels in June 1963 and during the summers of 1961 and 1962 were given earlier, but such estimates do not provide a sufficiently firm basis for further estimation of standing crop and turnover rates during those time periods.

DISCUSSION

To provide a point of reference for interpretation of the dynamics of the estuarine fish population the critical data are compared in Table 11 with similar information derived from a fresh-water bluegill population studied by Gerking (1962, 1972). In terms of wet weight and protein weight, the midsummer standing crops of the two populations were remarkably similar. By contrast, the pinfish population consisted of nearly nine times the number of individuals, and the estimated food consumption rate was between three and four times as great. The bluegill population inhabited a small self-contained lake along with several competing species, and the estimates included only the larger size classes in the population (i.e., over 90 mm). The bluegill population was known to be slow-growing, and it was apparently adjusted to maximum utilization of the locally produced invertebrate food supply.

TABLE 11
Comparison of standing crop and food consumption of the estuarine pinfish population with similar data on the bluegill population of Wyland Lake, Indiana

Comparison	Pinfish Population[*]	Bluegill Population[**]
Standing crop		
Number of individuals	16,900	1,975
Wet weight (kg)	65.4	72
Protein weight (kg)[†]	11.6	12.2
Daily food consumption		
Total food consumed (kg, wet wt)	12.6	4.0
Protein consumed (kg)	1.5	.38

[*]July, 1973 data are taken as the most representative of the average pinfish population. All data in table are given on a per hectare basis.
[**]The average summer values are employed in the comparison. Data are from Gerking, 1962.
[†]Protein weight was calculated at 6.25 times the organic nitrogen weight.

The pinfish population consisted entirely of small, fast-growing young-of-the-year individuals capable of exploiting a variety of food resources including plant and animal material as well as organic detritus (Darnell, 1970) and at extremely high rates of availability. Although a number of competing species were present, none achieved anywhere near the local abundance of the young pinfish which completely dominated the fauna of the weed bed.

The value (12.6 kg of food consumed/hectare/day) must stand as one of the highest ever recorded in the literature of aquatic populations. Whether one accepts the use of mean size of field-captured individuals to estimate field growth rates, the live-cage study at least indicates that the values are not unreasonable. Obviously such feeding and growth rates could not be long sustained on the basis of local production alone, but must rest, in large measure, upon imported organic matter. It is well known that weed beds in flowing waters act as sediment and plankton traps (Reif, 1939), but the *Zostera* bed must be especially efficient in this regard. Each blade of the *Zostera* plant consists of a long strap coated with filamentous algae which brushes the tidal water flowing past. The twice-daily tide results in four periods of estuarine tidal flow (one on each rise and one on each fall). Once trapped, the suspended matter that has not been caught on the *Zostera* "brushes" would tend to sink to the bottom due to the lower flow rates prevailing within the bed. Williams and Murdoch (1966) determined that the annual gross primary production for the euphotic zone of the nearby Beaufort Channel is 225 g C/m^2. They noted the presence of quantities of suspended organic detritus in the channel waters, and they pointed out that the channel obtains nutrients from forests and agricultural lands in the drainage basin of the Newport River, wastes from upstream fish processing plants, and raw sewage from communities totaling several thousand people. The grasses, filamentous algae, and diatoms of the bed itself undoubtedly exhibit high rates of primary production, and when supplemented by the waterborne exogenous organic material the bed is capable of supporting a remarkable fish population.

The respiration of pinfish from Texas populations has been investigated (Wohlschlag and Cameron, 1967; Wohlschlag, Cameron, and Cech, 1968). Although they worked with larger sizes of fish than we did, their findings are of particular interest in light of the present investigation. The Texas studies clearly demonstrated that the scope for routine activity (*sensu* Fry, 1957) decreases with temperature, especially for summer-acclimated fish. Temperature, thus, imposes a metabolic limitation on the larger fish which undoubtedly explains the retreat of larger pinfish from the warm shallow weed beds. Considering the energy which must be expended by the smaller fish in their daily search for food, however, they must be adapted for high-temperature activity. Temperature-related respiratory studies on the smaller size classes should prove highly instructive.

These researchers noted high variability in their respiration-weight data, and they postulated that such variability may be associated with behavioral changes (e.g., feeding, spawning, migrations) at seasons of low or rising temperatures. Our own data show high variability in individual growth rates, to some extent in the laboratory-feeding experiments, but to a remarkable extent in the live-cage studies. The same phenomenon is exhibited in the natural field population where modal size classes in weekly samples are quite evident in June but indistinguishable thereafter. We tentatively conclude that within any aggregation of pinfish, group metabolism becomes programmed so that a spectrum of feeding-growth responses will result. The observed variability probably reflects genetic-based metabolic differences exaggerated by a dominance-subordination behavioral pattern. This would tend to spread out the size classes and thus reduce intraspecific food competition, insuring more complete utilization of the available food resources. It is common practice in physiological and ecological investigations to deal with averaged data (and we have done the same), but if the above supposition is correct, data averaging may be masking an important ecological phenomenon.

Related to the above point is the finding that the pinfish feeds maximally in a group and will essentially starve to death when deprived of the stimuli associated with other fish feeding. We note that most of the respiration experiments mentioned above were carried out on isolated fish, and the few group experiments took place in the dark. Group experiments in the light should certainly be carried out for comparative purposes if the scope for activity of fishes in the natural population is to be estimated.

The above remarks imply that behavior may play an important role in the physiological processes of this estuarine fish. They also imply that the physiology may change dramatically during the life history of a given individual. How widespread these phenomena may be in other animals remains to be discovered. It must be apparent to all, however, that information derived from natural populations, and particularly from field-related experiments, will greatly aid in determining the relevance of laboratory investigations in the physiological ecology of estuarine organisms.

ACKNOWLEDGMENTS

This study was supported by NSF Grants G-10865, G-22561, and GB-813 awarded to R. M. Darnell, and the investigation was conducted at the Duke University Marine Laboratory, Beaufort, North Carolina. The following persons assisted in portions of the study: J. S. Balsano, G. Barrett, R. Bielski, R. Engstrom, J. Resko, and the late V. Fioravanti.

LITERATURE CITED

Caldwell, D. K.   1957.   The biology of the pinfish *Lagodon rhomboides* (Linnaeus).   Bull. Fla. St. Mus. Biol. Sci. 2: 77-173.

Carr, W. E. S. and C. A. Adams.   1973.   Food habits of juvenile marine fishes occupying seagrass beds in the estuarine zone near Crystal River, Florida.   Trans. Amer. Fish. Soc. 102: 511-40.

Darnell, R. M.   1958.   Food habits of fishes and larger invertebrates of Lake Pontchartrain, Louisiana, an estuarine community.   Publ. Inst. Mar. Sci., Univ. Texas 5:353-416.

_____.   1961.   Trophic spectrum of an estuarine community, based on studies of Lake Pontchartrain, Louisiana.   Ecol. 42:553-68.

_____.   1964.   Organic detritus in relation to secondary production in aquatic communities.   Verh. Intern. Verein. Limnol. 15: 462-70.

_____.   1970.   Evolution and the ecosystem.   Amer. Zool. 10:9-15.

Fry, F. E. J.   1957.   The aquatic respiration of fish.   In: The Physiology of Fishes (M. E. Brown, ed.), pp. 1-63.   New York: Academic Press.

Gerking, S. D.   1952.   The protein metabolism of sunfishes of different ages.   Physiol. Zool. 25:358-72.

_____.   1954.   The food turnover of a bluegill population.   Ecol. 35:490-98.

_____.   1955.   Endogenous nitrogen excretion of bluegill sunfish.   Physiol. Zool. 28:283-89.

_____.   1962.   Production and food utilization in a population of bluegill sunfish.   Ecol. Monogr. 32:31-78.

_____.   1972.   Revised food consumption estimate of a bluegill sunfish population in Wyland Lake, Indiana, U. S. A.   J. Fish. Biol. 4:301-8.

Gunter, G.   1945.   Studies on marine fishes of Texas.   Publ. Inst. Mar. Sci., Univ. Texas 1:1-190.

Hawk, P. B., B. L. Oser, and W. H. Summerson.   1954.   Practical Physiological Chemistry.   New York: McGraw-Hill.

Karsinkin, G. S.   1935.   Zur Kenntniss der Fischproduktivität der Gewässer.   II. Mitteilung. Erforschung der Physiologie der Ernährung des Spiegelkarpfens.   Kossino Limnol. Stat. 19:21-66.

Ketchum, B. H.   1967.   Phytoplankton nutrients in estuaries.   In: Estuaries (G. H. Lauff, ed.), pp. 329-35.   AAAS Publ. No. 83.

Lang, C. A.   1958.   Simple microdetermination of Kjeldahl nitrogen in biological materials.   Analyt. Chem. 30:1692-94.

Menzel, D. W.   1959.   Utilization of algae for growth by the angelfish. *Holacanthus bermudensis*.   Jour. du Conseil 24: 308-13.

_____.   1960.   Utilization of food by a Bermuda reef fish *Epinephelus guttatus*.   Jour. du Conseil 25:216-22.

Odum, E. P. and A. A. de la Cruz. 1967. Particulate organic detritus in a Georgia salt marsh-estuarine ecosystem. In: Estuaries (G. H. Lauff, ed.), pp. 383-88. AAAS Publ. No. 83.

Pandian, T. J. 1967a. Food intake, absorption, and conversion in the fish *Ophiocephalus striatus*. Helgoländer wiss. Meeresunters. 15.

_____. 1967b. Intake, absorption, and conversion of food in the fishes *Megalops cyprinoides* and *Ophiocephalus striatus*. Mar. Biol. 1:16-32.

_____. 1967c. Transformation of food in *Megalops cyprinoides*. I. Influence of quality of food. Mar. Biol. 1:60-64.

Reif, C. B. 1939. The influence of stream conditions of lake plankton. Trans. Amer. Micros. Soc. 58:398-403.

Tagatz, M. E. and D. L. Dudley. 1961. Seasonal occurrence of marine fishes in four shore habitats near Beaufort, N. C., 1957-60. USF&WS, Spec. Sci. Rept.-Fisheries 390:1-19.

Teal, J. M. 1962. Energy flow in the salt marsh ecosystem of Georgia. Ecol. 43:614-24.

Williams, R. B. and M. B. Murdoch. 1966. Phytoplankton production and chlorophyll concentration in the Beaufort Channel, North Carolina. Limnol. Oceanogr. 11:73-82.

Wohlschlag, D. E. and J. N. Cameron. 1967. Assessment of a low-level stress on the respiratory metabolism of the pinfish (*Lagodon rhomboides*). Contrib. Mar. Sci. 12:160-71.

_____, _____, and J. J. Cech, Jr. 1968. Seasonal changes in the respiratory metabolism of the pinfish (*Lagodon rhomboides*). Contrib. Mar. Sci. 13:89-104.

Factors controlling metabolic capacity adaption in marine invertebrates

R. C. Newell

The great complexity of environmental and other factors which affect the metabolism of invertebrates is well known and has been the subject of several reviews (Bullock, 1955; Prosser, 1955, 1958; Precht, 1958; Precht, Christophersen, and Hensel, 1955; Newell, 1970). In some cases a simple relationship has been established between changes in metabolism and those of one particular environmental factor such as temperature or salinity. Such responses may be defined as "acclimation." More often, however, many environmental influences may act together to produce a change in the metabolism of the intact organism. These latter responses have been distinguished as "acclimatization" (Prosser and Brown, 1961). In an effort to understand the nature of these factors, physiologists have commonly been forced by experimental design and analytical techniques to consider each parameter separately from the rest. This has often led to an overly simplistic view of factors controlling metabolism and, more important, has tended to make the results of laboratory studies less applicable to the environmental situation than one would wish.

Partly in an effort to overcome these difficulties, we have over the past few years made a fairly intensive study of factors affecting the metabolism of intertidal invertebrates. We have found the gastropod *Littorina littorea* a useful animal but more recently have been extending our work to include the supralittoral isopod *Ligia oceanica*, which has many advantages over *Littorina* as an experimental animal. In this review, however, I should like to

present a summary of the results of the analysis of data on
Littorina which I have gathered with colleagues at our laboratory
and to attempt to relate this to work which we have also carried
out on subcellular preparations of this animal. The analysis of
the data for the oxygen consumption of intact *Littorina* has been
carried out in collaboration with A. Roy (Newell and Roy, in press)
and suggests one way in which the complex relationship between
environmental factors and invertebrate metabolism may be interpreted
The second problem which poses difficulties for the environmental physiologist is the extent to which studies made at the subcellular and biochemical level can be related to the whole organism.
It is often easier from an experimental point of view to study the
metabolic rate of mitochondrial preparations or even the reaction
rates of isolated enzyme systems than the metabolism of the whole
animal, but real difficulties are encountered in relating the result
of such studies to the intact organisms. We do have, however, considerable data which suggest the way in which subcellular events
may control the responses of the metabolism of intact *Littorina* to
temperature. These suggest that it would be worthwhile to make an
integrated study of the effect of environmental factors on the
metabolism of not only intact invertebrates, but also of their subcellular components and isolated enzyme systems.

FACTORS AFFECTING THE OXYGEN CONSUMPTION OF INTACT ORGANISMS

The data obtained on *Littorina* amounted to a total of 697 different observations on the rate of oxygen consumption of animals of
different sizes at four seasons of the year and at various exposure
temperatures. Of these, 395 of the rates were obtained on active
animals and 302 were for quiescent animals. Regression lines were
then fitted to the data relating log oxygen consumption to log dry
tissue weight and these clearly showed that not only the level but
also the slope was influenced by activity level (Ac), day of the
year (Da), and exposure temperature (Te). The multiple regression
procedure which we have used (Newell and Roy, in press; Fredette,
Planté, and Roy, 1967; Roy, 1969) then yielded the following general
equation which related log metabolism (Lm) to log dry weight (Ld),
activity (Ac), exposure temperature (Te), day of the year (Da), as
well as their mutual interactions:

$$Lm = -0.003873 \tag{1}$$
$$+0.3029. \quad \text{log dry weight, mg (Ld)}$$
$$+0.6286. \quad \text{Activity (Ac)}$$
$$+0.03359. \quad \text{Exposure temperature (Te)}$$
$$-0.4441. \quad 10^{-4} \; Te^3$$
$$+0.8539. \quad 10^{-6} \; Te^4$$
$$+0.007743. \quad Ld.Te$$
$$+0.001123. \quad Ac.Da$$
$$-0.8523. \quad 10^{-4} \; Te.Da$$

The multiple correlation coefficient of this multiple regression is 0.9491. This suggests that most of the relevant factors influencing metabolism in *Littorina littorea* have been taken into account, although the use of day of the year may summarize a complex of other environmental factors including environmental temperature and day length. From Equation (1) it is possible to calculate the effect of body size (Ld), time of the year (Da), and exposure temperature (Te) on metabolism.

The Influence of Body Size

It is well known from the studies of Zeuthen (1947, 1953), Hemmingsen (1950, 1960), and many other workers that the slope of the common regression line relating log metabolism to body weight of protozoans through to the larger poikilotherm metazoans is approximately 0.75. There are, however, many instances where significant deviations from this mean value occur in response to temperature or aerial and aquatic conditions. Rao and Bullock (1954), for example, reviewed the occurrence of such deviations in the earlier literature and showed that the slope (b) varies directly with exposure temperature in some invertebrates, such as the sand crab *Emerita* and the amphipod *Talorchestia*. In contrast, the reverse relationship is found in *Megascolex* (Saroja, 1959), *Uca* (Vernberg, 1959), *Arion* (Roy, 1969), and in several barnacles (Barnes and Barnes, 1969). The curves obtained for *Littorina* collected on 5 May 1970 and measured at various exposure temperatures between 5°C and 35°C can be calculated from Equation (1) and are shown in Figure 1. These illustrate several features of importance in an interpretation of the influence of environmental factors on the metabolism of this animal. First, the slopes of the lines for active and quiescent animals both vary directly with exposure temperature. The values for the slopes are similar to those reported by Toulmond (1967a, b) for several species of littorinids, and the influence of exposure temperature thus resembles the results for *Emerita* and *Talorchestia* cited by Rao and Bullock (1954). Second, the influence of exposure temperature is more pronounced on large animals than small ones; and third, exposure temperature results in a greater increase of metabolism at the extremes of the temperature range than at intermediate values. To put in another way, the temperature coefficients for metabolism of active and quiescent *Littorina* are rather low at intermediate environmental temperatures.

The Influence of Time of the Year

Analysis of the data obtained on *Littorina* also shows that seasonal changes in metabolism occur and are complicated by the effects of exposure temperature and activity. In general, there is a reduction in the metabolism with advancing season from February to July and in this respect the response is a typical acclimatory

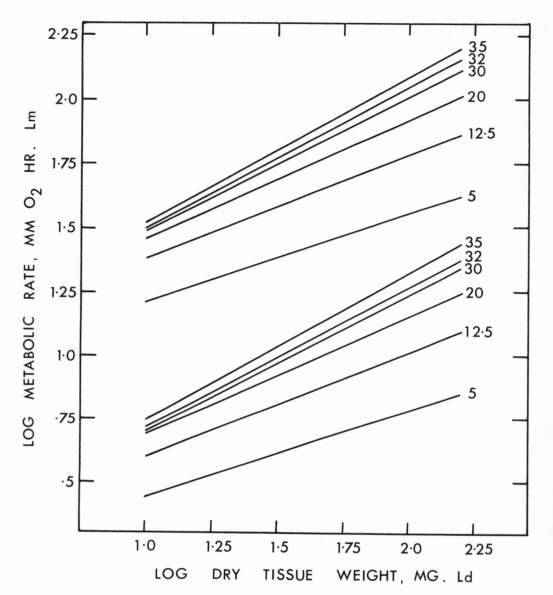

Fig. 1.   Graphs showing the relationship between log metabolism
(Lm) and log dry tissue weight (Ld) in mg in *Littorina
littorea* collected in May 1970 and measured at exposure
temperatures (Te) between 5°C and 35°C. Regression lines
calculated from Equation (1).   (Newell and Roy, unpublished
data.)

one. Superimposed upon such a suppression, however, are changes in the slopes of the regression lines relating log metabolism (Lm) and Activity (Ac) from 0.6915 in February to 0.7690 in May, 0.8049 in June, and 0.8431 in July 1970. It can be shown that in February the rate of oxygen consumption for active animals would have been 4.9 times that for a quiescent one of the same size and at the same exposure temperature, whereas this value increases to 7.0 in July. Thus, the "scope for activity" (Fry, 1957) is maximal during the summer months and cannot necessarily be considered to have a fixed value for a particular organism.

The Influence of Exposure Temperature

The influence of exposure temperature on the metabolism of intact *Littorina* is of particular interest because it can be most directly related to work which we have carried out on subcellular preparations of this and other intertidal animals. There is also a wealth of data on the influence of exposure temperature on the metabolism of organisms from many different environments.

The relationship between exposure temperature and log metabolism is not a simple rectilinear one. The curves are of sigmoid form with a region of relative temperature-insensitivity over parts of the environmental temperature range and correspond to the following expression which is derived from Equation (1):

$$Lm = a + (0.03359 + 0.07743.Ld - 0.00008523.Da).Te \qquad (2)$$
$$- 0.4441 \times 10^{-4}Te^3 + 0.8539 \times 10^{-6}Te^4$$

The sigmoid relationship is, however, modified by both body size (Ld) and time of the year (Da). The metabolism of large animals is somewhat more affected by exposure to temperature than that of small animals and the metabolism of all size ranges is suppressed with the passage of time from February to July. Figure 2 shows the rate:temperature curves of *Littorina* of 160 mg dry tissue weight during February, May, June, and July. The progressive alteration in the form of the curve from a relatively temperature-dependent one early in the year to one which has a pronounced region of temperature-insensitivity in the summer is in accordance with our earlier observations on this and other intertidal animals (Newell, 1969; Newell and Pye, 1970a, b, and 1971). It follows from this that the temperature coefficient (Q_{10}) for metabolism is at a minimum at the normal environmental temperature of the animal as is shown in Figure 3. Very similar types of rate:temperature relationships have been reported in many intertidal invertebrates, including the sea urchin *Strongylocentrotus purpuratus* (Ulbricht and Pritchard, 1972), the crab *Callinectes sapidus* (Leffler, 1972), the isopod *Ligia* (Wieser, 1972; Newell and Armitage, unpublished data), barnacles (Barnes and Barnes, 1969), as well as in a spider *Lycosa* (Moeur and Eriksen, 1972) and many other organisms (Newell, 1970, 1973a, b). However, in some subtidal organisms, and in those which burrow and thus evade temperature stress, more temperature-dependent rates of metabolism are obtained (Mangum and Sassaman, 1969).

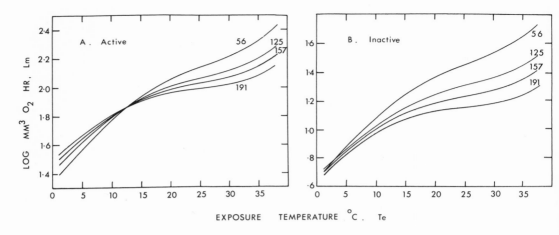

Fig. 2. Graphs showing the effect of exposure temperature ($\underline{Te}$) on log metabolism ($\underline{Lm}$) in *Littorina littorea* collected in February (day 56), May (day 125), June (day 157), and July (day 191) 1970. Data for a hypothetical animal of 160 mg dry tissue weight ($\underline{Ld}$ = 2.2). A. Active animal; B. Inacti animal. Derived from Equation (1). (Newell and Roy, unpublished data.)

The interaction between body size ($\underline{Ld}$) and exposure temperatur ($\underline{Te}$) in influencing the metabolism of *Littorina* is illustrated in Figure 4, which shows the rate:temperature curves for animals of different dry tissue weight collected in June 1970. It is evident that in small winkles the temperature insensitivity is more marked and over a greater thermal range than in large animals. This is very similar to results which we have obtained on the shore crab *Carcinus maenas* (Newell, Ahsanullah, and Pye, 1972; Wallace, 1972).

The Influence of Starvation

The influence of starvation on the metabolism of invertebrates has been systematically investigated in only a comparatively few instances. These have been recently reviewed by Marsden (1973), Newell and Bayne (1973), and Newell (1973b). Roberts (1957) showed that in the crab *Pachygrapsus* the metabolism falls to approximately 60% of its initial rate following 23 days of starvation at 16°C, although the most rapid decline occurred within the first 7 days. In *Uca* the major decline in respiration also occurs within 7 days (Vernberg, 1959) and much the same type of response has been obtained in the crab *Paratelphusa* (Rajabi, 1961). In both this cra and in *Carcinus* (Marsden, Newell, and Ahsanullah, 1973), the metabo lism of small crabs is suppressed to a greater extent in large ones The active rate of oxygen consumption in some invertebrates may be more affected by starvation than the standard rate of quiescent animals. Thus, in the shrimp *Crangon*, the active rate declines towards the standard rate over a period of 8 days' starvation at

10°C(Hagerman, 1970) and a similar result has been obtained in the mussel *Mytilus edulis* (Thompson and Bayne, 1972; Bayne, Thompson, and Widdows, 1973). It is clear, therefore, that either seasonal or laboratory-induced starvation may have a profound effect on the level of metabolism in intertidal invertebrates, and, moreover, there is increasing evidence of an interaction between body size, activity level, and starvation. Further, it has recently been shown that the metabolism of small starved *Carcinus* is less dependent upon temperature than that of large starved crabs (Marsden, Newell, and Ahsanullah, 1973). This finding implies that exposure temperature may also be involved in a complex interaction with metabolism.

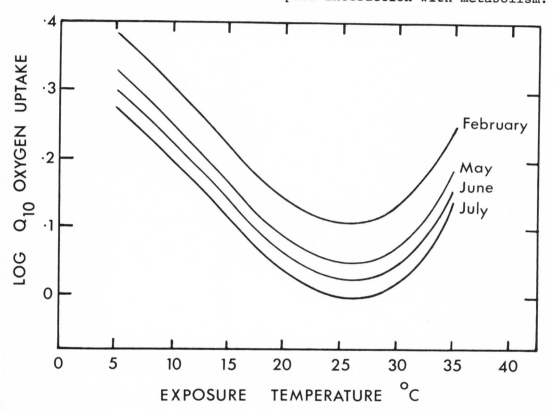

Fig. 3.   Graphs showing the effect of exposure temperature (Te) on the temperature coefficient (Q_{10}) for metabolism in *Littorina* collected in February (day 56), May (day 125), June (day 157), and July (day 191) 1970. Data for a hypothetical animal of 40 mg dry tissue weight (Ld = 1.6). Derived from Equation (1). (After Newell and Roy, in press.)

Unfortunately, we have not yet taken starvation into account quantitatively in our laboratory studies on factors affecting the metabolism of intertidal invertebrates, so this factor does not enter into the multiple regression Equation (1) for *Littorina* metabolism. However, K. B. Armitage and I have recently carried out an extensive series of measurements on the aerial metabolism of quiescent *Ligia* in relation to body size (Ld), exposure temperature (Te), acclimation temperature (Ta), days of starvation (measured also in terms of lipid reserves), and photoperiod. We expect to report our final analysis of the results shortly, but preliminary inspection shows a clearly sigmoid rate:temperature curve which is induced in response to warm acclimation and whose level is controll by nutritional conditions.

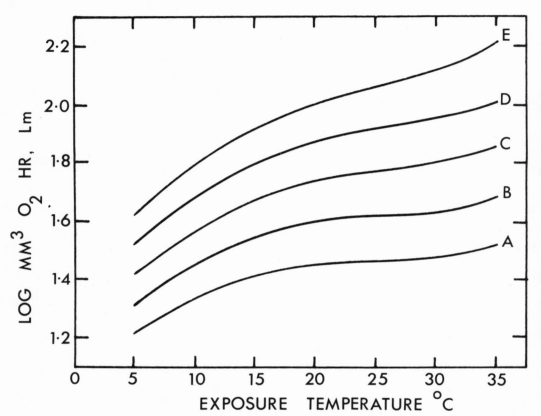

Fig. 4. Graphs showing the effect of exposure temperature on log metabolism (Lm) in active *Littorina* collected in June 1970 (day 157). Data for animals of various dry tissue weights A = 10 mg, B = 20 mg, C = 40 mg, D = 80 mg, E = 160 mg. Derived from Equation (1). (Newell and Roy, unpublished data.)

THE RELATIONSHIP OF SUBCELLULAR ACCLIMATORY RESPONSES TO THOSE IN THE INTACT ORGANISM

From this brief review of factors which we can now relate quantitatively to metabolism in *Littorina*, it is clear that the control of metabolism in this and other intertidal invertebrates is likely to be very complex. This in itself makes the interpretation of the responses of the intact animal in terms of its component parts extremely difficult, particularly as we need to know a great deal more about (1) metabolic pathways and the conditions under which they operate, (2) naturally occurring substrate levels in the cell, (3) intracellular pH values, and (4) other details of the intracellular "milieu," before we can devise suitable in vitro conditions for our experiments. Nevertheless, it does seem relevant to examine what fundamental features of the metabolic rate:temperature curves can be simulated at a subcellular level. The first of the two most obvious features of the rate:temperature curves for the metabolism of Littorina and many other invertebrates is the variability of the temperature-dependence from rates in small starved Carcinus (Marsden, Newell, and Ahsanullah, 1973) and in several other intertidal invertebrates (Newell and Northcroft, 1967; Newell, 1969; Newell and Pye, 1970a, b, 1971) which are essentially independent of exposure temperature, through typically sigmoid curves as described above for Littorina and several other invertebrates to more temperature-dependent rates in other animals. The second feature which may be demonstrable at a subcellular level is the seasonal or temperature-induced modification of the form of the rate:temperature curves with the establishment of temperature-insensitive regions over the normal environmental temperature range during the summer (see Fig. 2).

Rates of oxygen consumption which are very similar to those obtained in the intact animals can be demonstrated in mitochondrial preparations extracted from Littorina and suspended in pasteurized seawater to which known amounts of substrate have been added (Newell and Pye, 1971). Figure 5 shows the rate of oxygen consumption of active and inactive specimens of an arbitrary-sized winkle of 30 mg dry protein weight. Also shown is the rate:temperature curve for a crude homogenate containing 30 mg dry protein weight in pasteurized seawater. These results for the intact animal can then be compared with data obtained with mitochondrial preparations in the presence of various concentrations of pyruvate plus 0.001 mM malate. In order to relate these values quantitatively to those of the intact animal, the rates of oxygen consumption are expressed per unit protein in the original homogenate from which the mitochondria were extracted. The results surprised us at the time for it had been customary to measure mitochondrial respiration in the presence of high (saturating) levels of substrate. Yet our data clearly indicated that virtually temperature-independent reaction rates quantitatively similar to those of the intact quiescent animal were obtained at substrate levels as low as 0.01 mM. Reversion towards a sigmoid rate:temperature curve was obtained at 0.03 to 0.1 mM whereas reaction rates similar to those found in intact active Littorina were obtained at 0.5 mM. This suggests that substrate availability to the mitochondrial enzyme systems may be reduced during quiescence or starvation and that relatively low rates of oxygen consumption would occur under these conditions. The active

metabolism, however, may be associated with a rise in available
substrate level toward 0.5 mM. Since these measurements were made,
sigmoid rate:temperature curves for the respiration of mitochondria
from poikilotherms in the presence of physiological substrate concentrations have been commonly reported in the literature (Davison,
1971; Johnson, Newell, and Hanson, 1972; Johnson and Newell, 1973).
     These data give support to an important series of studies on
the rates of reaction of a variety of regulatory enzymes isolated
from poikilotherms and measured at a range of physiological substrate concentrations (for reviews, see Somero and Hochachka, 1969;
Somero, 1969; Hochachka and Somero, 1971; Hochachka, 1973). It has
been found that at low substrate concentrations within the normal
physiological range of 0.01 to 0.1 mMoles in King Crab leg muscle,
the enzyme-substrate affinity of several key regulatory enzymes is
at a minimum at the normal environmental temperature of the organism.
Above this temperature the affinity decreases and thereby offsets
the thermodynamic tendency of the reaction rate to increase. Also
of considerable significance in connection with the sigmoid rate:
temperature relationships described above for intact animals is the
fact that the enzyme-substrate affinity also decreases toward low
temperatures (Fig. 6) which would account for the sharp decline in
respiration at low temperatures noted in the intact organisms and
in many mitochondrial preparations. The high temperature coefficient
toward the upper lethal temperature of the intact organism, however

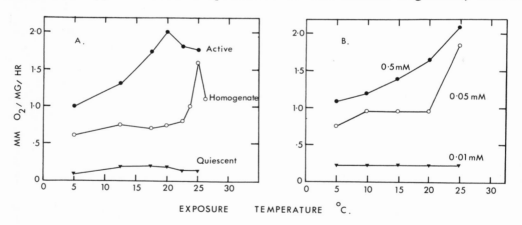

Fig. 5. Graphs showing the effect of exposure temperatures on the
         oxygen consumption of A. Active and inactive *Littorina* of
         30 mg dry protein weight collected in June 1970. The rate
         of oxygen consumption of a homogenate of 30 mg dry protein
         weight suspended in pasteurized seawater is also shown.
         B. The oxygen consumption of mitochondrial suspensions in
         pasteurized seawater in the presence of 0.001 mM malate
         plus various concentrations of pyruvate. Data expressed
         per unit of protein in the original homogenate from which
         the mitochondria were extracted. (After Newell and Pye,
         1971.)

Fig. 6. Graph showing the effect of temperature on the Michaelis constant ($\underline{Km}$) of an idealized regulatory enzyme (e.g., glucose-6-phosphate dehydrogenase, or lactage dehydrogenase). Also shown is the rate:temperature curve which would be expected to result if metabolism were limited by an enzyme with these kinetic properties. (Based on Hochachka and Somero, 1968; Somero, 1969.)

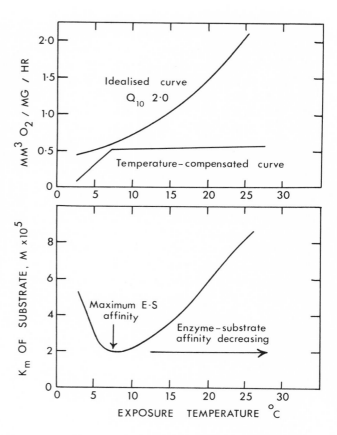

would need to be accounted for in terms of some other component, such as disruption of the mitochondrial membrane systems.

It is also possible to account for the sigmoid rate:temperature curves of many marine invertebrates, as well as the variation in reaction rate from the standard rate of quiescent animals to the active rate, in terms of membrane phenomena related to either membrane-bound enzymes or the structure of the membrane itself. The recent data of Thorhaug (1971) on the effect of temperature on the bioelectric potential across the membrane of the alga *Valonia* show striking similarities with the rate:temperature curves for the metabolism of isolated mitochondria from marine and other poikilotherms and suggest that other processes than those involved in metabolism may exhibit sigmoid rate:temperature relationships. The concept that reaction rates may be diffusion-limited is not new (Bělehrádek, 1935; 1957) but more recently McLaren (1965) has drawn attention to the possible importance of biophysical rather than biochemical phenomena in the control of reaction rates. It would be possible to envisage that changes in the entry of substrates through the mitochondrial membranes, rather than changes in their free intracellular concentration, may account for the increase of the metabolic rate of active animals.

Changes in the temperature range over which metabolism is rendered relatively independent of temperature may be associated with the induction of different regulatory enzymes or isozymes. Lactate dehydrogenase isozymes are known to be induced in response to thermal acclimation in salmonids (Hochachka, 1965, 1967) while in the King Crab there is a temperature-induced interconversion between two forms of pyruvate kinase (Somero, 1969). Both these and several other enzymes have the necessary kinetic properties to render reaction rates virtually independent of temperature, and a stepwise adjustment of the metabolic rate:temperature curve in response to thermal acclimation might be expected to occur. Interestingly, we have recently found that in homogenates of *Littorina* which had been acclimated to a variety of temperatures for two weeks, there appear to be only five different rate:temperature curves irrespective of the temperatures to which the animals had been acclimated (Fig. 7). This strongly implicates one or more regulatory isozyme systems and suggests, moreover, that changes in the sigmoid nature of the rate:temperature curve for the metabolism of intact *Littorina* reflect the induction of such isozymes in response to changes in environmental conditions.

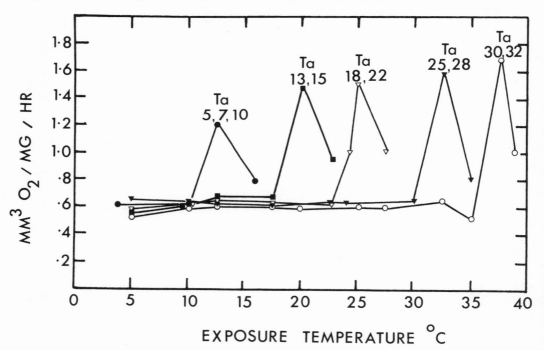

Fig. 7. Graphs showing the different rate:temperature curves obtained with homogenates of *Littorina littorea* acclimated to a variety of temperatures at intervals of 2 to 3°C between 5 and 32°C. Note that despite the use of eleven different acclimation temperatures, only five types of response curves were obtained. This suggests a stepwise acclimatory pattern in response to a change in environmental conditions. (After Newell, 1973a).

CONCLUSION

The summary which I have given of our work on *Littorina* suggests that many factors may interact to affect metabolism in this animal, and it seems likely that in other intertidal organisms it may also be possible to interpret "acclimatization" phenomena in terms of interactions between a variety of environmental factors. Such responses are, however, the sum of the metabolic processes occurring within the cell and it is only by detailed quantitative comparisons between the metabolism of intact animals and those of subcellular components that biochemical events can be related to those occurring in whole organisms. Such data as we have on *Littorina* suggest that variations in the level of metabolism may be associated with membrane-dependent phenomena such as availability of substrate to the mitochondrial enzyme systems, although an ATP-coupled feedback may also be implicated (Newell, 1973a). Seasonal changes in the thermal range over which metabolism is relatively unaffected by temperature may, however, be associated with the induction or repression of a limited set of key regulatory enzymes.

ACKNOWLEDGMENTS

It is a pleasure to acknowledge discussions with colleagues who have helped clarify the ideas expressed here. In particular I am grateful to Professor A. Roy of the Université de Montréal, Canada, for his help in the analysis of the data for intact *Littorina*, and to Dr. V. I. Pye for carrying out much of the experimental work with me. Most of the data presented in this paper has been obtained in collaboration with one or both of these colleagues.

LITERATURE CITED

Barnes, H. and M. Barnes. 1969. Seasonal changes in the acutely determined oxygen consumption and effect of temperature for three common cirripedes, *Balanus balanoides* (L), *Balanus balanus* (L) and *Chthamalus stellatus* (Poli). J. Exp. Mar. Biol. Ecol. 4:35-50.

Bayne, B. L., R. J. Thompson, and J. Widdows. 1973. Aspects of temperature acclimation in *Mytilus edulis*. In: Effects of Temperature on Heterothermic Organisms (W. Wieser, ed.). Heidelberg: Springer-Verlag. (In press.)

Bělehrádek, J. 1935. Temperature and living matter. Protoplasma Monogr. 8.

_____. 1957. Physiological aspects of heat and cold. Ann. Rev. Physiol. 19:59.

Bullock, T. H. 1955. Compensation for temperature in the metabolism and activity of poikilotherms. <u>Biol</u>. <u>Rev</u>. 30:311-42.

Davison, T. F. 1971. The effect of temperature on oxidative phosphorylation in isolated flight muscle sarcosomes. <u>Comp</u>. <u>Biochem</u>. <u>Physiol</u>. 38B:21-34.

Fredette, V., C. Planté, and A. Roy. 1967. Numerical data concerning the sensitivity of anaerobic bacteria to oxygen. <u>J</u>. <u>Bacteriol</u>. 93:2012-17.

Fry, F. E. J. 1957. The aquatic respiration of fish. In: <u>The Physiology of Fishes</u>, Vol. 1 (M. E. Brown, ed.), pp. 1-63. New York: Academic Press.

Hagerman, L. 1970. The oxygen consumption of *Crangon vulgaris* (Fabricius) (Crustacea, Natantia) in relation to salinity. <u>Ophelia</u> 7:283-92.

Hemmingsen, A. M. 1950. The relation of standard (basal) energy metabolism to total fresh weight in living organisms. <u>Rep</u>. <u>Steno</u>. <u>Hosp</u>. <u>Copenh</u>. 4:7-58.

_____. 1960. Energy metabolism as related to body size and respiratory surfaces and its evolution. <u>Rep</u>. <u>Steno</u>. <u>Hosp</u>. <u>Copenh</u>. 9:7-110.

Hochachka, P. W. 1965. Isoenzymes in metabolic adaption of a poikilotherm. Subunit relationships in lactic dehydrogenases of goldfish. <u>Arch</u>. <u>Biochem</u>. <u>Biophys</u>. 111:96-103.

_____. 1967. Organisation of metabolism during temperature compensation. In: <u>Molecular Mechanisms of Temperature Adaptation</u> (C. L. Prosser, ed.), pp. 117-203. Washington, D. C.: Americ Association for the Advancement of Science Publication No. 84.

_____. 1973. Basic strategies and mechanisms of enzyme adaptation to temperature. In: <u>Effects of Temperature on Heterothermic Organisms</u> (W. Wieser, ed.). Heidelberg: Springer-Verlag. (In press.)

_____ and G. N. Somero. 1968. The adaptation of enzymes to temperature. <u>Comp</u>. <u>Biochem</u>. <u>Physiol</u>. 27:659-68.

_____ and _____. 1971. Biochemical adaptation to the environment. In: <u>Fish Physiology</u> VI (W. S. Hoar and D. J. Randall, eds.), pp. 99-156. New York: Academic Press.

Johnson, L. G. and R. C. Newell. 1973. Temperature compensated oxygen consumption of subcellular preparations from vertebrate poikilotherm, homoiotherm and tumor tissues. In: <u>Effects of Temperature on Heterothermic Organisms</u> (W. Wieser, ed.). Heidelberg: Springer-Verlag. (In press.)

_____, _____, and R. D. Hanson. 1972. Temperature relationships of mitochondrial respiration during chicken development. <u>Comp</u>. <u>Biochem</u>. <u>Physiol</u>. 42B:693-701.

Leffler, C. W. 1972. Some effects of temperature on the growth and metabolic rate of juvenile blue crabs, *Callinectes sapidus* in the laboratory. <u>Mar</u>. <u>Biol</u>. 14:104-110.

Mangum, C. P. and C. Sassaman. 1969. Temperature sensitivity of active and resting metabolism in a polychaetous annelid. <u>Comp</u> <u>Biochem</u>. <u>Physiol</u>. 30:111-16.

Marsden, I. D. 1973. Influence of starvation on temperature rela-
tionships in *Carcinus maenas*. In: Effects of Temperature on
Heterothermic Organisms (W. Wieser, ed.). Heidelberg:
Springer-Verlag. (In press.)

_____, R. C. Newell, and M. Ahsanullah. 1973. The effect of
starvation on the metabolism of the shore crab, *Carcinus
maenas*. Comp. Biochem. Physiol. 45A:195–213.

McLaren, I. A. 1965. Temperature and frog eggs. A reconsideration
of metabolic control. J. Gen. Physiol. 48:1071–79.

Moeur, J. E. and C. H. Eriksen. 1972. Metabolic responses to
temperature of a desert spider, *Lycosa (Pardosa) carolinensis*
(Lycosidae). Physiol. Zool. 45:290–301.

Newell, R. C. 1969. Effect of fluctuations in temperature on the
metabolism of intertidal invertebrates. Am. Zoologist 9:293–
307.

_____. 1970. Biology of Intertidal Animals. London: Elek Books.

_____. 1973a. Environmental factors affecting the acclimatory
responses of heterotherms. In: Effects of Temperature on
Heterothermic Organisms (W. Wieser, ed.). Heidelberg:
Springer-Verlag. (In press.)

_____. 1973b. Factors affecting the respiration of intertidal
invertebrates. Am. Zoologist 13:513–28.

_____, M. Ahsanullah, and V. I. Pye. 1972. Aerial and aquatic
respiration in the shore crab, *Carcinus maenas* (L). Comp.
Biochem. Physiol. 43A:239–52.

_____ and B. L. Bayne. 1973. Temperature and metabolic acclimation
in marine invertebrates. Neth. J. Sea Research. (In press.)

_____ and H. R. Northcroft. 1967. A re-interpretation of the
effect of temperature on the metabolism of certain marine
invertebrates. J. Zool. Lond. 151:277–98.

_____ and V. I. Pye. 1970a. Seasonal changes in the effect of
temperature on the oxygen consumption of the winkle *Littorina
littorea* (L) and the mussel *Mytilus edulis* L. Comp. Biochem.
Physiol. 34:367–83.

_____ and _____. 1970b. The influence of thermal acclimation on
the relation between oxygen consumption and temperature in
Littorina littorea (L) and *Mytilus edulis* L. Comp. Biochem.
Physiol. 34:385–97.

_____ and _____. 1971. Quantitative aspects of the relationship
between metabolism and temperature in the winkle *Littorina
littorea* (L). Comp. Biochem. Physiol. 38B:635–50.

_____ and A. Roy. 1973. A statistical model relating the oxygen
consumption of a mollusc *(Littorina littorea)* to activity,
body size and environmental conditions. Physiol. Zool.
(In press.)

Precht, H. 1958. Concepts of temperature adaptation of unchanging
reaction systems of cold-blooded animals. In: Physiological
Adaptation (C. L. Prosser, ed.), pp. 50–78. Washington, D. C.:
American Physiological Society.

_____, J. Christophersen, and H. Hensel. 1955. Temperatur und
Leben. Berlin: Springer-Verlag.

Prosser, C. L. 1955. Physiological variation in animals. Biol. Rev. 30:229-62.

_____. 1958. The nature of physiological adaptation. In: Physiological Adaptation (C. L. Prosser, ed.), pp. 167-80. Washington, D. C.: American Physiological Society.

_____ and F. A. Brown. 1961. Comparative Animal Physiology, 2nd ed. Philadelphia: Saunders.

Rajabi, K. G. 1961. Studies on the oxygen consumption in tropical poikilotherms. VI. Effect of starvation on the oxygen consumption of the freshwater field crab Paratelphusa sp. Proc. Ind. Acad. Sci. 54:276-80.

Rao, K. P. and T. H. Bullock. 1954. Q_{10} as a function of size and habitat temperature on poikilotherms. Amer. Nat. 87:33-43.

Roberts, J. L. 1957. Thermal acclimation of metabolism in the crab Pachygrapsus crassipes Randall. I. The influence of body size, starvation and molting. Physiol. Zool. 30:232-42.

Roy, A. 1969. Analyse des facteurs de taux de métabolisme chez la limace Arion circumscriptus. Revue Can. Biol. 28:33-43.

Saroja, K. 1959. Studies on the oxygen consumption in tropical poikilotherms. II. Oxygen consumption in relation to body size and temperature in the earthworm Megascolex mauritii, when kept submerged under water. Proc. Ind. Acad. Sci. 49:183-93.

Somero, G. N. 1969. Enzymic mechanisms of temperature compensation immediate and evolutionary effects of temperature on enzymes of aquatic poikilotherms. Am. Nat. 103:517-29.

_____ and P. W. Hochachka. 1969. Isoenzymes and short-term temperature compensation in poikilotherms: Activation of lactate dehydrogenase isoenzymes by temperature decreases. Nature (London) 223:194-95.

Thompson, R. J. and B. L. Bayne. 1972. Active metabolism associate with feeding in the mussel Mytilus edulis L. J. Exp. Mar. Biol. Ecol. 9:111-24.

Thorhaug, A. 1971. Temperature effects on Valonia bioelectric potential. Biochim. Biophys. Acta. 225:151-58.

Toulmond, A. 1967a. Consommation d'oxygène, dans l'air et dans l'eau chez quatre Gastéropodes du genre Littorina. J. de Physiologie 59:303-4.

_____. 1967b. Étude de la consommation d'oxygène en fonction du poids, dans l'air et dans l'eau, chez quatre espèces du genre Littorina (Gastéropoda, Prosobranchiata). C. R. Acad. Sci. Paris 264:636-38.

Ulbricht, R. J. and A. W. Pritchard. 1972. Effect of temperature on the metabolic rate of sea urchins. Biol. Bull. 142:178-85.

Vernberg, F. J. 1959. Studies on the physiological variation between tropical and temperate zone fiddler crabs of the genus Uca. II. Oxygen consumption of whole organisms. Biol. Bull. 117:163-84.

Wallace, J. C. 1972. Activity and metabolic rate in the shore crab, Carcinus maenas (L). Comp. Biochem. Physiol. 41A: 523-33.

Wieser, W. 1972. Oxygen consumption and ammonia excretion in
    Ligia beaudiana M-E. Comp. Biochem. Physiol. 43A:869-76.
Zeuthen, E. 1947. Body size and metabolic rate in the animal
    kingdom. C. r. Trav. Lab. Carlsb. Sér. Chim. 26:15-161.
_____. 1953. Oxygen uptake as related to body size in organisms.
    Q. Rev. Biol. 28:1-12.

Respiratory adaptations in *Limulus polyphemus* (L).

K. Johansen and J. A. Petersen

Limulus polyphemus L., the North American representative among the four living species of horseshoe crabs, is of special interest to respiratory physiologists engaged in problems of environmental adaptation. *Limulus* also occupies a unique phylogenetical position (Størmer, 1952). The morphology and the ventilatory mechanism of its gills are exceptional among water-breathing animals. Its hemocyanin-containing blood shows the rare quality of a reversed Bohr shift (Redfield, Coolidge, and Hurd, 1926), and its heart and blood vascular system are markedly specialized (Patten and Redenbaugh, 1899).

Tolerant to diverse environmental conditions, *Limulus* is found in coastal and estuarine waters along the east coast of North America from Nova Scotia to Yucatan (Robertson, 1970). Their migration to shallow water during spawning entails intertidal air exposure which may result in body temperatures as high as 40°C (Fraenkel, 1960). Their habit of burrowing in the often muddy substrate in shallow water may expose them to hypoxic water conditions.

The objectives of the present investigation include the assessment of gas exchange and blood gas transport at variable ambient conditions in unrestrained animals free to move or bury in sand. The data obtained allow an alternative explanation of how a reversed Bohr shift may be of physiological significance in *Limulus* during normal behavior in its natural habitat.

MATERIALS AND METHODS

Twelve specimens of the horseshoe crab, *Limulus polyphemus*, were used in the present investigation. The animals were transported by air from the east coast of the United States to the Friday Harbor Laboratories in Washington State where they were kept in running seawater at 10 to 15°C.

Gas exchange was measured in closed respirometers at 10, 15, 20, and 25°C. Measurements were done with animals in seawater only or with sand (sterilized by boiling) which partly filled the respirometers in order to allow the animals to bury. The water in the respirometer, thoroughly stirred prior to sampling, was periodically sampled and analyzed for O_2 and CO_2 tension as well as pH. The samples were analyzed on a Radiometer BMS 3 gas analyzer and a PHM 71 acid-base analyzer with appropriate electrodes. The solubility coefficients for O_2 in seawater were taken from Harvey (1966). The O_2 uptake rates were computed on a weight basis with reference to total live weight or to the actual weight of live metabolizing tissues by subtracting the weight of the harder chitinous parts. The efficiency in gas exchange was assessed by sampling arterial and venous blood (1) via indwelling polyethylene catheters or (2) by direct percutaneous needle puncture into the heart or pericardium for arterial blood or into the large longitudinal venous sinuses accessible from the base of the tail spine (telson) close to the anus. Samples of expired water were obtained through a catheter fastened to the exoskeleton close to the telson and immediately behind the most posterior pair of gill plates. Catheters for blood sampling were implanted by placing needles in the heart or the venous sinuses. Following placement of the catheters, the needles were withdrawn and the catheters were secured in place. Sampling of blood was done while animals were actively walking or swimming and when they were digging into or buried in the substrate. Blood samples were also withdrawn at intervals during air exposure. Blood was analyzed immediately following sampling using the Radiometer equipment, since stored blood coagulated rapidly and had to be discarded.

The respiratory rate, expressed by the frequency of the waving motion of the gill plates bearing the gill books, was recorded as a pressure change caused by each respiratory stroke in the catheter placed behind the gills. The pressure changes were measured by a Statham strain gauge manometer (P23bb) and recorded continuously on a Brush mark 220 recorder.

Earlier published data by Redfield, Coolidge, and Hurd (1926), and Redfield and Ingalls (1933) on the oxygen dissociation curves of the hemocyanin-containing blood of *Limulus* were used in the analyses and computations of the data obtained.

RESULTS

Gill Surface Area

The gills of *Limulus* are unique among water-breathing animals. Five pairs of abdominal appendages fused medially are gill-bearing. Each appendage or gill plate has about 150 gill lamellae arranged in a configuration like leaflets in a book. The most anterior pair of abdominal appendages is fused and modified and forms an operculum which covers the gill plates when they are adducted (the end expiratory position).

As part of the present study the total surface area of the gill leaflets or lamellae was determined in four different specimens. Figure 1 depicts the relationship of total surface area to live weight (left ordinate) of the animals; a 20% increase in surface area is shown for a live weight increase of 100 g. The relationship between the total number of lamellae (right ordinate) and the gill surface area is also apparent from Figure 1. The surface areas computed on a weight bases are comparable to values reported for molluscs (Pelseener, 1935) and for decapod crustaceans (Gray, 1957). The chitinous, parchmentlike support of the gill filaments suggests they may withstand collapse when exposed to air and thus be effective for gas exchange also in air (see later).

Fig. 1. Relationships of gill area to live weight and number of gill leaflets in *Limulus polyphemus*.

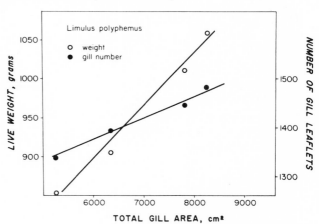

Oxygen Uptake (VO$_2$)

Eight specimens were used for this part of the study. Figure 2 shows the relationship of the oxygen uptake rate to the water temperature for *Limulus* without a substrate to bury in (filled circles). The animals went through periods of rest and activity during the period of measurement. No quantitative method for measurement of activity allowed a separation of standard from active O$_2$ uptake. The animals were preconditioned to the various temperatures used for at least several hours and usually overnight.

Fig. 2. Oxygen uptake in
   relation to water temperature for *Limulus*.
   •:   active animals;
   o: buried animals.

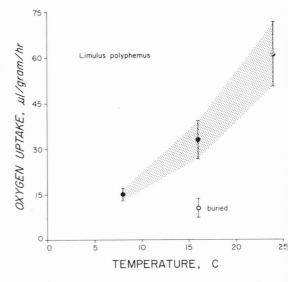

Activity increased markedly as the temperature was increased and
the difference between the minimum and maximum O_2 uptake at the
respective temperatures increased with temperature. At 8 to 10°C
the animals were nearly motionless. The Q_{10} for the average O_2
uptake between 15 and 25°C was 1.84. At 16°C the O_2 uptake was also
measured when the animals had dug into the sand and remained buried
and quiescent except for their ventilatory activity.

   The minimum O_2 uptake recorded at 16°C without sand was higher
than the average value for O_2 uptake in the buried condition. Also
the O_2 uptake in the buried condition at 16°C was considerably lower
than the O_2 uptake at 8°C although the animals at the lower temperature appeared nearly motionless. These findings suggest that the
buried condition itself is associated with a lowered O_2 uptake.

   Figure 3 depicts the O_2 uptake values for one specimen of
Limulus at 16°C during a 60-min period of quiescence alternating wit
digging into or out of the sand. The lowest O_2 uptake was obtained
during the buried condition. Digging could increase the O_2 uptake
more than fivefold.

Fig. 3. Changes in O_2 uptake
   of *Limulus* during digging
   and burying in sand.

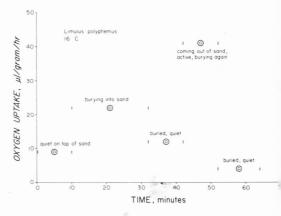

Data on oxygen uptake rates at falling ambient O_2 availability demonstrate that *Limulus* at 16°C and 25°C is able to maintain an unchanged O_2 uptake down to ambient water PO_2 of 50 to 60 mm Hg. Below this level, the O_2 uptake usually fell, but could also increase in animals which reacted by restlessness and increased activity when the water became hypoxic. At 8 to 10°C the animals were able to maintain their O_2 uptake down to 30 to 40 mm Hg.

Blood Gas Tensions and pH

More than 100 pairs of arterial and venous blood samples were withdrawn via catheters or direct needle puncture from 12 specimens of *Limulus*. Blood was sampled during activity, rest, and when the animals were buried in sand. Blood was also sampled prior to and at successive stages during air exposure and after return to aerated water. The samples were analyzed for PO_2, PCO_2 and pH. Although the sampling procedures were organized to permit immediate analysis, many samples had to be discarded due to coagulation of the blood.

The values obtained were variable and probably reflected differences in the state of locomotor activity as well as of ventilatory activity. The sampling of blood from an animal with a sluggish circulation and a large blood volume (approximately 20% of live tissue weight estimated by drainage, most of which is located in spacious venous sinuses) makes it difficult to obtain reproducibility between repetitive samples. The sampling site also inevitably differs from one specimen to another and this contributes additionally to the variability of the measured values. There was, however, no doubt that certain discernible trends in the blood gas picture attended behavior like burrowing, motor activity on the surface, or air exposure.

When the animals were confined in tanks at 18 to 20°C in normoxic water ($PO_2 > 120$) with no sand or other substrate to dig or bury in, their activity ranged from brief periods of rest to continuous moderate activity. Arterial O_2 tension (PaO_2) at such conditions averaged 77 mm Hg for twelve sets of samples. The corresponding venous tensions (PvO_2) averaged 20 mm Hg. No significant a-v differences in blood pH or PCO_2 were detectable. Overall blood pH ranged from 7.40 to 7.55. Blood PCO_2 averaged 2.9 mm Hg ranging upward to 4.1.

After voluntary burrowing, the animals were completely buried. In this condition they were completely quiescent. Both in the buried state and in the resting unburied condition, a marked reduction in ventilatory activity occurred. These combined changes resulted in a marked reduction in both arterial and venous O_2 tensions and a concurrent rise in blood PCO_2. Arterial PO_2 averaged 10 mm Hg with corresponding venous values of 5 mm Hg. Blood PCO_2 rose to an average of 6.5 mm Hg with a maximum value of 10.5 mm Hg. Blood pH as low as 7.1 was measured. No clear-cut a-v differences in PCO_2 or pH were apparent.

During air exposure the blood gas values changed conspicuously
Initially following exposure, both arterial and venous PO_2 dropped
precipitously while blood PCO_2 increased. In three experiments,
the arterial values while the animals were in water dropped from
more than 60 mm Hg to values between 7 to 12 mm Hg within the first
hour of air exposure. Corresponding venous values were 14 to
28 mm Hg before exposure to air, dropping to values between 2 to
10 mm Hg after exposure. Attending this reduction in PO_2 there was
a rise in PCO_2 to values ranging from 6.2 to 11.5 mm Hg depending
upon duration of exposure.

After the initial decline in blood O_2 tensions, continued air
exposure resulted in increasing arterial as well as venous O_2
tensions. Blood PCO_2, however, also continued to rise and levelled
off at values between 11 and 14 mm Hg after exposure lasting up to
48 hrs. Arterial tensions returned and leveled off at 21 to
27 mm Hg, while the venous tensions were more variable but rose to
an average of 10 mm Hg. These blood gas values and a tolerance to
air exposure lasting up to 48 hrs attest to an obvious ability of
Limulus to use its gills for O_2 absorption under both conditions.
A few experiments during which the air-exposed animals became rest-
less and attempted to move about gave unchanged values for arterial
O_2 tensions around 25 mm Hg, while the venous tensions dropped to
values below 5 mm Hg. These arterial gas tensions suffice to
nearly saturate the hemocyanin with oxygen. Air breathing in
Limulus was coupled with a general elevation of blood PCO_2. This
condition is common among water breathers able to subsist temporarily on air breathing during intertidal exposure.

O_2 Extraction from the Ventilatory Current. Effectiveness in Gas
Exchange, PO_2 Gradients Across the Gills. Transfer Factor for O_2

Among the expressions used below, the effectiveness in O_2
uptake by the gills was introduced by Hughes and Shelton (1962) in
their research on fish respiration. Also from work on fishes, the
concept of a transfer factor (diffusion capacity) for oxygen across
gills was introduced by Randall, Holeton, and Stevens (1967) and
Piiper and Baumgarten-Schumann (1968).

% Extr = Oxygen extraction from water, %.

$$\% \text{ Extr} = \frac{PiO_2 - PeO_2}{PiO_2} \times 100.$$

% Ew   = Effectiveness in oxygen removal from water, %.

$$\% \text{ Ew} = \frac{PiO_2 - PeO_2}{PiO_2 - PvO_2} \times 100.$$

ΔPO_{2G} = PO_2 gradient across gill exchange surfaces, (mm Hg).

$$\Delta PO_{2G} = (\frac{PiO_2 + PeO_2}{2}) - (\frac{PaO_2 + Pv\ O_2}{2}).$$

TO_2 = Transfer factor for oxygen, $(\frac{\text{ml } O_2/\text{kg/min}}{\text{mm Hg}})$.

$$TO_2 = \frac{VO_2}{\frac{1}{2}(PiO_2 + PeO_2) - \frac{1}{2}(PaO_2 + PvO_2)}.$$

The sampling of expired water revealed a very high percentage of O_2 extraction in *Limulus*. In normoxic water at 18 to 20°C when the animals were unburied and moderately active, O_2 extraction varied from about 60% to 85%, the higher values correlating with a less vigorous ventilatory activity. When buried and the ventilatory rate had slowed down, extraction was very high, often exceeding 80%. When exposed to decreasing ambient O_2 tensions, the O_2 extraction declined, but remained as high as 40 to 50% even at water PO_2 levels between 30 to 50 mm Hg.

In four experiments it was possible to sample expired water simultaneously with arterial and venous blood samples from half-buried or buried animals. Arterial and venous blood PO_2 values were generally low during such periods of inactivity. The O_2 extraction ranged from 60% to 81% and averaged 71%. The effectiveness in O_2 removal from water (Ew) ranged from 62% to 86%, averaging 75%. The average PO_2 gradient from water to blood across the respiratory exchange surface was very high, averaging 82 mm Hg. This gradient is expressive of the effectiveness of convection on both sides of the exchange surface and the magnitude of the diffusion barrier separating water and blood. Effective convection and/or a small diffusion barrier correlate with a small ΔPO_{2G}. The transfer factor for oxygen across the gills was low and averaged $0.00213 \frac{ml\ O_2/kg/min}{mm\ Hg}$.

Due to technical difficulties, expired water and blood samples could not be obtained simultaneously from active animals. Samples obtained successively gave an average value for effectiveness in O_2 uptake by the gills of 70%. The PO_{2G} was approximately 50 mm Hg and the transfer factor for O_2: $0.011 \frac{ml\ O_2/kg/min}{mm\ Hg}$.

Ventilatory Activity. Responses to Gas Composition of the Water, Motor Activity and Burrowing

Each respiratory movement consists of abduction of the gill plates, i.e., raising them from their ventral resting position (inspiration), followed by an adduction, i.e., actively bringing them back during expiration which forces the water to pass over the gill lamellae and exit posteroventrally. During each respiratory cycle, abduction of the operculum precedes the next gill plate, succeeded in turn by the next posterior and so on, resulting in the passage of the inspiratory water current over successive gill plates. Adduction or expiration is also successive, but in the reverse direction (Waterman and Travis, 1953). The pressure changes recorded with each respiratory movement provide a continuous record of the respiratory rate, while the amplitude of the pressure

pulses provides an indication of the force of each respiratory
movement.

Continuous recording of respiratory movements from undisturbed
animals revealed that the most common ventilatory pattern was one
of periodic breathing. Figure 4, recorded from a resting animal in
well-aerated water at 15°C, shows that the bursts of breathing can
be separated by complete respiratory rest (bottom tracing) or superimposed on a steady slower respiratory rhythm (top tracing). In
the latter case, the bursts of more intense breathing are characterized by (1) a change to a frequency four to six times higher than
the steady rhythm, (2) the amplitude of the breathing movements
gradually increasing to a peak, and (3) the amplitude again gradually declining back to that of the steady rhythm. The same amplitude changes prevail when breathing is periodic with complete
respiratory pauses (bottom tracing).

Temp, 15°C

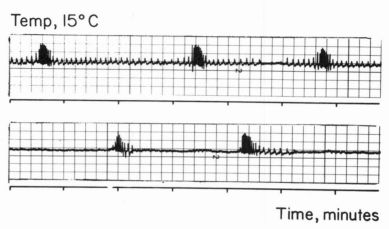

Time, minutes

Fig. 4. Continuous tracings of respiratory movements in *Limulus*
showing periodic changes in ventilatory rhythm.

Figure 5 demonstrates the relationships between the frequency of
respiratory movements (left bottom ordinate), the relative duration
of respiratory rest (right ordinate), and the O_2 tension of the
water (top left ordinate) plotted against time.

Initially when ambient water O_2 tension was about 50 mm Hg,
the respiratory rate varied between 22 and 24 beats per minute. At
this time about 2 min of each 10 min showed complete respiratory
inactivity. At time 100 min, the water O_2 tension was gradually
increased up to full air saturation. In response to this, the
respiratory rate declined on the average about 27%, while the time
occupied by respiratory quiescence increased from 20 to 65%. Later
when water PO_2 again was lowered to about 40 mm Hg, the respiratory
rate increased to about 24 per minute when breathing became
continuous.

Other experiments revealed that the respiratory rate recorded
in well-aerated water could be further lowered when the water PO_2

was increased by O_2 equilibration. These data suggest that the breathing rhythm in normoxic water, as well as in water of lowered O_2 tension, is driven by an oxygen-sensitive feedback mechanism (a hypoxic stimulus).

Fig. 5. Relationships between respiratory rate, the relative duration of respiratory rest, and the O_2 tension of the water.

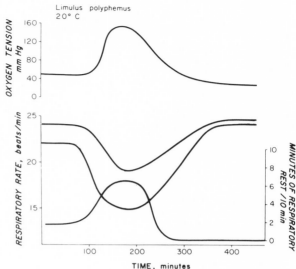

Figure 6 shows a situation where a continuous breathing rhythm at rest is interrupted by a brief episode of motor activity. The activity resulted in an increased amplitude as well as an increase in the rate of the breathing movements.

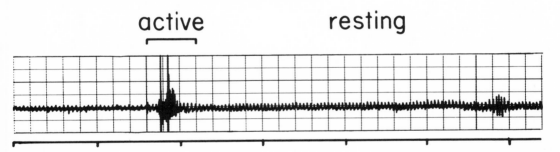

Fig. 6. Change in ventilatory movements of *Limulus* caused by a brief period of motor activity.

Figure 7 shows two traces of ventilatory movements in a buried, quiescent animal (top tracing), and in the same animal resting on the bottom. The buried condition shows typical periodic breathing with complete respiratory arrest between the bursts of respiratory movements. The animal resting on the bottom shows continuous respiratory activity with superimposed bursts of higher frequency and amplitude.

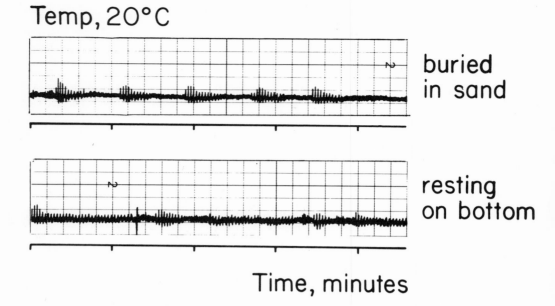

Temp, 20°C

buried in sand

resting on bottom

Time, minutes

Fig. 7.   A comparison of ventilatory rhythm in a buried and resting *Limulus*.

DISCUSSION

The rate of O_2 uptake is complexly dependent upon so many ambient and behavioral conditions that a comparison of the data on *Limulus* with those from other invertebrates obtained by other investigators has limited value. We are of the opinion that the now common practice of separating O_2 uptake rates into standard and active values can lead to serious mistakes if one has no way of measuring activity as actual work output.

A salient feature of the O_2 uptake measurements was the fact that the uptake in the completely buried animals was markedly lower than that of the inactive unburied animals kept at the same tempera-ture. This finding contests the validity of the term basal or standard metabolism as has been done for vertebrates, for instance, diving species (Pickwell, 1968). The recordings of ventilatory activity also indicate that the animal hypoventilates when buried. Similarly, if ventilation volume is computed on the basis of the measured values for O_2 uptake and O_2 extraction from the ventilatory current, the calculated ventilation for the unburied condition is nearly four times that for the buried animals. The possibility that the lower O_2 uptake in the buried condition simply reflects a reduced O_2 uptake due to a lowered O_2 availability in the water drawn over the respiratory organs from the water layer adjacent to the sand (i.e., oxygen confirmer) can be discounted. Our experi-ments, in contrast to earlier reports (Maloeuf, 1937), show *Limulus* to be a so-called regulator of O_2 uptake down to water oxygen tensions of 40 to 50 mm Hg. Furthermore, the O_2 tension of the water in the respirometers remained higher at all times when buried animals were compared to active animals.

It has long been advocated that the gill epithelium of decapod crustaceans represents a very large diffusion barrier and hence results in very low circulating O_2 tensions in the blood leaving the gills (Redmond, 1955; Larimer, 1964). Similar conditions have been surmised to exist in molluscs (Redmond, 1962) and also allegedly in *Limulus*. This so-called low-tension transport of O_2 has come to be generalized and regarded as a principle governing the O_2 transport of a number of invertebrates (Newell, 1970; Jones, 1972).

If gas transport, as a rule, should be of the low tension type, it would be an insult not only to the design(er) of respiratory organs, but also to the evolution and significance of respiratory pigments as O_2 carriers since their oxygen-binding capacity would never be put to full use. The low tension principle has recently been contested in experiments on at least five species of decapod crustaceans (Johansen, Lenfant, and Mecklenburg, 1970; McMahon and Wilkens, 1972; Taylor, Butler, and Sherlock, 1973) and one mollusc, *Cryptochiton stelleri* (Petersen and Johansen, 1973). These experiments utilized more direct methods for measuring blood gas tensions and techniques permitting sampling from undisturbed animals. In all cases, much higher arterial O_2 tensions have been measured than reported earlier; these have resulted in near full or full saturation of the arterial blood. Moreover, circulating venous O_2 tensions have been demonstrated high enough to leave a considerable venous O_2 reserve when animals are inactive, e.g., *Cancer magister* (Johansen, Lenfant, and Mecklenburg, 1970), or when an intertidal animal like the sea cradle, *Cryptochiton stelleri*, is submerged under conditions optimal for gas exchange in water (Petersen and Johansen, 1973). This situation leaves these two animals with a reserve in O_2 transport to meet increased demands in case of motor activity *(Cancer magister)* or a reduced efficiency in gas exchange during air exposure at low tide *(Cryptochiton)*.

The great variability in blood gas tensions obtained on *Limulus* cannot be expressive of a large diffusion barrier in the gills, since arterial O_2 tensions at times could exceed 100 mm Hg (average for twelve measurements on active animals were 77 mm Hg). Rather, the variability must reflect a close dependency of circulating blood gas levels on the ventilation and/or perfusion of the gills. Thus, there is a large reduction in blood gas tensions when the animals become inactive, and an even more marked reduction when the animals are buried, reflecting a reduced ventilatory activity. The sharp reduction in blood O_2 tensions during air exposure probably results from a derangement of the ventilation perfusion matching. Continued air exposure does, however, raise arterial O_2 tensions to give nearly full O_2 saturation, which attests to an ability of *Limulus* to breathe air directly. The continued increase in blood

PCO_2 levels during air exposure results from loss of contact with water which has a high CO_2 solubility.

The consequence of the recorded changes in blood O_2 and CO_2 tension for internal gas transport can only be appreciated when discussed in the context of the O_2-dissociation properties of *Limulus* hemocyanin.

The hemocyanin of *Limulus* has been studied extensively because it has a reversed or positive Bohr shift, resulting in increased O_2 affinity at lowered pH (increased PCO_2) (Redfield, Coolidge, and Hurd, 1926; Redfield and Ingalls, 1933). *Limulus* shares this rare quality with a few gastropods, notably the prosobranch, *Busycon* (Redfield, Coolidge, and Hurd, 1926).

In vertebrate blood, the regular Bohr shift is present at variable magnitudes and in O_2 transport provides the advantage of increasing the O_2 utilization (i.e., turnover to tissues) for a given a-v PO_2 difference or, expressed differently, allowing the unloading of O_2 in the tissues at a higher capillary PO_2 than does blood with a smaller or no Bohr shift. The significance of the Bohr shift in vertebrate O_2 transport is linked with the presence of an arteriovenous difference in pH. The metabolism of invertebrates is generally too low to provide significant arteriovenous differences in pH or the difference is minimized by the ease with which CO_2 dissolves in ambient water. Bohr shifts of most invertebrate bloods are small or nonexistent (Redmond, 1963). A notable exception is represented by cephalopods, which do show significant a-v differences in pH and PCO_2 and for which a large Bohr shift is crucially important for gas transport (Redfield and Goodkind, 1929; Johansen and Lenfant, 1966).

The Bohr shift could potentially be of importance in the gas transport of invertebrates if a change in ambient conditions or behavior would shift both arterial and venous blood pH. This possibility has formed the basis for attempts to reason the physiological importance of the reversed Bohr shifts seen in a few invertebrates. Redmond (1955) has suggested in the case of *Busycon* that a steepening of the dissociation curve in its operational PO_2 interval would improve the O_2 loading potential more than it would reduce the unloading potential if loading occurred at the steepest portion of the curve. Such conditions could allegedly arise if *Busycon* were exposed to periodic reductions in O_2 availability associated with an increased ambient PCO_2 from, for instance, fermentation in a muddy substrate. Redmond's hypothesis, although theoretically tenable, has not been supported by any measurements of environmental parameters, or of circulating arterial and venous blood gases and pH.

The present results which have been obtained on *Limulus* invite another attempt to rationalize a potential usefulness of a reversed Bohr shift based on actual measurements. Figures 8 and 9 illustrat how two acts of normal behavior in *Limulus*, i.e., burying and intertidal air exposure, will shift the operational PO_2 interval of the oxyhemocyanin binding from the upper knee of the dissociation curve to its steepest portion, while simultaneously the blood PCO_2 will increase.

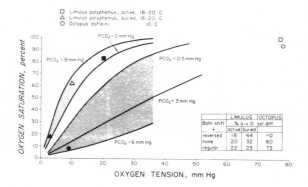

Fig. 8. **In vivo** blood gas parameters superimposed on **in vitro** dissociation curves, showing the blood O_2 utilizations in *Limulus* and *Octopus*. Insert shows the effects of an alteration of their Bohr shifts. Open symbols indicate arterial blood samples; tilled symbols, venous samples. Open square to the far right indicates arterial PO_2 of active *Limulus*; the open circle, arterial PO_2 of *Octopus*.

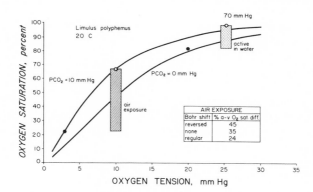

Fig. 9. A comparison of blood O_2 utilization in *Limulus* during activity in water and air exposure. Insert shows the effect of an altered Bohr shift. Dotted rectangle marked 70 mm Hg expresses O_2 utilization from circulating blood when *Limulus* is active in water. 70 mm Hg refers to the actual arterial PO_2. Shaded rectangle to the left indicates O_2 utilization in air-exposed *Limulus*.

In the case of the burying *Limulus*, the shift in operational PO_2 range and blood PCO_2 results from the aforementioned hypoventilation associated with the buried condition. Whereas O_2 utilization in an active animal occurs at high arterial O_2 tensions, the hypoventilation associated with burying will bring arterial PO_2 down into the steepest portion of the dissociation curve. The retention of CO_2 as a result of hypoventilation will, because of the reversed Bohr shift, further steepen the in vivo dissociation curve. The table insert on Figure 8 gives average values for O_2 utilization in the active unburied *Limulus* with a reversed Bohr shift compared with the calculated utilization had there been (1) no Bohr shift or (2) a regular Bohr shift of a magnitude simila to the reversed shift. To the active animal, the Bohr shift is seen to be of little consequence for O_2 turnover, whereas in the buried condition, considerable advantage accrues from the presence of a reversed Bohr shift. Conditions in a cephalopod, *Octopus dofleini*, are included for comparison, since O_2 transport in this animal is crucially dependent on a large regular (negative) Bohr shift (Lenfant and Johansen, 1965; Johansen and Lenfant, 1966). *Octopus* could simply not exist if its blood possessed a reversed Bohr shift.

Figure 9 similarly illustrates how the intertidal air exposure of *Limulus* will alter the conditions for gas exchange such that the reversed Bohr shift affords an advantage in O_2 turnover to the tissues. The reduced arterial O_2 tensions occurring initially on air exposure, and the retention of CO_2 as an obligate result of air-exposing the gills, are the principal factors responsible for the change in O_2 turnover.

The above proposed solution to the paradox of a reversed Bohr shift cannot be declared valid until confirmed by further experimentation. To do this, additional data are required on blood gas levels in undisturbed animals exposed to conditions which mimic those of the natural environment of *Limulus*.

The extraction of O_2 from the respiratory current in *Limulus* was very high and much higher than reported for most other invertebrates (Hazelhoff, 1938; Johansen, Lenfant, and Mecklenburg, 1970; Petersen and Johansen, 1973). The results obtained by using the indirect method for measurement of ventilation measured as the rate and amplitude of gill plate movement suggest that O_2 extraction is closely correlated with ventilation. A comparison of directly measured ventilation values in the crab, *Cancer magister* (Johansen, Lenfant, and Mecklenburg, 1970), with values computed for *Limulus* of approximately similar weights at the same temperature and simila activity levels discloses that ventilation volumes in *Limulus* are four to five times less than in *Cancer*. The average O_2 extraction rate for twelve specimens of *Cancer* was 18%, which is approximately one-fourth to one-fifth of the values obtained on *Limulus*. The effectiveness in gas exchange (Ew) in *Limulus* was also very high compared to other invertebrates (Johansen, Lenfant, and Mecklenburg 1970) and similarly was due to low ventilation volumes.

When the average PO_2 gradients (ΔPO_{2G}) across the gills in half-buried or buried animals are compared with values from active specimens, very high gradients are seen in the buried animals and these gradients decrease to approximately one-half in the active animals. Values for moderately active decapod crustaceans, such as *Cancer magister* (Johansen, Lenfant, and Mecklenburg, 1970), and the cephalopod, *Octopus dofleini* (Johansen and Lenfant, 1966), are

60 mm Hg and 65 mm Hg, respectively. These values fall between those for the resting and active *Limulus*. The transfer factor for O_2 similarly increased about sixfold going from the buried to the active condition in *Limulus*. These changes attest to the major importance of ventilation and, hence, convection of the respiratory surfaces for efficiency in gas exchange and the level of circulating blood gases.

The control of respiratory movements (i.e., ventilation) in *Limulus* has earlier been studied by Hyde (1894, 1906) and by Waterman and Travis (1953). Since the neurological basis for the rhythmic ventilatory movements is discussed in the papers by Hyde, only information on alteration of respiratory activity from changes in normal behavior or ambient gas composition will be discussed here.

Hyde (1894) reported a respiratory rate of about 27 beats per minute in undisturbed *Limulus* at 24C. She also found that long periods of quiescence were associated with apneusis or periodic breathing. By provoking increased activity (exercise) of restrained animals, the frequency of gill plate movements increased. Hydrogen-saturated seawater induced ventilatory slowdown and arrest. Waterman and Travis (1953) reinvestigated the respiratory responses to alterations in ambient gas composition of *Limulus*. Their results, which appear to represent the current teaching on the subject, are completely discordant with ours. They claim that deoxygenation of the water evokes no compensatory change in the breathing rate whatsoever in *Limulus* and that the breathing rate is directly proportional to the oxygen content of the medium. They claim that this finding is consonant with earlier data indicating that *Limulus* shows no regulation of O_2 consumption over wide ranges of ambient PO_2 (Maloeuf, 1937). All of these findings are totally at variance with our results on *Limulus*.

The experimental procedures of Waterman and Travis (1953) include recordings by kymographic methods of gill plate movements in animals restrained ventral side up. In another series gill movements were visually observed while animals were transferred from air-saturated water to nitrogen- or CO_2-equilibrated water. Such drastic changes are completely alien to any normal situation and not likely to explain physiological mechanisms operating under normal conditions.

Our data offer evidence from undisturbed animals that *Limulus* possesses a negative feedback control of ventilatory movements resulting in compensatory increase in ventilatory frequency when water PO_2 is lowered gradually (Fig. 5). This finding is in accord with earlier studies on crustaceans (Lindroth, 1938; Larimer, 1961).

ACKNOWLEDGMENTS

This study was supported by Grant G.B. 7166 from the National Science Foundation.

LITERATURE CITED

Fraenkel, G. 1960. Lethal high temperatures for three marine invertebrates: *Limulus polyphemus*, *Littorina littorea* and *Pagarus longicarpus*. Oikos 11:171-82.

Gray, I. E. 1957. A comparative study of the gill areas of crabs. Biol. Bull. 112:34-42.

Harvey, H. W. 1966. The Chemistry and Fertility of Sea Waters. Cambridge, Mass.: Cambridge University Press.

Hazelhoff, E. H. 1938. Über die Ausnützung des Sauerstoffs bei verschiedenen Wassertieren. Z. vergl. Physiologie 26:306-27.

Hughes, G. M. and G. Shelton. 1962. Respiratory mechanisms and their nervous control in fish. Adv. Comp. Physiol. Biochem. 1:275-364.

Hyde, I. H. 1894. The nervous mechanism of the respiratory movements in *Limulus polyphemus*. J. Morph. 9:431-48.

_____. 1906. A reflex respiratory centre. Amer. J. Physiol. 16: 368-77.

Johansen, K. and C. Lenfant. 1966. Gas exchange in the cephalopod, *Octopus dofleini*. Amer. J. Physiol. 210:910-18.

_____, _____, and T. A. Mecklenburg. 1970. Respiration in the crab, *Cancer magister*. Z. vergl. Physiologie 70:1-19.

Jones, J. D. 1972. Comparative Physiology of Respiration. London: E. Arnold Ltd.

Larimer, J. L. 1961. Measurement of ventilation volume in decapod Crustacea. Physiol. Zool. 34:158-66.

_____. 1964. The patterns of diffusion across the crustacean gill membranes. J. Cell. Comp. Physiol. 64:139-48.

Lenfant, C. and K. Johansen. 1965. Gas transport by the hemocyanin containing blood of the cephalopod, *Octopus dofleini*. Amer. J. Physiol. 209:991-98.

Lindroth, A. 1938. Atmungsregulation bei *Astacus fluviatilis*. Arkiv. f. Zool. 30:1-7.

Maloeuf, N. S. R. 1937. Studies on the respiration of animals. II. Aquatic animals with an oxygen transporter in their internal medium. Z. vergl. Physiol. 25:29-46.

McMahon, B. R. and J. L. Wilkens. 1972. Simultaneous apnoea and bradycardia in the lobster, *Homarus americanus*. Can. J. Zool. 46:585-96.

Newell, R. C. 1970. Biology of Intertidal Animals. London: Logos Press Ltd.

Patten, W. and W. A. Redenbaugh. 1899. Studies on *Limulus*. II. The nervous system of *Limulus polyphemus*, with observations upon the general anatomy. J. Morphology 16:91-200.

Pelseneer, P. 1935. Essay d'éthologie zoologique. Brussels: Acad. Roy. Belg. Classe Sci. Publ. Foud. Aqathon Potter.

Petersen, J. A. and K. Johansen. 1973. Gas exchange in the giant sea cradle, *Cryptochiton stelleri*. J. Exp. Mar. Biol. Ecol. In press.

Pickwell, G. V. 1968. Energy metabolism in ducks during submergence asphyxia: Assessment by a direct method. Comp. Biochem. Physiol. 27:455-85.

Piiper, J. and D. Baumgarten-Schumann. 1968. Effectiveness of O_2 and CO_2 exchange in the gills of the dogfish (*Scyliorhinus stellaris*). Resp. Physiol. 5:338-49.

Randall, D. J., G. F. Holeton, and E. D. Stevens. 1967. The exchange of oxygen and carbon dioxide across the gills of the rainbow trout. J. Exp. Biol. 6:339-48.

Redfield, A. C., T. Coolidge, and A. L. Hurd. 1926. The transport of oxygen and carbon dioxide by some bloods containing hemocyanin. J. Biol. Chem. 69:475-509.

_____ and R. Goodkind. 1929. The significance of the Bohr effect in the respiration and asphyxiation of the squid, *Loligo pealei*. J. Exp. Biol. 6:340-49.

_____ and E. N. Ingalls. 1933. The oxygen dissociation curves of some bloods containing hemocyanin. J. Cell. Comp. Physiol. 3:169-202.

Redmond, J. R. 1955. The respiratory function of hemocyanin in Crustacea. J. Cell. Comp. Physiol. 46:209-47.

_____. 1962. The respiratory characteristics of chiton hemocyanins. Physiol. Zool. 63:304-13.

_____. 1963. Bohr effect: Absence in a Molluscan Homocyanin. Science 139:1294-95.

Robertson, J. D. 1970. Osmotic and ionic regulation in the horseshoe crab, *Limulus polyphemus* (Linnaeus). Biol. Bull. 138: 157-83.

Størmer, L. 1952. Phylogeny and taxonomy of fossil horseshoe crabs. J. Paleontol. 26:630-40.

Taylor, E. W., P. J. Butler, and P. J. Sherlock. 1973. The respiratory and cardiovascular changes associated with the emersion response of *Carcinus maenas* (L.) during environmental hypoxia at three different temperatures. J. Comp. Physiol. In press.

Waterman, T. H. and D. F. Travis. 1953. Respiratory reflexes and the flabellum of *Limulus*. J. Cell. Comp. Physiol. 41:261-90.

The extraction of oxygen by estuarine invertebrates

C. P. Mangum and L. E. Burnett

A few years ago we were attracted for several reasons to the problem of measuring oxygen extraction rates, or the fractional removal of oxygen from the incoming ventilatory stream. Generally, the movement of water across sites of respiratory exchange with the aquatic environment is caused by one of two kinds of mechanisms, ciliary or muscular. One question that has been with us for a long time concerns the adaptive determinants of flow rates in different groups of animals. For example, ciliary flow is usually associated with a straining device such as the lamellibranch gill which serves a dual function, i.e., the water current brings the animal's food as well as its oxygen. Indeed, most studies of ciliary flow have concluded that the rate-determining factor, in the teleological sense, is the animal's nutritional requirement rather than its respiratory demand (Jørgensen, 1966).

Direct methods of estimating rates of oxygen extraction have either involved some physical imposition on the animal's sensibilities, such as catheterization of the siphon (a highly refined sense organ) or else they have entailed discrete and therefore discontinuous sampling of incurrent and excurrent fluid. In the first case, the experience is clearly perceived by the animal and one must wonder whether it results in any sort of feedback. In the second, the oxygen regime at respiratory surfaces is portrayed acurately by a feasible number of samples only if the flow rate does not change very much in time. This lack of large-scale fluctuation on a short-term basis is implicitly assumed to be true of ciliary flow, whose control

mechanism is unknown although evidence of central nervous input into
the molluscan gill seems to be growing (Aiello, 1960). In contrast,
if flow is intrinsically rhythmic in character, then oxygen extraction is almost certain to be poorly described by discontinuous samples unless an enormous number are subjected to time series analysis
a procedure that is not widely used by biologists. Spontaneously
rhythmic currents are generated by muscular mechanisms in most lower
metazoans, where the control center lies in endogenous pacemakers,
and probably also in the holothurians (R. C. Newell, personal communication). As an alternative, oxygen extraction rates have been
estimated less directly from separate measurements of ventilatory
flow and oxygen uptake. This procedure is both tedious and complicated, due to rather arduous calibration and computation procedures
(Dales, Mangum, and Tichy, 1970).

In the recent past, accurate measurement of the PO_2 of very
small volumes of aqueous fluids has become routine. These measurements can be made by anaerobically procuring small samples, a necessarily discontinuous procedure, or they can be in situ, in which
case they may be continuous. We initiated our studies on a group of
marine invertebrates displaying different pumping mechanisms by making the measurements shown in Table 1, most of which derive from

TABLE 1
Oxygen extraction rates in marine and estuarine invertebrates exposed to
well-oxygenated conditions

Species	Conditions	Incurrent PO_2 (mm Hg)	N	% Extraction (± S.E.)
Porifera				
Halichondria bowerbanki	20°C, running seawater	146	7	1.5 ± 1.0
Haliclona loosanoffi	22°C, running seawater	132	11	4.0 ± 1.0
Annelida				
Diopatra cuprea	22°C, natural tubes, still water	125	12	53.7 ± 6.0
	20°C, natural tubes, running seawater (Mangum, Santos, and Rhodes, 1968)	140–145	78	39
Glycera dibranchiata	14°C, aerated aquarium	155–160	11	52.1 ± 7.3
	22°C, running seawater (Hoffmann and Mangum, 1970)	140–145		58
Nereis virens	22°C, glass tubes, running seawater	146	15	22.0 ± 3.0

TABLE 1 (cont.)

Species	Conditions	Incurrent PO_2 (mm Hg)		% Extraction (± S.E.)
Pectinaria gouldi	22°C, natural tubes in sand, running seawater	146	14	9.0 ± 1.0
Mollusca				
Anadara ovalis	22°C, running seawater	157	16	5.0 ± 1.0
Busycon carica	18°C, aerated aquarium	159	7	31.5 ± 7.2
Dinocardium robustum	18°C, mixed aquarium	128	3	9.8 -12.8
Modiolus demissus	20°C, attached, running seawater	140	9	6.3 ± 0.4
Mytilus edulis	20°C, attached, running seawater	140	6	8.5 ± 1.0
Crustacea				
Upogebia affinis	22°C, burrowed in sand, running seawater	144	5	2.0 ± 1.0
Urochordata				
Ciona intestinalis	22°C, running seawater	153	14	4.0 ± 1.0

discontinuous samples injected into the PO_2 microelectrode chamber of our Radiometer blood gas machine (BMS1, equipped with Acid-Base Analyzer PHM71). The exceptions are in situ measurements from the mollusca *Busycon carica* and *Dinocardium robustum*, and the annelids *Diopatra cuprea* (Mangum, 1973) and *Glycera dibranchiata* (see below).

In general these estimates do not differ very much from those in the literature. If anything, those for animals with ciliary currents are lower than previous reports (Jørgensen, 1966). Among species with muscular ventilation, the only surprise is the very low figure for the ice cream cone worm, *Pectinaria gouldi*. To our knowledge, details of the function and pattern of the current in this species have not been reported; this worm may be an interesting exception to the annelid rule.

PASSIVE FLOW: ITS RESPIRATORY FUNCTION

We were particularly interested in the sponges and the annelids because of the intriguing suggestion by Vogel and Bretz (1972) that passive flow might aid ventilation of their respiratory surfaces. If fluid is moved along an unmoving surface bearing separate incurrent and excurrent openings to a tube below that surface, and if the

excurrent opening is elevated only slightly above the incurrent ope
ing, then fluid flow through the tube is caused by its boundary lay
properties. Previously it was pointed out (Hoffmann and Mangum, 19
that the model of passive flow is not applicable to annelids. The
tube or burrow is blocked by longitudinal muscle contractions that
maintain setal anchoring and cuticular contact at more or less regu
lar intervals along the length of the body, known as the points
d'appui. Mechanistically, this occlusion is an inevitable conseque
of the mutually excitatory and inhibitory reflex pathways existing
between adjacent sets of body wall muscles. Teleologically, the wor
plugs its tube because it must work with the resistance of the burrow walls in order to maintain position at rest and to move substan
tial volumes of fluid during a burst of pumping activity. Finally,
we pointed out that the importance of passive ventilation in sponge
needs quantitative evaluation, especially since current dogma holds
that actively generated currents in these animals are nutritive
rather than respiratory, a point with which Vogel and Bretz (1972)
agreed.

With Dr. Vogel's help, we have attempted to make such a quantitative evaluation. We set up several species in systems that meet
physical requirements for high velocity passive flow (Vogel, 1974),
and then made the measurements shown in Table 2, by anaerobically

TABLE 2
Oxygen extraction rates in marine and estuarine invertebrates under control
and passive flow conditions

Species	Conditions*	Incurrent PO_2 (mm Hg)	N	% Utilization (± S.E.)
Porifera				
Halichondria bowerbanki	Control	146	7	1.5 ± 1.0
	Passive flow	146	8	1.5 ± 1.0
Haliclona loosanoffi	Control	146	11	4.0 ± 1.0
	Passive flow	146	17	4.0 ± 1.0
Annelida				
Nereis virens	Control (bur- rowed in sand)	140	5	50.0 ± 2.0
	Passive flow	140	6	50.0 ± 2.0
Crustacea				
Upogebia affinis	Control (bur- rowed in sand)	144	5	3.0 ± 2.0
	Passive flow	144	6	1.0 ± 1.0

removing 200 μl samples from a site a few millimeters below the
excurrent aperture and comparing their PO_2 with that of samples take
simultaneously from the incurrent stream.

We could not detect a reduction in oxygen extraction rates unde
conditions physically conducive to passive flow either in the sponge

in which Vogel (1974) has shown that its magnitude is impressive, or in the polychaete *Nereis virens*, in which passive flow is blocked (Table 2). Had we continued sampling burrow fluid from the crustacean *Upogebia affinis*, we probably would have achieved the results of a significantly lower oxygen extraction rate under passive flow conditions when the number of observations (N) became sufficiently great to reduce the error. But in this species, oxygen extraction rates under both control and passive flow conditions are so low that the endeavor did not seem worthwhile.

We were prepared to concur that passive flow, whose velocity reaches very high values under the right conditions (Vogel, 1974), may be quite important for sponge nutrition but not for respiration. And then it occurred to us that we might have overlooked the most important condition of its importance: low oxygen, particularly in the estuary. Most estimates of oxygen extraction rates have been confined to animals exposed to air-saturated conditions, and the oversight is not entirely without reason. Flourishing populations of epifaunal ciliary pumpers are usually associated with well-oxygenated intertidal habitats like wharf pilings. But the distribution of estuarine species is somewhat distinctive. In the Chesapeake Bay system, for example, where the intertidal zone is small and hard substrata are scarce, these animals are most prolific at several meters depth below the halocline on oyster reefs and gorgonian skeletons. It is here, in a greatly attenuated two-layer transport system such as the estuary, that low oxygen conditions developing along with summer stratification (Carpenter and Cargo, 1957; Biggs, 1967) are very much a part of the normal living conditions of benthic species.

We chose several of these species and made in situ measurements of oxygen extraction with a PO_2 hypodermic microelectrode whose signal was amplified by a Beckman Model 160 Physiological Gas Analyzer. The microelectrode was gently positioned with a micromanipulator in the excurrent siphon while incurrent PO_2 was monitored simultaneously with a macroelectrode (Yellow Springs Instrument Co. Model 5420). PO_2 was altered by bubbling nitrogen gas (Matheson Gas Products) or air through the water in a closed aquarium, and mixing it with submersible magnetic stirrers.

We quickly learned that the notion of continuous rates is quite inapplicable to most species, even when both incurrent and excurrent channels appear to be fully open (Fig. 1). Consequently, our data are computed where possible from the integral or estimate of the area under a continuous trace of PO_2 as a function of time (Yellow Springs Instrument Co. Models 80 and 81 recorders). We found, however, that the excurrent stream of many lamellibranch molluscs changes direction with such frequency that continuous traces spanning more than a few minutes are not reliable. Even though measurement was continuous, these data are given as discrete points which were noted simultaneously with visual confirmation that the microelectrode was in the appropriate position a few millimeters within the siphon and not touching its inner walls. Oxygen extraction rates in the soft-shelled clam *Mya arenaria* increase exponentially as the animal encounters low oxygen levels, but they never reach very impressive proportions before the clam ceases pumping (Fig. 2A). At least one

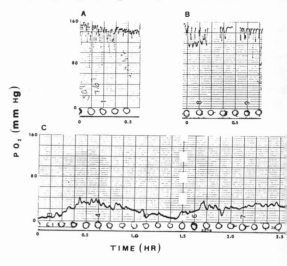

Fig. 1. Record of PO_2 in excurrent channels of A) *Dinocardium robustum* (incurrent PO_2=160 mm Hg), B) *molgula manhattensis* (incurrent PO_2=160 mm Hg), and C) *Adocia tubifera* incurrent PO_2=45 mm Hg) in mixed aquaria at 19°C.

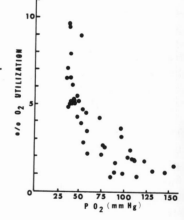

Fig. 2A The fractional extraction or utilization by *Mya arenaria* of oxygen available in the incurrent ventilatory stream at 16°C and 20 o/oo.

very common natural event elevates those rates, however, The clam is exquisitely sensitive to vibration and other forms of mechanical stimuli, to which it responds by briefly withdrawing its siphons, thereby shutting off flow across its gills. After a period of sipho retraction lasting only a few seconds, oxygen extraction is enhanced to the point that respiratory function of the current is clear. The data in Figure 2B were obtained after inducing siphon retraction by gently tapping the side of the aquarium, a mode of stimulation that must occur frequently in the natural habitat.

Possibly the most interesting lamellibranch in the present context is the oligohaline clam *Rangia cuneata*. Its anaerobic metabolism, which has been studied in detail by Awapara and co-workers, differs in important respects from classical vertebrate glycolysis.

Fig. 2B. The fractional extraction or utilization of oxygen by *M. arenaria* after brief periods of siphon retraction. Incurrent PO$_2$= 86 mm Hg.

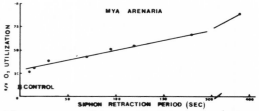

Perhaps its most bizarre feature is that anaerobic pathways continue to operate under well-oxygenated conditions (Chen and Awapara, 1969). Yet its rate of aerobic metabolism is quite typical for a lamellibranch mollusc of its body size. Figure 3A shows the effect of declining oxygen levels on aerobic metabolism before and after a 6-hr period of anoxic exposure. The aerobic metabolism before the anoxic exposure resembles that of *Mya arenaria* (Fig. 2C).

Fig. 2C. Oxygen uptake by *M. arenaria* under declining oxygen conditions.

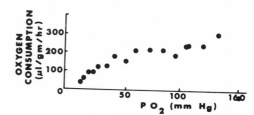

Fig. 3A. Oxygen consumption as a function of PO$_2$ in *Rangia cuneata* at 2 o/oo and 22°C before (o) and after (□) a 6-hr period of anoxia.

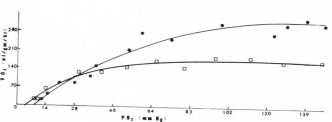

The data were calculated from the slope of the line describing oxygen depletion in a closed container (Yellow Springs Instrument Co. Model 5420 Polarographic Electrode). Since the exterior of the shell had been coated a week previously with paraffin wax (Kushins and Mangum, 1971), oxygen uptake cannot be attributed to microorganisms inhabiting the hard parts. In this particular example, the rate of aerobic metabolism was decreased by the anoxic experience, but that result is not consistent. When we analyzed paired observations on nine animals, we found no significant overall change (Mangum and Van Winkle, 1973). More important, the oxygen extraction rate is typically lamellibranch and minuscule (Fig. 3B). Unlike the results for the more polyhaline *M. arenaria*, the values for *R. cuneata* begin to reach respectably respiratory levels at low PO$_2$ values known to occur in the habitat for several months of the year (Carpenter and Cargo, 1957; Biggs, 1967).

We are left with the paradox of an animal that removes only a tiny fraction of the oxygen molecules available at its gill, although it burns quite a lot, and at the same time it is believed to operate anaerobic pathways at a high level (Chen and Awapara, 1969). While the relative magnitudes of contributions from aerobic and anaerobic

pathways to the overall energy budget cannot at present be gauged,
we are unable to reconcile these rates with the concept of an essen
tially anaerobic organism. Instead, we should at least consider th
possibility that the key to understanding the anomaly lies in the
oligohaline habitat. In the Chesapeake Bay system, *R. cuneata* is

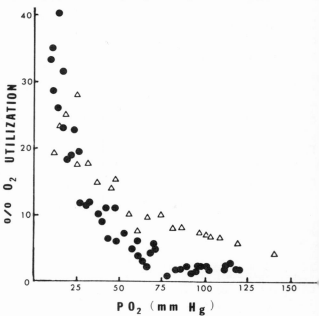

Fig. 3B. The fractional
extraction or utilization
by *R. cuneata* of oxygen
available in the incur-
rent ventilatory stream
and at 2 o/oo and 14°C
under declining (o) and
rising (△) oxygen condi-
tions.

restricted to salinities below about 12 o/oo, and it becomes most
abundant below 5 o/oo. One of its chief anaerobic pathways involve:
the conversion of pyruvate to alanine (Stokes and Awapara, 1968);
regulation of free amino acid content is, of course, the major means
of osmotic adjustment in molluscan tissue. Moreover, Anderson
(1974) has shown that the clam is hyperosmotic to its dilute enviror
ment. This reaction may very well be more crucial in opposing dimir
tion of the free amino acid pool in this very rigorous habitat than
in the energy budget.

The hard-shell clam *Mercenaria mercenaria* yields a somewhat
different result. In this case we were able to record excurrent PO
continuously, so the data in Figure 4A were computed from integrals

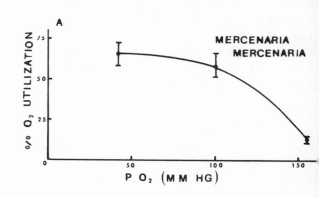

Fig. 4A. Oxygen extraction
rates at 19°C in *Mercenaria
mercenaria*.

Oxygen extraction reaches high values at intermediate PO_2 values and essentially ceases at low ones. Since we found that siphon retraction elevates these rates in *M. arenaria*, high values may be related to the lower frequency of pumping periods than in the other clams. Diving observations lead to the belief that pumping is more nearly continuous in nature than in the laboratory, but we have observed animals in nature only under well-oxygenated conditions.

The estuarine sponge *Adocia tubifera* (Fig. 4B) has much higher oxygen extraction rates than its close relative *Haliclona loosanoffi*

Fig. 4B. Oxygen extraction rates at 19°C in *Adocia tubifera*.

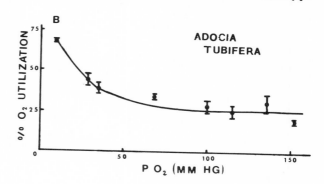

(Table 1). The rates do not change markedly until the animal encounters low PO_2, when they ascend to values more commonly associated with muscular mechanisms of ventilation. The same picture emerges for the mesohaline tunicate *Molgula manhattensis* (Fig. 4C), in

Fig. 4C. Oxygen extraction rates at 19°C·in *Molgula manhattensis* and *Styela plicata*.

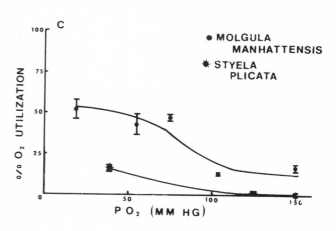

which the data suggest a rather abrupt change at intermediate PO_2 from lower to higher rates. At this point, we wondered whether higher values might not be an inevitable outcome of the computation procedure rather than a distinctive feature of estuarine animals. So we made another set of measurements on a more euhaline tunicate, *Styela plicata* (Fig. 4C), and reassured ourselves that the low rates obtained previously (Table 1) do in fact exist.

To summarize, under well-oxygenated conditions animals with ciliary mechanisms of generating water currents extract a fraction of oxygen from incurrents that is too small to reconcile with the concept of flow rate determination in response to respiratory demands. Consequently, adaptation of habit or body architecture to meet physical conditions requisite for high velocity passive flow cannot be understood in respiratory terms. However, in species characteristically reaching maximum abundance on the bottom of meso and oligohaline regions of estuaries, where summer stratification produces very low oxygen environments, the oxygen extraction rate is appreciably elevated at lower PO_2's. While the generalization is severely limited by the number of species studied, this adaptation of at least several conspicuously dominant estuarine species enables them to cope with a special feature of their habitat other than its more obvious and better known osmotic properties. And in those estuarine species whose design meets physical requirements, passive as well as actively generated flow may make important contributions to respiration.

THE COST OF MUSCULAR VENTILATION

In a group of animals with muscular mechanisms of generating ventilatory flow, we took advantage of the very convenient rhythmicity of annelid ventilation to estimate its cost to the animal. When the microelectrode is inserted into the tube or burrow of an infaunal annelid, the rhythmic change in PO_2 reflects the alternating bursts of pumping activity and rest. The chronology of the rhythm and, therefore, changes in oxygen content at the site of gas exchange vary in different species. Our particular example is from the bloodworm, *Glycera dibranchiata* (Fig. 5); the pattern in the onuphid polychaete *Diopatra cuprea* is quite different (Mangum, 1973)

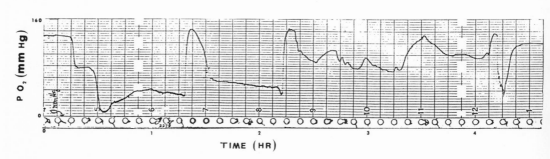

Fig. 5.   Record of PO_2 in the head shaft of a glass tube inhabited by *Glycera dibranchiata*.   Incurrent PO_2 = 160 mm Hg. Temperature 14°C.

The animal experiences wildly unstable oxygen conditions, although they are predictably unstable. Microenvironmental oxygen levels are not especially low, however; at high tide when the ventilatory stream in unimpaired, oxygen in the microhabitat is high enough to permit functioning of bloodworm hemoglobin in gas transport. External PO_2 falls to levels where bloodworm hemoglobin operates as an oxygen reservoir only during the resting phase at high tide and then of course at low tide when the worm cannot ventilate (Mangum and Carhart, 1972).

When PO_2 changes are recorded in a closed container, a similar rhythmicity appears. Therefore, derivatives of the trace yield a simple, direct measurement of active and resting levels of aerobic metabolism at different oxygen concentrations. Identification of the two phases of oxygen uptake can be confirmed by making independent recordings of the product of activity, which is fluid flow (Mangum and Sassaman, 1969; Hoffmann and Mangum, 1970). In some annelids, aerobic respiration shuts down at a PO_2 well above zero and a switchover to anaerobic pathways presumably occurs. This phenomenon has also been encountered in some twenty-three species representing seven animal phyla (Prasada Rao and Ganapati, 1968; Mangum, 1970; Kushins and Mangum, 1971; Sassaman and Mangum, 1972; Mangum and Van Winkle, 1973), and it is not ordinarily a premonition of death. Many species live for days after it occurs. We are naturally intrigued by the question as to why the switch occurs, in the adaptive sense, when oxygen remains for the taking. An obvious answer might be that it happens when production of the ventilatory stream is no longer economical, i.e., in modern economic terms, when the marginal cost of the product equals the marginal revenue gained, or when the volume of oxygen that can be extracted from the ventilatory stream falls below that expended by the animal in generating the current. Testing this hypothesis in rigorous econometric fashion would be extraordinarily difficult because it rests on the projection of variable costs per production unit (Samuelson, 1955), or the measurement of different levels of metabolic and muscular activity at low PO_2's near the aerobic shutdown point where it is hard enough to distinguish active from resting metabolism. But we can project fixed costs or resting metabolism, and average costs of the activity at different oxygen levels. Under well-oxygenated conditions, the muscular activity involved in a ventilatory burst requires a four- to sevenfold metabolic increase in the terebellid polychaete *Amphitrite ornata*, a three- to fourfold increase in the onuphid *Diopatra cuprea*, and a two- to tenfold increase in the bloodworm *Glycera dibranchiata*, depending on the temperature (Coyer and Mangum, 1973). At lower PO_2's, the increase in aerobic respiration is reduced, but contributions from anaerobic pathways still cannot be evaluated.

Under well-oxygenated conditions, the oxygen extraction rate in the onuphid polychaete *Diopatra cuprea* is among the lowest known in animals with muscular mechanisms of ventilation. The water current in this species is exceptional in that it plays an important role in the biology of feeding (Mangum, Santos, and Rhodes, 1968). The current carries chemical stimuli which, if they consist of suprathreshold concentrations of a glycoprotein or any one of several amino acids, elicit a complex response that culminates in ingestion (Mangum and Cox, 1971). The rate-determining requirement is therefore nutritional and not respiratory, a fairly unusual feature among animals with muscular ventilation. But at low oxygen concentration the volume of flow and, thus, of oxygen available at the gills declines precipitously until it is only slightly higher than and not significantly different from the volume expended in the activity of procuring it. In Figure 6, the volume of oxygen available at each of three concentrations of environmental oxygen was computed from estimates of flow made by Dales, Mangum, and Tichy (1970), and the level of active metabolism was measured as described by Mangum and Sassaman (1969). We believe that the proximity of the intersec of these two curves to the shaded area of the abscissa, which spans the region in which aerobic shutdown occurs, is more than coinciden

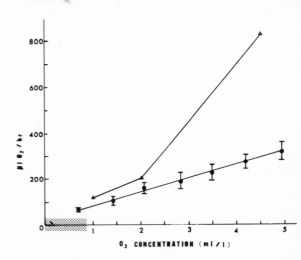

Fig. 6. The cost of ventilation in *Diopatra cuprea*. Circles (●) indicate oxygen consumption in 1.5 gm worm and triangles (Δ) show volume of oxygen in incurrent ventilatory stream as a function of oxygen concentration at 17°C. Shaded area spans range of oxygen concentration at which aerobic shutdown occurs. Vertical bars represent standard error.

The argument is formulated in terms of understanding the adaptive significance of aerobic shutdown; its form is therefore teleological and not causal. At present we have no insight into the nature and operation of the switch. The adaptive importance of the response, however, is emphasized by evidence that it does not occur at a series of temperatures lower than 17.5°C, where production costs are reduced and revenue, or oxygen content, is enhanced (Mangum, 1972). One might suppose that the response would be similarly contingent upon body size, but within a species flow rate is not related in a systematic way to a relatively small range of body weights. Values of the correlation coefficient r for linear and logarithmic regression lines describing the relationship of these two variables (Dales, Mangum, and Tichy, 1970) are not signif cant (P > .05). While the correlation should certainly be tested over a wider range of body size, it is not apparent that the expres sion of flow rates in body size units would be a useful addition.

In contrast, oxygen extraction in the terebellid polychaete *Amphitrite ornata* is more typical of animals making currents by muscular activity. The data in Figure 7 were estimated from integrals of continuous in situ measurements with the PO₂ microelectrode positioned in the head shafts of natural tubes. The shape of that curve, interestingly enough, resembles the shape of a curve describing the quantitative participation of the animal's hemoglobin in aerobic respiration at the same temperature (Mangum, unpublished data). Figure 7B shows the volume expended in active and resting metabolism down to an oxygen concentration where the two cannot be distinguished, and the volume of oxygen available as computed from the data in Figure 7A. Alternatively, if we compute the volume of oxygen available from flow measurements (Coyer and Mangum, 1973), the result is lower but not significantly so. In any event, the curve in Figure 7B describing the rate of decline of procured oxygen has a different shape from that in *D. cuprea*, and it predicts a net profit in oxygen essentially down to anoxia. In fact, we have never encountered the phenomenon of aerobic shutdown in *A. ornata*, regardless of temperature, and we know that the worm continues to engage

Fig. 7A. The fractional extraction or utilization by *Amphitrite ornata* of oxygen available in the incurrent ventilatory stream at 20°C and 31 o/oo. Vertical bars represent standard error.

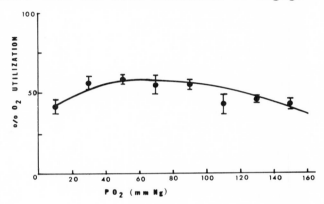

Fig. 7B. The cost of ventilation in a 3.37 gm *Amphitrite ornata* at 20°C. Oxygen consumption indicated during active (●) and resting (o) phases of the rhythmic ventilation cycle as a function of oxygen concentration. Closed squares (■) are used where the two can no longer be distinguished. Triangles (Δ) show volume of oxygen entering the animal, computed from data in 7A. Open square (□) shows volume of oxygen in ventilatory stream, computed from flow rates (Coyer and Mangum, 1973). Vertical bars represent standard error.

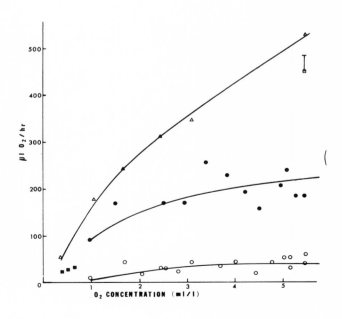

in bursts of muscular ventilation at external PO$_2$'s lower than 10 mm Hg. At least in these two examples then, the phenomenon of aerobic shutdown seems to have a rational basis in respiratory economy.

REGULATION OF OXYGEN EXTRACTION

At a constant PO$_2$, changes in oxygen extraction rates such as those observed in animals generating ciliary currents (Fig. 1) may be due to alterations in fluid flow either inside or outside the gill, or changes of either term in the expression $\dot{V}_w/\dot{Q}$ for the ventilation perfusion ratio, although we would be reluctant to assign it a numerical value for the reasons discussed below. Similarly, when the external PO$_2$ is reduced, the failure of oxygen extraction rates to reflect this reduction precisely constitutes evidence for a compensatory response. Low oxygen compensation may be affected by increased ventilation and/or circulation, or it may be due to enhanced transport of oxygen by a respiratory pigment. While a compensatory response in oxygen extraction by a vertebrate is usually explained solely in terms of fluid flow, there is some evidence that in the terebellid polychaete *Amphitrite ornata* it is partly attributable to the mode in which its hemoglobin operates (Mangum, unpublished data). Here, as in most invertebrates, our knowledge of circulatory dynamics is so primitive that we cannot assess the relative importance of gill perfusion and hemoglobin oxygenation. But we can evaluate the changes under low oxygen conditions in (1) ventilation or (2) both gill perfusion and hemoglobin oxygenation. When measurements of the volume of water pumped through the tube were made according to the procedure described by Dales, Mangum, and Tichy (1970), variation between the eleven individuals studied was so great that changes would be obscured by presenting the results in absolute values. When the data are analyzed as paired observations, however, it is clear that an initial reduction the environmental oxygen level to 3.1 ml/l brings about a transient increase in flow of about 25% over the control value (Fig. 7B). This response, achieved by an increase in the amplitude of muscular movements rather than a change in parameters of the spontaneous rhythm itself, persists for only a few hours and then it diminishes until the rate is indistinguishable from the control. Further reductions to 1.9, 1.1, and 0.5 ml/l cause no further changes, although one of the eleven worms did attempt to emerge from its tube under essentially anoxic conditions. There is no mechanism increasing the ventilatory rate at low oxygen levels. Therefore, the failure of oxygen uptake and extraction to show a linear decline in the face of a decreasing PO$_2$ gradient across the gills (Mangum, unpublished data) as well as a decreasing volume of oxygen available (Fig. 7B) implicates a compensatory response by the circulatory system.

Flow across the gills of lamellibranch molluscs has often been computed from the clearance rate of small particles in suspension (Jørgensen, 1966). The method probably yields highly unrealistic values due to incomplete particle retention. When *Mya arenaria* and *Rangia cuneata* are allowed to clear prodag (Anderson, 1972) at air saturation, the estimates of flow (vol/time) are an order of magnitude lower than those required to explain the oxygen uptake and oxygen extraction figures. In contrast, the estimates made volumetrically for *A. ornata* (Fig. 7B) are not significantly lower than those required to explain oxygen uptake and oxygen extraction.

If we assume that particle retention is not oxygen-dependent, however, the response of clearance rate to low oxygen levels may permit identification of the compensatory system. In *M. arenaria*, the estimated rate at PO_2 = 140 mm Hg is 344 ± 32 (N = 11) ml/hr, and at PO_2 = 30 mm Hg it is only 15% of that value, or 52 ± 2 (N = 7) ml/hr. In *R. cuneata*, the rates are 68 ± 9 (N = 5) ml/hr at PO_2 = 105 mm Hg, and 7 ± 0 (N = 10) ml/hr at PO_2 = 28 mm Hg. In both species the clearance rate varies directly with PO_2, while the oxygen extraction rate varies inversely (Figs. 2A and 3B). This result suggests that the compensatory responses enabling the clams to maintain a relatively stable rate of oxygen uptake (Figs. 2C and 3A) occur within the gill, a conclusion that also has been reached from different kinds of evidence for the mussel *Mytilus edulis* (Bayne, 1971).

ACKNOWLEDGMENTS

Our work was supported by a research grant from the Biological Oceanography Program of the National Science Foundation (NSF GA-34221). It was undertaken in part while the senior author was a William and Mary Faculty Research Fellow. We are grateful for the use of facilities belonging to the Marine Biological Laboratory, Woods Hole, Massachusetts, and the Duke University Marine Laboratory, Beaufort, North Carolina. The measurements of active and resting metabolism in annelids were made in part by C. Sassaman and P. E. Coyer, and those of oxygen consumption in *Rangia cuneata* by L. J. Kushins. We are grateful to P. W. Hochachka for calling our attention to the problem of aerobic respiration in *Rangia*.

LITERATURE CITED

Aiello, E. 1960. Factors affecting the ciliary activity on the gill of *Mytilus edulis*. Physiol. Zool. 33:120-35.

Anderson, G. E. 1972. The effects of oil on the gill filtration rate of *Mya arenaria*. Va. J. Sci. 23:45-47.

Bayne, B. L. 1971. Ventilation, the heart beat and oxygen uptake by *Mytilus edulis* L. in declining oxygen tension. Comp. Bioch Physiol. 40A:1065–85.

Biggs, R. B. 1967. The sediments of Chesapeake Bay. In: Estuaries (G. H. Lauff, ed.), pp. 239–60. Washington, D. C.: AAAS Publ. No. 83.

Carpenter, J. H. and D. G. Cargo. 1957. Oxygen requirement and mortality of the blue crab in the Chesapeake Bay. Ches. Bay Inst. Tech. Rep. 13:1–22.

Chen, C. and J. Awapara. 1969. Effect of oxygen on the end-products of glycolysis in *Rangia cuneata*. Comp. Biochem. Physiol. 31:395–401.

Coyer, P. E. and C. P. Mangum. 1973. Effects of temperature on active and resting metabolism in polychaetes. In: Temperatur Adaptation in Heterothermic Organisms (W. Wieser, ed.). Heidelberg: Springer-Verlag. In press.

Dales, R. P., C. P. Mangum, and J. C. Tichy. 1970. Effects of changes in oxygen and carbon dioxide concentrations on ventilation rhythms in onuphid polychaetes. J. Mar. Biol. Assoc. U. K. 50:365–80.

Hoffmann, R. J. and C. P. Mangum. 1970. The function of coelomic cell hemoglobin in the polychaete *Glycera dibranchiata*. Comp. Biochem. Physiol. 36:211–28.

_____ and _____. 1972. Passive ventilation in benthic annelids? Science 176:1356.

Jørgensen, C. B. 1966. The Biology of Suspension Feeding. New York: Pergamon Press.

Kushins, L. J. and C. P. Mangum. 1971. Response to low oxygen conditions in two species of the mud snail *Nassarius*. Comp. Biochem. Physiol. 39A:421–35.

Mangum, C. P. 1970. Respiratory physiology in annelids. Amer. Scient. 58:641–47.

_____. 1972. Temperature sensitivity of metabolism in offshore and intertidal onuphid polychaetes. Mar. Biol. 17:108–14.

_____. 1973. Evaluation of the functional properties of inverte-brate hemoglobins. Neth. J. Sea Res. In press.

_____ and J. A. Carhart. 1972. Oxygen equilibrium of coelomic cel hemoglobin from the bloodworm *Glycera dibranchiata*. Comp. Biochem. Physiol. 43A:949–57.

_____ and C. D. Cox. 1971. Analysis of the feeding response to chemical stimulation in the onuphid polychaete *Diopatra cuprea* (Bosc). Biol. Bull. 140:215–29.

_____, S. L. Santos, and W. R. Rhodes. 1968. Distribution and feeding in the onuphid polychaete *Diopatra cuprea* (Bosc). Mar. Biol. 2:33–40.

_____ and C. Sassaman. 1969. Temperature sensitivity of active and resting metabolism in a polychaetous annelid. Comp. Bioch Physiol. 30:11–16.

_____ and W. Van Winkle. 1973. The response of aquatic inverte-brates to declining oxygen conditions. Amer. Zool. 13:529–41.

Prasada Rao, D. V. G. and P. N. Ganapati. 1968. Respiration as a function of oxygen concentration in intertidal barnacles. Mar. Biol. 1:309-10.

Samuelson, P. A. 1955. Economics: An Introductory Analysis. New York: McGraw-Hill.

Sassaman, C. and C. P. Mangum. 1972. Adaptations to environmental oxygen levels in infaunal and epifaunal sea anemones. Biol. Bull. 143:657-78.

Stokes, T. M. and J. Awapara. 1968. Alanine and succinate as end-products of glucose degradation in the clam Rangia cuneata. Comp. Biochem. Physiol. 25:883-92.

Vogel, S. 1974. Flow outside and inside the sponge Halichondria. In preparation.

_____ and W. L. Bretz. 1972. Interfacial organisms: Passive ventilation in the velocity gradient near surfaces. Science 175:210-11. Also, Ibid. 176:1356.

Adaptations to extreme environments

F. J. Vernberg and W. B. Vernberg

An organism must be able either to compensate or endure changes in any of the ambient environmental parameters to survive. For convenience one can view this organismic-environmental relationship as a dynamic equilibrium. When one of the components of the environment becomes too extreme, the organism cannot compensate and the organismic-environmental equilibrium is destroyed and death results. Less harsh environmental conditions, however, may drastically alter the equilibrium characteristics, but the organism survives. Species and, in some cases, different populations of the same species exhibit great diversity in their response to fluctuations in environmental complexes. The physiological ecologist seeks to analyze this diversity, not only to characterize the limits of toleration for various biotic groupings but also to understand the functional basis that enables organisms to compensate for environmental change.

Typically when an organism is exposed to an environmental gradient, such as temperature, it can survive intermediate exposures, but the extremes cause death. The middle portion of this gradient in which the organism can withstand prolonged exposure is called the biokinetic zone (zone of capacity adaptation). In contrast, the extreme portions of the gradient which result in death are referred to as the lethal zone (zone of resistance adaptation). These zones are graphically represented in Figure 1. Obviously the lethal zone represents an extreme environment. The lethal zone cannot be easily characterized since duration of exposure in some unit of time and intensity of the environmental factor must be considered as well as certain intrinsic organismic factors. For example, 50% of all adult fiddler crabs (genus *Uca*) from the tropics die within 35 min when exposed to 7°C, but at 15°C significant levels of mortality are not reached for a two-week period. Thus, both temperatures are lethal, but are both equally extreme? Can and should the term "extreme environment" be expressed quantitatively?

Fig. 1. The lethal and biokinetic zones (from Vernberg and Vernberg, 1970).

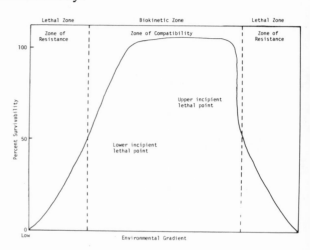

The response of a species to an environmental extreme may differ with stage of its life cycle, and larval and adult stages frequently occupy different positions on an environmental gradient. However, it is not possible to predict with a high degree of reliability which stage is most resistant. The following example illustrates this point. The adult stage of most species of fiddler crabs survives a salinity range from 0 to 60 o/oo but the first stage zoeae die at salinities below 20 o/oo. In contrast, zoeae of certain tropical species survive 5°C for 48 hrs, but the adults die within 35 min at 7°C. An extreme environment for tropical fiddler crabs depends on the stage of development: adults are euryhaline and the larvae are stenohaline, and the adults are stenothermal and the larvae eurythermal.

The physiological state of an organism determines both the lethal influence of an environmental extreme and the type of compensatory adaptation within the biokinetic zone. For example, starved isopods died more quickly than well-fed individuals when both groups were exposed to high temperature, and fiddler crabs which were exposed to food demonstrated a higher degree of compensatory metabolic response to decreased temperature than nonfed individuals (Vernberg, 1959b). Many other variables can be cited which may influence the organism's physiological response to extreme environments, such as body size, sex, molting, previous exposure to environmental conditions, and season (see Vernberg and Vernberg, 1970, and Vernberg and Vernberg, 1972, for a more detailed discussion of these factors).

A given environmental complex may have little effect on the survival of adult individuals but it could prevent the successful completion of the reproductive cycle, thereby resulting in the death of the population in terms of population continuity. Under these circumstances, the use of the term "extreme environment" must be carefully defined. The polychaete *Capitella capitata*, for example, survives for several days in water with an oxygen content below 2.9 ppm, but it will not feed and grow until the oxygen level is higher than this value. However, this species will not reproduce until the oxygen concentration is greater than 3.5 ppm (Reish and Barnard, 1960). Thus, an oxygen regime of 2.9 to 3.5 ppm is extreme in terms of the survival of the species but not to the individual organism. It is apparent that the term "extreme environment" must be carefully defined if it is to have scientific significance.

Although most of the examples presented above demonstrate how one environmental parameter has an extreme effect on organisms, one should keep in mind that the environment is the sum total of the interaction of many separate factors. These factors may act synergistically on organisms to produce an effect which is more extreme than that which results when studying each factor independently. Also, exposure to sublethal intensities of two environmental factors may be lethal to an organism (Vernberg, DeCoursey, and Padgett, 1974). To adequately describe an extreme environment or to predict when an environment will become extreme requires that physiological ecologists understand multiple-factor interaction.

In summary, we can conclude that the environmental-organismic interaction represents a complex dynamic equilibrium. Whenever the environment becomes extreme for sufficient time, this equilibrium is destroyed and the organism or the population perishes. However, an environment may show various degrees of extremeness depending on the numerous variables which influence the critical transition point between the zone of lethality and the biokinetic zone.

Another view of defining extreme environments is based on the comparative study of the various habitats available to organisms. An extreme environment (habitat) is characterized as having low species diversity and an environmental parameter (or parameters) in which intensity of expression is either high or low compared with its expression in other habitats. For example, the prolonged cold temperature of the polar intertidal zone and the high temperatures of hot springs are expressions of an extreme physical factor. The anaerobic muds of certain lakes and seas and the waters of tidepools supersaturated with oxygen are extremes of a chemical gradient in these diverse habitats. Not only extremes in physical and chemical factors result in a harsh environment, but also biotic parameters, such as overcrowding and intense predation, have a profound influence on the presence or absence of a species. Obviously organisms living in these areas are adapted to these environmental complexes and the environments are not extreme to them. However, few species occupy these habitats and thus in a relative sense these areas represent extreme environments.

On a broad geographical basis, there are physiological differences between organisms from different habitat types. Not unexpectedly, polar organisms are adapted to cold temperature, tropical animals to warmer temperatures, and many temperate zone species can withstand a wider thermal range. The data graphically presented in Figure 2 show the correlation between lethal limits and geographical distribution of marine fishes and fiddler crabs. Of significance, some polar animals die of heat death at temperatures below that which is the lower lethal level of tropical animals. Also, various functional processes, including respiration, reproduction, and growth, demonstrate this phenomenon of physiological adaptation (Vernberg, 1962, 1974). On this basis an extreme environment must be defined in terms of the range of environmental factors "normally" encountered by a species. Irregularly occurring environmental perturbations may represent an extreme environment. Many of these happenings are transitory and unpredictable, such as winter and/or summer thermal kills, the appearance of a warm counter-currents, and oil spills.

Fig. 2. The lethal limits of fiddler crabs and fish from different latitudes (fish data modified from Brett, 1970; fiddler crab data from Vernberg and Tashian, 1959; and Vernberg and Vernberg, 1967).

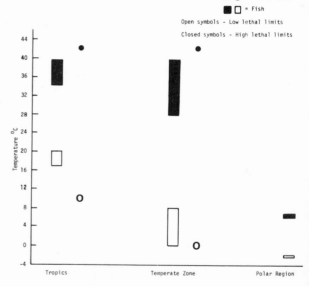

Certain unique habitats are found in restricted geographical regions which represent extreme environments. These are extreme in the sense that few species are adapted to survive the peculiar set of environmental conditions found there. For example, hot springs are the warmest habitats known to be occupied by active organisms; some algae live at temperatures of 85.2°C. Brues (1939) reported that the number of species decreases as the temperatures increase: fifty-seven species in springs with temperatures from 36 to 40°C, twelve from 41 to 45°C, and four from 56 to 50°C.

Another type of extreme environment is that which an animal encounters at the end of its geographical range. This paper will deal specifically with this problem as represented by fiddler crabs (genus *Uca*), a group of decapod crustaceans found in the north and south temperate zones and in the tropics. New data on both resistance and capacity adaptations of larval fiddler crabs are compared with previously published data on adults.

MATERIALS AND METHODS

Animals

Five species of *Uca* representing the tropical and temperate zones were studied. Tropical species were *Uca rapax* and *U. thayeri* from Puerto Rico, while the temperate zone species were *U. minax*, *U. pugilator*, and *U. pugnax* from Georgetown, South Carolina, and from North Carolina. A detailed description of their distribution was given in a previous paper (Vernberg, 1959a).

Ovigerous females were collected and maintained individually in finger bowls in the laboratory under conditions of constant temperature (25°C) and salinity (30 o/oo) in B.O.D. incubators. They were exposed to a controlled photoperiod of 14 hrs light and 10 hrs dark. Each morning the bowl was checked for free-swimming first stage zoeae. Only freshly hatched zoeae were used experimentally. During the course of the experiments to be described below, the zoeae were fed *Artemia* and fertilized *Arbacia* eggs. In all, about 250 ovigerous females were used. This was necessary to insure that a representative sample of a population was used rather than being restricted by the limited genetic variation that might result if zoeae from only one female was studied.

Resistance Adaptation

The thermal-salinity tolerance of each species was determined by the same procedure. Fifteen to twenty-five zoeae were placed in a flask containing seawater of the desired salinity. Each flask had been immersed in a constant temperature bath for a sufficient period of time to allow the seawater in the flask to reach the desired experimental temperature before the animals were introduced. The temperature was read frequently by means of a telethermometer (Yellow Spring Instrument Company). Low salinity water was obtained by diluting seawater with distilled water and determining the salinity by means of a standard hydrometric technique. At the end of a test period the animals were removed and their condition noted. An animal was considered dead if under microscopic examination there was no visible movement of appendages and no discernible heart beat. In a few cases an animal was judged dead by these criteria when viewed immediately after the termination of an experiment, but recovered with time. Very infrequently an animal initially considered alive was dead upon subsequent examination. Therefore, it was necessary to make observations immediately and after a period of time (4 to

8 hrs). Each animal was only used once. From one hatch of larvae, some zoeae were used for temperature-salinity tolerance studies, and in almost all cases, another sample was used for rearing studies. This last group served as a control to indicate the viability of the hatch. Although most of the studies on temperate zone species was done during the period May through September, some observations on animals collected in late March were made to determine if seasonal differences occurred.

Combinations of the following temperatures and salinities were used: salinities, 10 o/oo, 20 o/oo, 35 o/oo, and 50 o/oo at temperatures of 34°, 36°, 38°, and 40°C. These combinations may be encountered in the animal's natural environment. Additional observations were made at low temperatures. During the course of an experiment, flasks containing zoeae were withdrawn at intervals of time to give an estimate of the duration of exposure necessary to produce 50% mortality. The 50% mortality level, referred to as the LD_{50}, was calculated for each temperature-salinity combination and for each species.

Capacity Adaptation

To determine the influence of temperature and salinity on larval growth, 100 first stage zoeae were isolated and placed in finger bowls. Daily, they were fed *Artemia* and fertilized *Arbacia* eggs and transferred to a clean finger bowl. Since the number of molts and living animals were recorded each day, the time to reach specific life cycle stages and the mortality could be determined.

RESULTS AND DISCUSSION

Resistance Adaptations

The points at which 50% of the animals survive (LD_{50}) determined at 38°C and 40°C and at different salinities are represented in Figures 3 and 4.

At 38°C the following trends were noted for all species. The lowest survival values were obtained at 10 o/oo and the maximum values at 35 o/oo. Although a curve somewhat resembling a bell shape resulted when plotting salinity against LD_{50} (in time), only in one species, *thayeri*, was the survival level at 50 o/oo as low as the value at 10 o/oo.

Two distinct patterns of response were noted: one in which the survival times at 35 o/oo and 20 o/oo were similar, and the second in which the greatest survival was reached at 35 o/oo and no plateau was observed. *Uca pugilator*, *pugnax*, and *rapax* are examples of the first type of response while *minax* and *thayeri* fit the second type of response. No apparent broad zoogeographical significance can be attached to this observation since tropical and temperate zone animals showed each type of response. However, habitat preference of the adult could play some role in the larval response. *Uca pugilator*, *pugnax*, and *rapax* have somewhat similar habitat requirements while *minax* and *thayeri* require similar substrate conditions.

Fig. 3. The influence of various combinations of salinity on the mortality of zoeae of fiddler crabs determined at 30°C (after Vernberg, 1969).

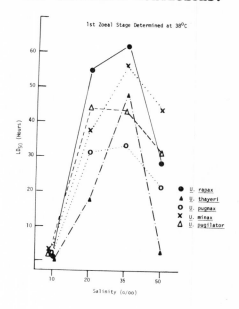

Fig. 4. The influence of various combinations of salinity on the mortality of zoeae of fiddler crabs determined at 40C (after Vernberg).

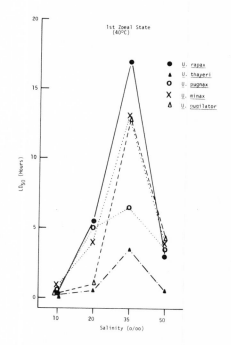

The species exhibiting the greatest resistance to reduced salinity (*minax*) is the species that consistently penetrates the greatest distance up the estuary with some populations living in fresh water. Species differences were observed when comparing the three species from the temperate zone. *Uca pugilator* was the least tolerant of reduced salinity. This species is restricted to a sandy substrate habitat, particularly protected sand beaches, and, in general, most of these sites are near high salinity waters or near an area in which there is fairly rapid mixing of water with the higher saline water of the incoming tide. Hence, it could be expected that larvae of *pugilator* would not come into contact with 10 o/oo salinity water very frequently or for a long period of time.

To escape the lethal influence of high temperature and low salinity, ovigerous fiddler crabs go to the water when their embryos are ready to hatch. The time of hatching is restricted to the cooler evening hours between 7:00 and 10:00 P.M. (Hyman, 1920; Gray, 1942).

At a salinity of 50 o/oo all of the species, with the exception of *thayeri*, survived a period of time which would extend over several tidal cycles. Thus, larvae trapped in unfavorable conditions would have a physiological safety factor enabling them to survive until the tide changed. Although the magnitude of tidal change in Puerto Rico is slight, this change does influence the region in which fiddler crab burrows are found.

The same general relationships were observed at 40°C or at 38°C except that the survival times were much shorter (Fig. 4). One major difference was the appearance of only one response pattern and the lack of a plateau between 20 o/oo and 35 o/oo, while two types of response were observed at 38°C. It is not unusual to find a shift in patterns of response with temperature change (Fry, 1957). This might indicate a different mechanism is responsible for death.

Seasonal differences in survival values were observed at certain salinity-temperature combinations for the two species of temperate zone crabs investigated (*pugilator* and *pugnax*) (Table 1).

TABLE 1

The LD_{50} values of *Uca* determined at different seasons of the year, salinities and temperatures

Species and Season	Experimental Conditions			
	38°C			
pugnax	10 o/oo*	20 o/oo	35 o/oo	50 o/oo
Spring	30 min	31 1/2 hrs	33 1/2 hrs	15 hrs
Summer	90 min	31 1/2 hrs	40 1/2 hrs	22 hrs
pugilator	10 o/oo*	20 o/oo	35 o/oo	50 o/oo
Spring	20 min	47 1/2 hrs	50 hrs	33 hrs
Summer	45 min	44 hrs	43 1/2 hrs	31 3/4 hr
	40°C			
pugnax	10 o/oo*	20 o/oo	35 o/oo	50 o/oo
Spring	< 15 min	2 hrs	3 hrs	2 hrs
Summer	30 min	5 hrs	6 1/2 hrs	3 1/2 hrs
pugilator	10 o/oo*	20 o/oo	35 o/oo	50 o/oo
Spring	< 15 min	2 1/2 hrs	16 hrs	< 1 hr
Summer	20 min	1 hr	12 3/4 hrs	4 1/4 hrs

*o/oo = salinity in parts per thousand.

This response is in contrast to that observed for tropical species. In these species there was no seasonal shift although ovigerous females were collected throughout the year. At 10 o/oo summer zoeae of both species survived longer at both 38°C and 40°C than spring animals. Summer *pugnax* tended to survive longer than spring animals, but the same trend was not found for *pugilator* in all combinations of salinity and temperature. At 38°C and 40°C *pugilator* spring animals were slightly more resistant than summer zoeae at 20 o/oo and 35 o/oo. Although the functional significance of these differences is obscure at present, the data indicate that field acclimation phenomena influence the response of early stage larvae. The mode of action is not indicated in this study, but it is likely the environment acts on the reproductive physiology of adult fiddler crabs and/or any phase of embryological development up to the time of hatching.

No simple correlation between temperature-salinity survivalship and geographical distribution was apparent when comparing these five species of fiddler crabs. *Uca rapax* from the tropics was most thermally resistant, whereas *thayeri*, another tropical species, was least resistant at 40°C and the third least resistant species at 38°C. However, Patel and Crisp (1960) reported a distinct relationship between the seasonal breeding period, the upper thermal limit, and the geographical distribution of barnacles. Animals from higher latitudes had lower temperature limits and those species breeding during warmer months had the highest upper thermal lethal limits. But if we compare two closely related species, *rapax* and *pugnax*, we also find a similar correlation.

Comparative data on the salinity-temperature lethal limits at lower temperature (34°C and 36°C) are available for *rapax* and *pugnax* (Figs. 5 through 8). These two species were selected because they are closely related taxonomically and ecologically, and one is from the tropics while the other is from the temperate zone. At one time these two species were considered subspecies of one species. *U. rapax* was called *U. pugnax rapax* and *U. pugnax* was *U. pugnax pugnax* (Tashian and Vernberg, 1958). A thermal increase from 34°C to 36°C resulted in a decrease in survival for both species. The tropical species survived longest at 36°C but the species difference was much greater at 38°C. The LD_{50} at 38°C and 35 o/oo was 62 hrs for *rapax* and 33½ hrs for *pugnax*, while at 36°C these values were 90 hrs and 78 hrs, respectively. When comparing these two closely related species, the tropical species is more thermally resistant at a salinity of 35 o/oo (Fig. 9).

At a low temperature of 5°C zoeae of *rapax*, *pugnax*, and *pugilator* could survive exposure for 12 hrs. At 15°C *pugilator* and *pugnax* not only survived long exposures but also they molted. This response is unlike that reported by Passano (1960) and Miller and Vernberg (1968). They found this temperature inhibited molting in the adult, and Passano suggested that temperature limited the northward spread of these species in Massachusetts.

Fig. 5. Survival of larval *Uca rapax*
at different salinities when
determined at 34°C.

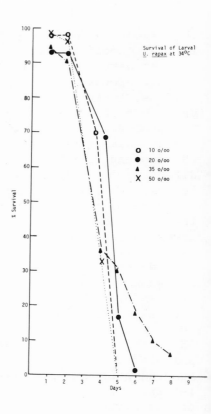

Fig. 6. Survival of larval *Uca pugnax*
at different salinities when
determined at 34°C.

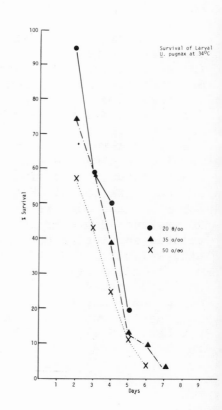

Fig. 7. Survival of larval *Uca rapax*
at different salinities when
determined at 36°C.

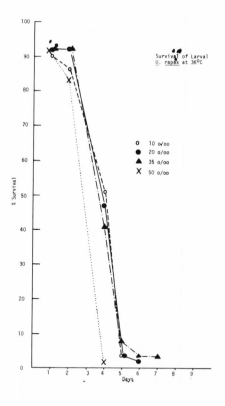

Fig. 8. Survival of larval *Uca pugnax*
at different salinities when
determined at 36°C.

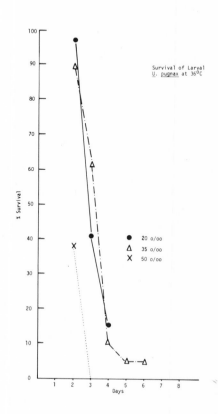

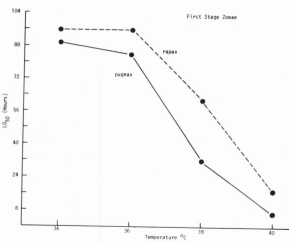

Fig. 9.   Comparative LD$_{50}$
values for zoeae of a
tropical fiddler crab
(Uca rapax) and a tem-
perate zone species
(Uca pugnax) determined
at 35 o/oo salinity.

   Thus, distinct differences exist between the capabilities of
adults and larvae to survive extreme environments. Neither stage
is consistently more resistant than the other to the environmental
parameters used in this study (temperature and salinity). The
adults are more tolerant to salinity stress than zoeae while in
contrast the zoeae of tropical fiddler crabs are more resistant to
low temperature than the adults. Low temperature limits the northward distribution of tropical fiddler crabs (Vernberg and Tashian,
1959).

Growth Studies

   Uca pugilator develop to the megalops stage faster than *pugnax*
at 20 o/oo and 30 o/oo and 25°C (Fig. 10). However, at 20°C and
30 o/oo *pugnax* tend to develop faster than *pugilator* while the
values for *minax* fall within the range for *pugnax* (Fig. 11). Salinities between 20 o/oo and 30 o/oo did not appreciably alter the development rate of *pugnax* although greater variation in time was observed
at the higher salinity. At salinities below 20 o/oo, no larvae
developed megalops.
   Temperature influences the rate of development in a predictable
manner: development is faster at 25°C than at 20°C. Marked differences between temperate zone species was not noticeable at 20°C

Fig. 10.   Development times to
megalopa stage from first stage
zoeae of fiddler crabs determined
at different salinities and at
25°C.   px = *Uca pugnax*; pt =
Uca pugilator.

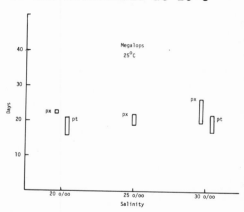

although *pugilator* tended to develop most slowly. However, at 25°C *minax* was much slower to reach the megalops stage than the other two species from Carolina. The tropical *rapax* responded similarly to *minax*. Generally it is expected that tropical species develop more slowly than temperate zone species at a common temperature (Vernberg, 1962).

Fig. 11. Development times to megalopa stage from first stage zoeae of fiddler crabs determined at different temperatures and at a salinity of 30 o/oo. px = *Uca pugnax*; pt = *Uca pugilator*; mx = *Uca minax*; rx = *Uca rapax*.

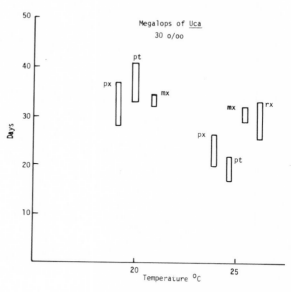

Developmental times to the crab stage vary both with temperature and species. Low temperature decreases the developmental rate (Fig. 12). This response is best illustrated by *pugilator* in that the range of development times is for 44 to 52 days at 20°C, 27 to 34 days at 25°C, and 21 to 35 days at 30°C. At 25°C marked interspecific differences were observed: *minax* and *rapax* developed more slowly than did *pugilator*.

Populations of *pugilator* from different geographical areas developed at different rates (Fig. 13). At 25°C, zoeae from Massachusetts developed into the crab stage faster than individuals from the Carolinas or Florida. Florida animals showed a wider range of development time than the populations from Carolina.

Fig. 12. Development times to crab stage from first stage zoeae of fiddler crabs determined at different temperatures and at a salinity of 30 o/oo. px = *Uca pugnax*; pt = *Uca pugilator*; rx = *Uca rapax*; mx = *Uca minax*.

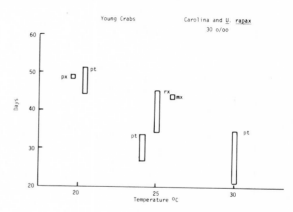

The observation that the population from the colder waters (Massachusetts) developed faster than animals from warmer water is consistent with the general theory of climatic adaptation; cold-adapted animals are functionally more active than closely related species (or populations of the same species) from warmer environments (see review of Vernberg, 1962). Differences in developmental rate have been reported for some geographically separated populations of marine animals (Bayne, 1965, *Mytilus edulis* from Wales and Denmark; and Dehnel, 1955, *Thais emarginata* from Alaska and California) but not all (Scheltema, 1967, *Nassarius obsoletus*).

Fig. 13. Development times to crab stage from first stage zoeae of *Uca pugilator* from different latitudes. Zoeae were reared at a salinity of 30 o/oo and at different temperatures. Mass. = Massachusetts; Fla. = Florida; and Carolina = South and North Carolina.

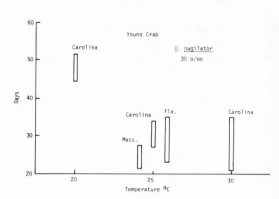

Seasonal differences were also observed in developmental times in *pugilator* from Carolina. During the summer the zoeae reached the megalops stage between 17 to 22 days whereas during September the range was from 16 to 29 days. No data are available to explain this observation, but it is possible that the variability in eggs produced is a function of differential reproductive capability of the female late in the breeding season. It is conceivable that mechanisms geared for egg production become less efficient near the termination of reproduction.

As organisms reach the limits of their geographical range, they experience environmental extremes which either may be lethal to some stage in its life cycle or restrict some physiological function within the biokinetic zone. Data from the present paper, as well as those from previous studies, indicate the complex nature of the organismic-environmental interactions.

LITERATURE CITED

Bayne, B. L. 1965. Growth and the delay of metamorphosis of larvae of *Mytilus edulis* L. Ophelia 2:1-47.
Brett, J. R. 1970. Temperature and fish. In: Marine Ecology (O. Kinne, ed.), pp. 515-60. New York: Wiley-Interscience.
Brues, C. T. 1939. Studies on the fauna of some thermal springs in the Dutch East Indies. Proc. Am. Acad. Arts and Sci. 73: 71-95.

Dehnel, P. A. 1955. Rates of growth of gastropods as a function of latitude. _Physiol_. _Zool_. 28:115-44.

Fry, F. E. J. 1957. The lethal temperature as a tool in taxonomy. _Annee Biol_. 33:205-19.

Gray, E. H. 1942. Ecological and life history aspects of the red-jointed fiddler crab, _Uca minax_ (Le Conte), in the region of Solomons Island, Maryland. Chesapeake Biol. Lab. Publ. No. 51.

Hyman, O. W. 1920. The development of _Gelasimus_ after hatching. _J_. _Morph_. 33:484-525.

Miller, D. C. and F. J. Vernberg. 1968. Some thermal requirements of temperate and tropical zone fiddler crabs influencing geographic distribution. _Am_. _Zool_. 8:459-69.

Passano, L. M. 1960. Low temperature blockage of molting in _Uca pugnax_. _Biol_. _Bull_. 118:129-36.

Patel, B. and D. J. Crisp. 1960. Rates of development of the embryos of several species of barnacles. _Physiol_. _Zool_. 33: 104-19.

Reish, D. J. and J. L. Barnard. 1960. Field toxicity tests in marine waters utilizing the polychaetous annelid _Capitella capitata_ (Fabricius). _Pacific_ _Nat_. 1:1-8.

Scheltema, R. S. 1967. The relationship of temperature to the larval development of _Nassarius obsoletus_ (Gastropoda). _Biol_. _Bull_. 132:253-65.

Tashian, R. E. and F. J. Vernberg. 1958. The specific distinctness of the fiddler crab _Uca pugnax_ (Smith) and _Uca rapax_ (Smith) at their zone of overlap in northeastern Florida. _Zoologica_ 43: 89-92.

Vernberg, F. J. 1959_a_. Studies on the physiological variation between tropical and temperate zone fiddler crabs of the genus _Uca_. II. Oxygen consumption of whole organisms. _Biol_. _Bull_. 117:163-84.

_____. 1959_b_. Studies on the physiological variation between tropical and temperate zone fiddler crabs of the genus _Uca_. III. The influence of temperature acclimation on oxygen consumption of whole organisms. _Biol_. _Bull_. 117:582-93.

_____. 1962. Latitudinal effects on physiological properties of animal populations. _Ann_. _Rev_. _Physiol_. 24:517-46.

_____. 1969. Acclimation of intertidal crabs. _Am_. _Zool_. 9:333-41.

_____. 1974. Editor. _Physiological_ _Adaptation_ _to_ _the_ _Environment_. New York: Intext Publishers.

_____ and R. E. Tashian. 1959. Studies on the physiological variation between tropical and temperate zone fiddler crabs of the genus _Uca_. I. Thermal death limits. _Ecology_ 40:589-93.

_____ and W. B. Vernberg. 1967. Studies on the physiological variation between tropical and temperate zone fiddler crabs of the genus _Uca_. IX. Thermal lethal limits of southern hemisphere _Uca_ crabs. _Oikos_ 18:118-23.

_____ and _____. 1970. _The_ _Animal_ _and_ _the_ _Environment_. New York: Holt, Rinehart and Winston.

Vernberg, W. B., P. J. DeCoursey, and W. J. Padgett. 1974. Synergistic effects of environmental variables on larvae of *Uca pugilator* (Bosc). <u>Marine</u> <u>Biology</u> 22:307-12.

_____ and F. J. Vernberg. 1972. <u>Environmental</u> <u>Physiology</u> <u>of</u> <u>Marine</u> <u>Animals</u>. New York: Springer-Verlag.

An analysis of water-content regulation in selected worms

L. C. Oglesby

A major factor in the successful adaptation of soft-bodied marine worms to estuarine conditions is the ability to stabilize the volume of the body at or near some optimal level. Unrestricted inflow of water into the coelom and tissues after exposure to lower salinities may stretch the body wall to the point of serious impediment to proper functioning of the body-wall musculature. The worms may be rendered incapable of burrowing, crawling, performing respiratory movements, or feeding (e.g., Oglesby, 1965a, 1969b, 1973). In estuaries with even moderate tidal variations in salinity, it is likely that excessive swelling and shrinking with every semidiurnal or even diurnal tidal cycle could effectively prevent the colonization of the estuary by worms with poor control over body volume, even if the osmotic or ionic concentrations of their body fluids never fell to lethal levels (Oglesby, 1969b).

Control of body volume in soft-bodied worms is generally equated with control of the amount of water within the animal. Studies of water-content control have been approached in at least three different ways. The first has been the measurement of total water content. Oglesby (1969a) has summarized the measured water contents for about 50 species of marine and estuarine worms (polychaete annelids, sipunculans, and echiurans), all apparently adapted to high salinities. No correlations were found between water content and either the ecological distributions or the taxonomic relationships of the various species. However, there were so many differences among analyses with respect to maintenance salinity

and period of adaptation to that salinity, nutritional state,
reproductive state, etc., that even if strong correlations were
present, they could easily have been obscured by the (usually
unstated) variability of the data and the lack of equivalent
conditions for measurements.

A second, more informative, approach has been the determination
of steady-state water contents of worms after adequate acclimation
to a wide range of salinities. Such measurements have been made
for a number of nereid polychaetes (Bogucki and Wojtczak, 1964;
Krishnan, 1952; Mokrousova, in Ivleva and Popenkina, 1966; Oglesby,
1965a, 1970), the lugworm *Abarenicola pacifica* (Oglesby, 1973), and
the sipunculan *Themiste dyscritum* (Oglesby, 1968b). In all of
these species, there is a similar trend: the more dilute the
external medium becomes (i.e., the lower the salinity), the more
hydrated the worms become. The relationship between salinity and
water content varies among species: the more euryhaline a species,
the less steep the slope of its water-content curve. In other
words, the most euryhaline species have a steady-state water content
which varies the least with external salinity.

A third approach has been the study of responses after transfers from one salinity to another. When a marine or estuarine worm
is taken from high salinity, and placed in any lower salinity, it
will take up water by osmosis. The amount of water taken up
initially depends on at least two factors. One of these is the
magnitude of the transfer. The greater the difference between the
initial and final salinity, the more water is taken up. There are
numerous published examples (reviewed in Oglesby, 1969a). A second
factor is the species. For transfers under identical conditions,
some species will take up more water than others. Stenohaline
worms incapable of surviving in diluted seawater will osmotically
take up much more water than euryhaline species which can live in
highly diluted seawater for long periods, even when the transfers
are within the tolerable salinity range. The clearest examples
include several species of nereid polychaetes (reviewed by Oglesby,
1969a).

After a worm is transferred to a lower salinity, the initial
osmotic inflow apparently takes place through the body wall. Water
enters chiefly into the coelom and other extracellular spaces;
swelling of the tissues is proportionally not as great as swelling
of the coelom (Oglesby, 1965a, 1969a, 1970). Net uptake of water
is fairly rapid for the first 2 to 4 hrs after the transfer. This
initial phase is followed by a plateau period, lasting 4 to 8 hrs
or so, during which there is relatively little change in water
content (Oglesby, 1969a). Following this plateau period, two
different types of responses may occur, depending on the species.

In one response, there is no subsequent loss of water. The
worms remain swollen at the plateau level until death, or until
transferred to a higher salinity. This response is characteristic
of stenohaline marine species, and even of some estuarine species
if the salinity transfer is beyond a certain limited, tolerable

range. Examples include the polychaetes *Eudistylia vancouveri* (Hoar, 1966), *Cirriformia spirabrancha* (Dice, 1969), and *Glycera* sp. (Krishnamoorthi, 1962), and the sipunculan *Themiste dyscritum* (Oglesby, 1968b). In such species, if the transfer involves too great a salinity difference, death may occur at any time during the period of initial water inflow or during the plateau period, sometimes accompanied or caused by the body wall's bursting from the inflow of water.

In the second type of response, there is subsequent loss of water. The weight of the worms approaches the initial (lower) weight in the higher salinity, a process generally taking some one to three days. The initial weight is seldom fully restored. Rather, the weight (body volume, or water content) levels off at some higher value, depending upon the salinity and the species. The more euryhaline a species is, the less water it is likely to gain initially and the more water it is likely to lose subsequently. This latter phase of slow weight loss is usually termed "volume regulation" (Oglesby, 1969a), a process which must involve an active metabolic response by the worm. Examples include most of the nereid polychaetes and a number of other estuarine and euryhaline marine species (Oglesby, 1969a).

Regulation of body volume (water content) and regulation of osmotic concentration of the body fluids do not necessarily involve the same physiological processes, nor do they necessarily occur with the same time course after transfers (Oglesby, 1969a). Some species are able to regulate water content to a considerable extent, but are osmoconformers and are therefore unable to regulate the total osmotic concentration of the body fluids. Therefore, the term osmoregulation should not be used as a synonym for volume regulation (e.g., Bair and Peters, 1971; Croll and Viglierchio, 1969), as confusion and even error in discussions of facts and proposed mechanisms in regulation of salt and water balance may result.

If a worm were totally incapable of volume regulation, and if it were totally incapable of resisting the initial osmotic inflow of water, there should be a maximal amount of water it would take up osmotically after a given transfer. Probably no worm which is capable of even a slight amount of volume regulation attains this maximum value. The difference between this theoretical maximum value after a transfer, and the actual value for a worm fully adapted to the new salinity represents a certain amount of water which either did not enter the worm in the first place, or else was subsequently eliminated during volume regulation.

It would be of value to be able to estimate the maximum amount of water a worm theoretically would take up after a transfer to a lower salinity if it were totally incapable of regulating its body volume and water content (i.e., if it behaves as a simple osmometer). This theoretical maximum can then be compared with the amount of water actually taken up after the transfer. In this way it would be possible to assess quantitatively the efficiency of various worms in regulating water content. In this paper, such a theoretical formulation is developed, and applied to several species of worms for which adequate data on water content exist.

THE CASE OF NO REGULATION OF WATER CONTENT (THE "SIMPLE OSMOMETER"

By definition, the percentage water content $\underline{P}$ in a worm is:

$$P = 100 \left(\frac{W}{S + W} \right)$$

where $\underline{W}$ is the weight of water in the worm, and $\underline{S}$ is the weight of solids (dry weight). These solids would include both osmotically active particles as well as a much greater weight of osmotically inactive material.

Assume that the chloride concentration of the medium (express as $\underline{mM}$ Cl^-) is proportional to the osmotic concentration $\underline{C}$ of the medium, which, for an iso-osmotic osmoconformer, is also the same as the osmotic concentration of the internal fluids of the worm. Assume also that all the water in the worm is osmotically active, and that it is uncompartmentalized (i.e., there is no barrier to water exchange between intra- and extracellular fluids). Since th is a simple osmometer, there is no alteration in the number of osmotically active particles present in the body fluids following changes in the external salinity.

Such a worm, behaving as a simple osmometer (i.e., ready permeability of the body surface to water, no permeability of the body surface to salts or other solutes), when transferred from a high salinity to a low salinity will gain water by osmosis in dire proportion to the magnitude of the concentration difference. All the water taken up (excess water) is retained even after full adaptation to the new salinity. Therefore,

$$W_2 = \left(\frac{C_1}{C_2} \right) W_1$$

or:

$$W_2 C_2 = W_1 C_1 = \underline{k}$$

Therefore,

$$\underline{k} = WC, \text{ or } W = \underline{k}/C$$

where $\underline{k}$ is a constant relating the water content of the organism to the osmotic concentration of the medium and of the coelomic fluid.

By substitution of Equation (2) into Equation (1), we obtain:

$$P_2 = \frac{100 \ (k/C)}{S + (k/C)} = \frac{100 \ (k/S)}{C + (k/S)}$$

For example, a worm is transferred from $C_1 = 560$ $\underline{mM}$ Cl^- (100% SW) to $C_2 = 280$ $\underline{mM}$ Cl^- (50% SW). If this worm behaves as a simple osmometer, the water content per unit dry weight will be twice as great in the lower salinity as in the higher salinity, as shown by Equation (2):

$$W_{280} = \left(\frac{560}{280}\right) W_{560} = 2W_{560}$$

Using Equation (4), we can calculate the percent water content P_2 of the worm at the new low salinity C_2, $\underline{k}$ having been calculated from Equation (3):

$$\underline{k} = \left(W_{560}\right)\left(C_{560}\right)$$

If our worm contained 80% water ($W_1 = 80$, S = 20) when in $C_1 = 560$ $\underline{m}M$ Cl^-, $\underline{k}$ will be:

$$\underline{k} = (80)(560) = 44,800$$

After the transfer to $C_2 = 280$ $\underline{m}M$ Cl^-, the new water content P_2 would be, from Equation (4):

$$P_2 = \frac{100\left(\frac{44,800}{20}\right)}{280 + \left(\frac{44,800}{20}\right)} = 88.9\% \text{ water}$$

Equation (4) will generate a curve of percentage water content of a worm behaving as a simple osmometer, as a function of the external salinity (see the line marked $\beta = 1.00$ in Figure 1).

THE CASE OF REGULATION OF WATER CONTENT

If, however, the worm is capable of regulating its steady-state water content (that is, if it is not a simple osmometer), the amount of water ultimately retained after such a transfer will be less than that calculated by Equation (2), and the ultimate percentage water content will be less than that calculated by Equation (4). (For the purposes of the following treatment, it is irrelevant how this reduction in ultimate water content is brought about.)

β is defined as the proportion of the maximum excess water which is actually retained after a transfer, such that $\beta = 1.0$ in the case of no regulation by a simple osmometer (maximum excess water retention), and $\beta = 0.0$ in the case of complete regulation (no excess water retention: water content same in all salinities). It follows that, after a transfer from C_1 to C_2,

$$W_2 = \left(W_{2max} - W_1\right)\beta + W_1 \tag{5}$$

where W_{2max} is the total water retained by a simple osmometer as given by Equation (2). By substitution with Equation (2), Equation (5) becomes:

$$W_2 = \left[\left(\frac{C_1}{C_2}\right) W_1 - W_1\right]\beta + W_1$$

or,

$$W_2 = W_1\left[\frac{(C_1 - C_2)\beta + C_2}{C_2}\right]$$

and, by substitution for W_1 from Equation (3), we obtain:

$$W_2 = \left(\frac{k}{C_1}\right) \left[\frac{(C_1 - C_2)\ \beta + C_2}{C_2}\right]$$

and, by substitution into Equation (1), we obtain a relationship which provides the percentage water content P_2 after transfer to any different salinity C_2, when a certain proportion $(1-\beta)$ of the excess water is not retained:

$$P_2 = \frac{100\ (k/S)\ \left[\dfrac{(C_1 - C_2)\ \beta + C_2}{C_2}\right]}{C_1 + (k/S)\ \left[\dfrac{(C_1 - C_2)\ \beta + C_2}{C_2}\right]} \tag{6}$$

When $\beta = 1.00$, Equation (6) reduces to:

$$P_2 = \frac{100\ (k/S)}{C_2 + (k/S)} \tag{4}$$

which is the equation for the behavior of a simple osmometer (0% regulation of water content).

When $\beta = 0.00$, Equation (6) reduces to:

$$P_2 = \frac{100\ (k/S)}{C_1 + (k/S)} = 100 \left(\frac{W_1}{S + W_1}\right) \tag{1}$$

which is the equation for constant water content in all salinities (100% regulation of water content).

Let us return to the earlier example of the worm with $P_1 = 80\%$ water in $C_1 = 560$ mM Cl⁻, transferred to $C_2 = 280$ mM Cl⁻, and postulate that this worm is capable of regulating its water content to the extent that 75% of the excess water gain expected of a simple osmometer is not retained after full adaptation to the lower salinity (that is, $\beta = 0.25$). From Equation (2), we know that the maximum water uptake by a simple osmometer is:

$$W_{280max} = 2W_{560}$$

Therefore, from Equation (5),

$$W_{280} = (W_{280max} - W_{560})\ (0.25) + W_{560} = 1.25W_{560}$$

That is, the water content per unit dry weight of this worm in the lower salinity will be 1.25 times that in the higher salinity. Using Equation (6), we can calculate the percentage water content P_2 of this worm in the new low salinity:

$$P_2 = \frac{(100)\left(\dfrac{44,800}{20}\right)\left(\dfrac{(560-280)\ (0.25) + 280}{280}\right)}{560 + \left(\dfrac{44,800}{20}\right)\left(\dfrac{(560-280)\ (0.25) + 280}{280}\right)} = 83.3\%\ \text{water}$$

Thus, a worm behaving as a simple osmometer ($\beta = 1.00$) and having 80% water when in 560 mM Cl⁻ would be expected to have 88.9% water after adaptation to 280 mM Cl⁻, whereas a worm capable of 75%

regulation of its water content (β = 0.25) would be expected to have only 83.3% water when adapted to the lower salinity.

Equation (6) may be used to generate a series of curves of water content over a full range of external concentrations for different given β values from 0.00 (complete or 100% regulation) to 1.00 (no regulation: the simple osmometer). Figure 1 shows such a series of curves for a worm with P_1 = 80% water in C_1 = 560 mM Cl$^-$, with β values chosen for illustrative purposes as 1.00, 0.50, 0.25, 0.10, 0.05, and 0.00 (corresponding to 0%, 50%, 75%, 90%, 95%, and 100% regulation of water content, respectively).

Fig. 1. Theoretical water-content regulation curves calculated from Equation (6) for a worm with 80% water when adapted to 100% SW (560 mM Cl$^-$), for β values of 1.00 (no regulation), 0.50, 0.25, 0.10, 0.05, and 0.00 (complete regulation).

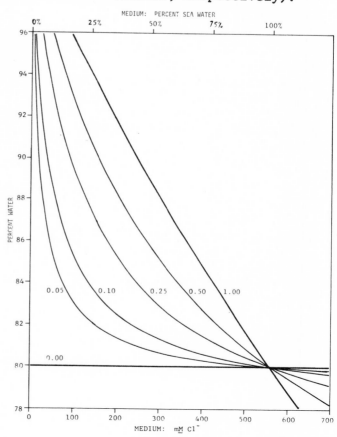

The family of curves in Figure 1 suggests that it requires rather low β values in order to effect any marked reduction in final water content after a transfer to a lower salinity. At lower β values (better regulation), the water-content curves become flatter at salinities greater than about 150 to 250 mM Cl$^-$, and then show a steep rise in slope at lower salinities.

By computing a series of curves for different β values, based on the actual water content of a worm when adapted to 100% SW, the actual water-content curve of that species can be compared with these theoretical curves in an attempt to find the β value which generates the best fitting curve. It would be expected that worms

with differing physiological capabilities for water-content regulation would have different β values. Determining β values will
permit direct comparisons of the abilities of different species to
control water content, using a quantitative measure. This analysis
will be performed for nine populations of seven species of polychaete annelids and sipunculans, using available published data for
steady-state water contents of worms in different salinities.

Water-Content Regulation in Osmoconformers

The first group of examples to be considered includes three
species in which there is limited or no regulation of the osmotic
concentration of the body fluids at any external salinity.
    1. *Nereis (Nereis) vexillosa* Grube. This nereid polychaete
is commonly found among rocks along the open sea coast and in
marine-dominated portions of certain estuaries from central California to Alaska. Figure 2 shows the water contents of worms from
Pt. Richmond in San Francisco Bay, adapted to salinities ranging
from 14% SW to 116% SW (Oglesby, 1965a). In Figure 2, theoretical
water-content regulation curves calculated from Equation (6) using
β = 0.35, 0.30, and 0.25, are shown.

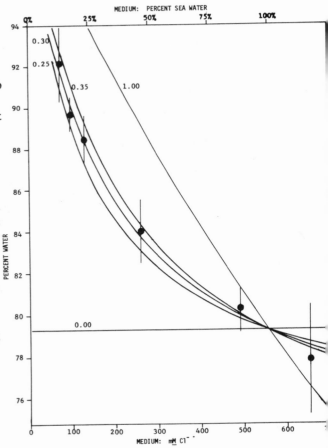

Fig. 2. Water content
of *Nereis vexillosa* from
Pt.Richmond,San Francisco
Bay (Oglesby,1965a),and
theoretical water-content
regulation curves calcu-
lated from Equation (6)
for β = 0.35, 0.30, and
0.25. In this and all
subsequent figures,
theoretical curves for
β = 1.00 (no regulation:
the simple osmometer)
and β = 0.00 (complete
regulation) are given
for comparative purposes
Water-content data are
given as the mean, with
vertical bars indicating
± one standard error.

In this figure and all subsequent figures, theoretical water-content regulation curves for β = 1.00 (no regulation) and β = 0.00 (complete regulation) are given for comparative purposes. The best fitting curve appears to be that for β = 0.30, suggesting that over the entire salinity range tested, *N. vexillosa* can regulate its water content and body volume such that 70% of the excess water gain expected in a simple osmometer is not retained after full adaptation following salinity transfers. The hyperbolic curve generated by Equation (6) fits the data closely, indicating that in this species, there is not a directly linear relationship between external salinity and body volume.

   2.   *Themiste* (formerly *Dendrostomum*) *dyscritum* (Fisher). This sipunculan is commonly found in burrows and crevices in intertidal rocks along the open coast from Pt. Conception, California, to central Oregon. It does not seem to penetrate into estuaries, perhaps due to lack of suitable substrate. Figure 3 shows the water contents of worms from Boiler Bay, Oregon, adapted to salinities ranging from 30% SW to 114% SW (Oglesby, 1968b); theoretical water-content regulation curves calculated for β = 0.65, 0.60, and 0.55 are shown. While the data are rather scattered, the best fitting curve appears to be that for β = 0.60, or 40% regulation, from the

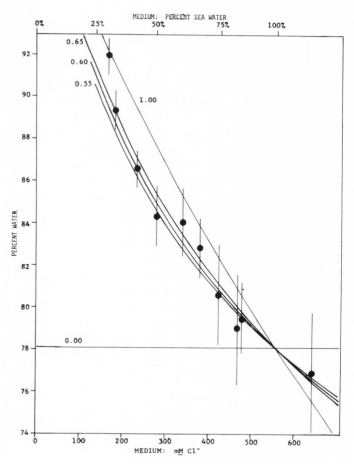

Fig. 3. Water content of *Themiste dyscritum* from Boiler Bay, Oregon (Oglesby, 1968b), and theoretical water-content regulation curves calculated from Equation (6) for β = 0.65, 0.60, and 0.55.

highest salinity tested down to 33% SW. The much higher value for the water content for those worms maintained in 30% SW, nearly to the β = 1.00 curve, suggests that water regulation has broken down almost completely in this low salinity. This species, both in the field and in the laboratory, is much less euryhaline than *N. vexillosa*, and it thus seems reasonable to find that *T. dyscritum* is less able to control its water content than is *N. vexillosa*.

A few data have been published for the water content of *Themiste zostericolum* (Chamberlain), a sipunculan very closely related to *T. dyscritum* but with a more southerly distribution (Hogue and Oglesby, 1972). Worms maintained in 125% SW were reported to contain 83.3±1.2% water (Gross, 1954), those maintained in 100% SW, 80 to 83% water (Peebles and Fox, 1933), and those maintained in 75% SW, 83.8±0.6% water (Gross, 1954). These data suggest that the water content of this stenohaline marine worm varies little over the salinity range 75 to 125% SW, and thus imply a considerable capacity for volume regulation after salinity transfers. However, Gross (1954) demonstrated that *T. zostericolum* in fact has a very limited capacity for volume regulation, certainly no greater than *T. dyscritum* (Oglesby, 1968b). It may be that these few data on water content of *T. zostericolum* are not correct.

Experiments by Adolph (1936) with the sipunculan *Phascolopsis gouldi* (Pourtalès) (as *Phascolosoma gouldii*) have been interpreted both by Adolph and by subsequent reviewers (e.g., Florey, 1966) as indicating that *P. gouldi* and all other sipunculans are merely simple osmometers. In fact, Adolph's experiments showed that *P. gouldi* is capable of a certain amount of volume regulation, also confirmed by Gross (1954). The present analysis shows that *T. dyscritum* is capable of preventing the retention of about 40% of the excess water gain expected in a simple osmometer, strongly supporting Gross's (1954) contention that sipunculans are not as osmotically simple as generally believed.

3. *Abarenicola pacifica* Healy and Wells. This lugworm (Polychaeta, family Arenicolidae) is commonly found in many estuaries and certain open coast areas from central California to British Columbia. Figure 4 shows the water contents of worms from Coos Bay, Oregon, adapted to salinities from 29% SW to 97% SW (Oglesby, 1973); theoretical water-content regulation curves calculated for β = 0.70, 0.65, and 0.60 are shown. While the data are rather scattered, the best fitting curve appears to be the one for β = 0.65, or 35% regulation, over the entire salinity range tested. This result suggests that the estuarine species *A. pacifica* has even less ability to regulate water content than does the open coast species *Themiste dyscritum*, a suggestion reinforced by the observation that several groups of lugworms had mean water contents above the β = 1.00 line (for no regulation). This surprising result is discussed by Oglesby (1973), who concludes from several kinds of evidence that lugworms are not as euryhaline as is generally believed, and that lugworms living in estuaries do not come into contact with low-salinity waters under normal conditions.

Fig. 4. Water content
of *Abarenicola
pacifica* from Coos
Bay, Oregon
(Oglesby, 1973),
and theoretical
water-content
regulation curves
calculated from
Equation (6) for
β = 0.70, 0.65, and
0.60.

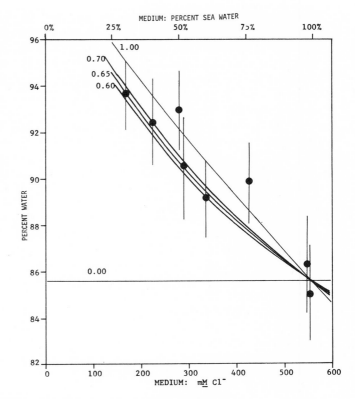

Water-Content Regulation in Osmoregulators

The second group of examples to be considered includes several
species which, to varying extent, show some ability to regulate the
osmotic concentration of the body fluids above that of the external
medium. Hyperosmotic regulation is found only in these worms when
they are adapted to salinities lower than about 25 to 35% SW; at
higher salinities, these worms are osmotic conformers.

1. *Nereis (Hediste) diversicolor* Müller. This nereid polychaete is perhaps the worm species most frequently used in studies
of salt and water balance. *N. diversicolor* is a dominant inhabitant
of estuaries and other brackish-water environments around the North
Atlantic Ocean, and is capable of withstanding salinities ranging
from fresh water up to at least 200% SW. However, it cannot
reproduce successfully in salinities lower than about 10% SW
(Bogucki, 1954, 1963; Smith, 1964b). Water contents have been
measured in three different populations from three different types
of brackish-water habitats. Figure 5 shows the water contents of
worms from a population in the River Wansbeck estuary in Northumberland, England, which were adapted to salinities ranging from 2% SW
to 99% SW (Oglesby, 1970). The theoretical water-content regulation
curve for β = 0.25, based on the external salinity, is shown by the
dashed line in Figure 5. This curve fits the data fairly well at
salinities greater than about 35% SW; that is, in the salinity range
where *N. diversicolor* is an osmoconformer. For worms maintained in

lower salinities, where they are hyperosmotic regulators, this
theoretical curve predicts much higher water contents than were
actually measured. The worms have a considerably better control
over body volume than predicted by a curve calculated on the basis
of external salinity. However, if the internal osmotic concentrati
(taken from Oglesby, 1970) is used in Equation (6), the resultant
curves (solid lines in Figure 5) closely approximate the actual
osmotic behavior of the worms at all salinities. Using internal
osmotic concentrations in Equation (6), theoretical water-content
regulation curves for β = 0.30, 0.25, and 0.20 are shown in Figure
The best fitting curve appears to be that for β = 0.25, or 75%
regulation, over the entire salinity range tested.

Fig. 5. Water content
of *Nereis diversi-
color* from the River
Wansbeck estuary,
Northumberland
(Oglesby, 1970), and
theoretical water-
content regulation
curves calculated
from Equation (6)
using internal
osmotic concentra-
tions (Oglesby,
1970), for β = 0.30,
0.25, and 0.20.
Dashed line indi-
cates theoretical
curve for β = 0.25,
calculated from
Equation (6) using
external osmotic
concentrations.

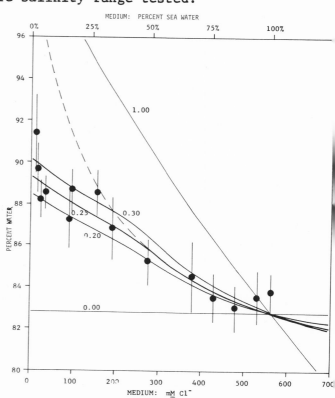

This analysis suggests that for the regulation of total water
content, the worm monitors the osmotic (or ionic) characteristics
of the internal body fluids, rather than those of the external
medium. For osmotic conformers such as *Nereis vexillosa*, *Themiste
dyscritum*, and *Abarenicola pacifica* in all tolerable salinities,
and for *N. diversicolor* when in salinities greater than 35% SW,
there would be no practical distinction between monitoring internal
osmotic concentration or external concentration, because internal
and external concentrations are approximately equal. But for an
osmoregulator such as *N. diversicolor* at salinities below 35% SW,
basing regulation of the water content on the internal osmotic

concentration confers a great reduction in water content over that achieved if the external osmotic concentrations were used. Thus, it can be argued that one of the adaptive advantages to a worm of being hyperosmotic in low salinities is that it assists in the prevention of undue osmotic swelling.

Figure 6 shows the water contents of a population of *N. diversicolor* from Kamyshovaya Bay on the Black Sea near Sebastopol (Mokrousova, in Ivleva and Popenkina, 1966), for worms adapted to a salinity range from 14% to 113% SW. Using internal osmotic concentrations taken from Oglesby (1970) in Equation (6), theoretical water-content regulation curves for β = 0.25, 0.20, and 0.15 are shown in Figure 6. The best fitting curve appears to be that for β = 0.20, or 80% regulation, over this entire salinity range.

Fig. 6. Water content of *Nereis diversicolor* from Kamyshovaya Bay, Black Sea (from Mokrousova, in Ivleva and Popenkina, 1966), and theoretical water-content regulation curves calculated from Equation (6) using internal osmotic concentrations (Oglesby, 1970), for β = 0.25, 0.20, and 0.15.

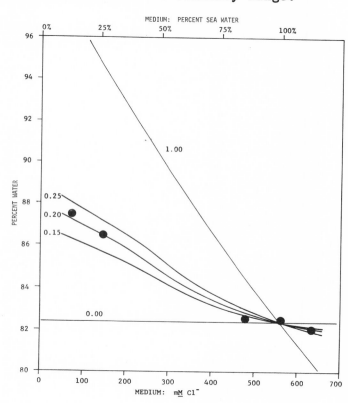

Figure 7 shows the water contents of a population of *N. diversicolor* from the Baltic Sea near Gdynia, where the worms live in constant and low salinities of about 20% SW (Bogucki and Wojtczak, 1964). These water-content data cover a range of salinities from 0.7% SW to 98% SW. Using internal osmotic concentrations taken from Oglesby (1970), which are similar to but more complete than those given by Bogucki and Wojtczak (1964), in Equation (6), theoretical water-content regulation curves for β = 0.15, 0.10, and 0.05 are shown in Figure 7. The best fitting curve appears to be that for β = 0.10, or 90% regulation, over this entire salinity range.

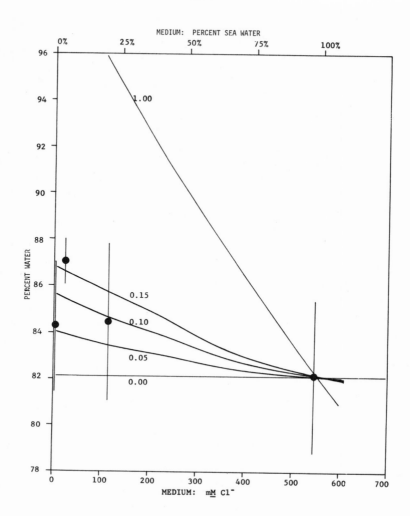

Fig. 7. Water content of *Nereis diversicolor* from Gdynia, Baltic
Sea (Bogucki and Wojtczak, 1964), and theoretical water-
content regulation curves calculated from Equation (6)
using internal osmotic concentrations (Oglesby, 1970), for
$\beta = 0.15$, 0.10, and 0.05.

The River Wansbeck and Kamyshovaya Bay populations of *N.
diversicolor* are similar in their abilities to regulate water content, as shown by their similar β values, 0.25 and 0.20, respectively. The Baltic Sea population has much lower water contents
when adapted to very low salinities, and the ability of this population to regulate water content is considerably greater than that
for the other two populations: only 10% of the excess water gain
expected in a simple osmometer is retained by this Baltic Sea
population. Perhaps the greater ability of the Baltic Sea population to regulate water content is related to a permanent habitat in
low salinities, as contrasted with the much higher, though variable,
salinities to which the River Wansbeck and Black Sea populations

are generally exposed. On the other hand, there has never been definitive proof of the frequently expressed idea (e.g., Muus, 1967; Bogucki and Wojtczak, 1964) that there are genetically determined differences in physiological abilities within a single polychaete species. As pointed out by Smith (1955), ontogenetic conditioning has not been excluded as a major factor in seemingly different osmotic behaviors displayed by different populations.

2. *Nereis (Hediste) limnicola* Johnson. Taxonomically and ecologically, this species is very closely related to *N. diversicolor*. *N. limnicola* is found in estuaries and freshwater lakes and streams from central California north to British Columbia and, unlike *N. diversicolor*, can reproduce successfully in fresh water (Smith, 1950). Figure 8 shows the water contents of a population from freshwater Lake Merced in California (Oglesby, 1965a), maintained over a salinity range from 5% SW to 130% SW. Using internal osmotic concentrations taken from Oglesby (1965a) in Equation (6), theoretical water-content regulation curves for β = 0.20, 0.15, 0.10, and 0.05 are shown. While some groups of worms examined in this experiment may have regulated at β values as high as 0.20 to 0.25, several groups of worms approximated constant water content in all salinities, as shown by some of the data falling near the curve for β = 0.00 (100% regulation). The best fitting curve for the entire experiment would seem to be that for β = 0.10, or 90% regulation, over the entire salinity range. This freshwater population of *N. limnicola* is at least as capable of preventing the retention of excess water gain after transfers as the Baltic Sea population of *N. diversicolor*.

Fig. 8. Water content of *Nereis limnicola* from Lake Merced, San Francisco (Oglesby, 1965a), and theoretical water-content regulation curves calculated from Equation (6) using internal osmotic concentrations (Oglesby, 1965a), for β = 0.20, 0.15, 0.10, and 0.05.

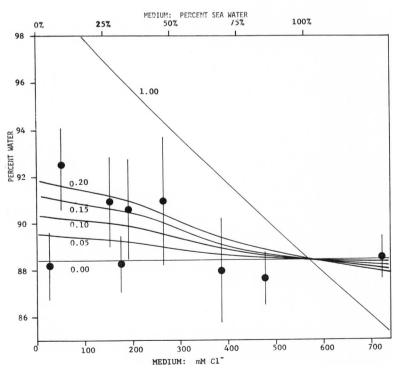

3. *Nereis (Neanthes) succinea* Leuckart. This nereid polychaete is a major estuarine species in the North Atlantic Ocean,
found in generally warmer waters and thus further south than
N. diversicolor. *N. succinea* has been introduced into several
estuaries in California, as well as into the inland salt lake, the
Salton Sea. In California, *N. succinea* is less euryhaline than
N. limnicola but more euryhaline than *N. vexillosa*. In San Francisco
Bay, where all three species are found, the habitat for *N. succinea*
is between the habitats of the other two species (Oglesby, 1965a).
Similarly, in northern Europe, *N. succinea* is less euryhaline than
N. diversicolor, and occupies a more seaward portion of those
estuaries where they both occur (Smith, 1963a). Figure 9 shows the
water contents of a population of *N. succinea* from near Hayward, in
San Francisco Bay, maintained over a salinity range from 2% SW to
131% SW (Oglesby, 1965a). Using internal osmotic concentrations
taken from Oglesby (1965a) in Equation (6), theoretical water-content
regulation curves for β = 0.35, 0.30, 0.25 are shown in Figure 9.
The best fitting curve appears to be that for β = 0.30, or 70%
regulation, over this entire salinity range.

Fig. 9. Water
content of
Nereis succinea
from near
Hayward, San
Francisco Bay
(Oglesby,
1965a), and
theoretical
water-content
regulation
curves calcu-
lated from
Equation (6)
using internal
osmotic con-
centrations
(Oglesby,
1965a), for
β = 0.35, 0.30,
and 0.25.

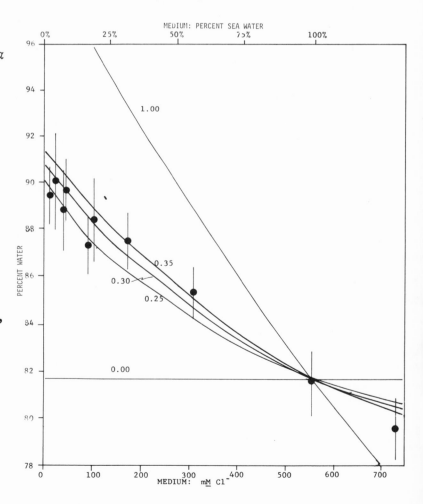

Both *N. vexillosa* (Fig. 2) and *N. succinea* (Fig. 9) have the same β value of 0.30. But since *N. succinea* osmoregulates when maintained in salinities lower than about 30% SW, whereas *N. vexillosa* is an osmoconformer, the water content and body volume of *N. succinea* when kept in low salinities are maintained at much lower levels than in *N. vexillosa*. Thus, *N. succinea* does not gain as much water as does *N. vexillosa* after transfers to lower salinities (Oglesby, 1965b). *N. succinea* has a higher β value than either *N. diversicolor* or *N. limnicola*, and also gains more water osmotically after transfers to lower salinities than do either of these two species (Oglesby, 1965b, 1968a; Smith, 1964a). The results of this comparative analysis of water-content regulation using β values calculated from Equation (6) for these four species in the genus *Nereis* are consistent with our information on other aspects of their osmotic behavior, and with their ecological distribution. Smith (1963a) stated that in European estuaries where they both occur, *N. succinea* occupies a habitat exposed to lower salinities than does *Nereis (Neanthes) virens* Sars. There are no published data on the osmotic concentration or the water content of *N. virens* (Oglesby, 1969a), but on the basis of the present analysis and discussion, it can be predicted that *N. virens* will show a β value which is at least as high as 0.30, and that it very likely does not hyperosmoregulate when kept in low salinities.

4. *Namalycastis indica* (Southern). This nereid polychaete has been recorded from a number of brackish-water locations in India (Wesenberg-Lund, 1958), but it has never been clearly stated to what overall salinity range this species is exposed naturally, or whether it ever is found naturally in a truly freshwater environment. Krishnan (1952) studied a population from "the brackish waters in the vicinity of Madras where it is abundantly distributed," and states that this species could be kept in the laboratory "for months in fresh water in a healthy condition." He also comments that *Namalycastis indica*, "though found in fresh water is able to live in waters of high salinity." Krishnan reported water contents in this species when "acclimated to sea-water," and when maintained in apparently fresh water; these percentages, recalculated from Krishnan's original data, are shown in Figure 10. Krishnan did not measure the osmotic or ionic concentrations of the body fluids, but in general those nereid polychaetes which show hyperosmotic regulation in low salinities have approximately the same level of hyperosmotic regulation, and differ only in the low critical external salinity at which osmoregulatory failure sets in (Oglesby, 1969a). Using Krishnan's two data points for water content, and assuming that the osmotic concentration of the body fluids of *Namalycastis indica* is similar to that reported for other euryhaline nereid polychaetes (Oglesby, 1965a, 1969a, 1970; Smith, 1955, 1957), it can be seen from Figure 10 that the water-content regulatory ability of *Namalycastis indica* may approximate a β value of 0.20, or 80% regulation. Thus, the ability of *Namalycastis indica* to regulate its body volume seems comparable to such other euryhaline nereids as *Nereis limnicola* and *N. diversicolor*.

198

Fig. 10. Water content
of *Namalycastis indica*
from Madras, India
(Krishnan, 1952), and
theoretical water-
content regulation
curves calculated
from Equation (6)
using internal osmotic
concentrations (for
Nereis diversicolor,
Oglesby, 1970), for
β = 0.25 and 0.20.

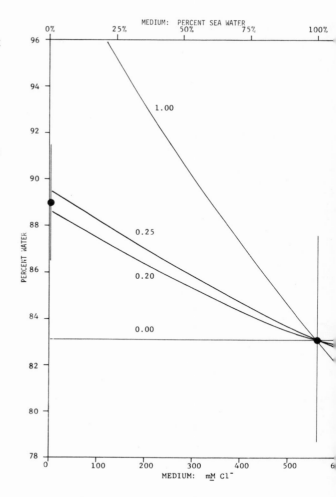

DISCUSSION

Table 1 provides a summary of the results of analyses of water
content regulatory ability of these several species of worms,
arranged in order of increasing β values (decreasing water-regulatory
ability). The resultant rank order for β values is comparable to
the rank ordering of these species on the bases of the following:
(1) penetration into estuaries and distribution within
estuaries (Oglesby, 1965a, 1969a, 1973; Smith, 1963a);
(2) tolerance of low salinities in the laboratory (Oglesby,
1965a, 1969a, this report);
(3) ability to regulate volume after salinity transfers
(Oglesby, 1965b, 1968a, 1969a, this report; Smith, 1964a); and
(4) differences in lower critical salinity for the maintenance
of hyperosmotic body fluids (Oglesby, 1965a, 1969a, 1970).

TABLE 1
Water-regulation ability of selected worms

Species	Source of population	Reference	Water content in 560 mM Cl^-	β "Best Fit"	Percent regulation
I. Osmoregulating Species					
Nereis limnicola	Lake Merced, San Francisco	Oglesby,1965	88.4%	0.10	90%
Nereis diversicolor	Gdynia, Baltic Sea	Bogucki & Wojtczak, 1964	82.1%	0.10	90%
Nereis diversicolor	Kamyshovaya Bay, Black Sea	Mokrousova, 1966	82.4%	0.20	80%
Namalycastis indica	Madras, India	Krishnan, 1952	83.1%	0.20	80%
Nereis diversicolor	River Wansbeck, England	Oglesby,1970	82.8%	0.25	75%
Nereis succinea	Hayward, San Francisco Bay	Oglesby,1965a	81.7%	0.30	70%
II. Osmoconforming Species					
Nereis vexillosa	Pt. Richmond, San Francisco Bay	Oglesby,1965b	79.3%	0.30	70%
Themiste dyscritum	Boiler Bay, Oregon	Oglesby,1968	78.1%	0.60	40%
Abarenicola pacifica	Coos Bay,Oregon	Oglesby,1973	85.6%	0.65	35%

Several types of mechanisms have been proposed as being involv
in volume regulation following transfers from one salinity to
another, and also in the maintenance of hyperosmotic body fluids.
Among the more important hypotheses are:

1. Prevention of entry of some or all of the osmotic water
inflow by means of lowered permeability to water. This was first
proposed for polychaetes by Jørgensen and Dales (1957), and was
confirmed for *N. diversicolor* by Smith (1964a, 1970a) and *N. limmi-
cola* (Oglesby, unpublished results) using deuterated and tritiated
water as tracers.

2. Expulsion of some or all of the excess water via the urine
following osmotic water uptake. This was originally proposed for
polychaetes by Beadle (1937). Experiments with sipunculans have
shown that interference with nephridial integrity prevents volume
regulation (Gross, 1954; Kamemoto and Nitta, 1964; Koller, 1939).
While it has frequently been suggested that the production of a
urine hypo-osmotic to the body fluids would be of great adaptive
advantage to worms, by conserving salts while assisting in the
expulsion of excess water, only *N. diversicolor* has been directly
shown to produce a hypo-osmotic urine, and then only when adapted
to very low salinities (Smith, 1970b).

3. Loss of salts from the body fluids, reducing the number of
osmotically active particles and thus lessening the gradient for
osmotic water influx. Such loss could be across the body wall, or
via the urine by means of nephridial activity, or it could occur
through sequestering of such particles within the body. Salt loss
across the body wall following transfers has been experimentally
demonstrated by Oglesby (1965b, 1968a, 1972) and Smith (1963b).
Excess salt elimination by the nephridia, that is, the production
of urine hyperosmotic to the body fluids, has been demonstrated in
the sipunculan *Themiste signifer* (Kamemoto and Larson, 1964).
However, any urinary loss of salts will decrease the number of
osmotically active particles in the body fluids. Iso-osmotic urine
has been measured in the sipunculan *Themiste dyscritum* (Oglesby,
1968b), and hypo-osmotic urine in *Nereis diversicolor* (Smith, 1970b
Gross (1954) postulated that osmotically active particles were
sequestered in the sipunculan *Themiste zostericolum* under certain
types of imposed osmotic stress. The accepted hypothesis for the
relationship of intracellular and extracellular fluids in most
animals is that the intracellular fluids always maintain iso-
osmoticity with the extracellular fluids. Intracellular iso-osmoti
regulation seems always to involve the intracellular binding and
release of such osmotically active particles as free amino acids
and certain other small molecular weight organic molecules, thus
preventing much osmotic swelling and shrinking of the cells
(Oglesby, 1969a).

The present analysis using Equation (6) to determine β values
does not require any assumptions about the relative importance of
these several mechanisms for regulating water content and body
volume. It seems clear from the experimental evidence just cited

that more than one mechanism is involved, very likely to different extents in different species. Rather, Equation (6) assesses the overall water-content regulatory ability of worms, and thus may prove of value in attempts to determine the nature and relative importance of mechanisms actually used.

SUMMARY

1. A formulation is developed to describe the ability of soft-bodied invertebrates, such as worms, to regulate water content and body volume in the face of varying environmental salinities.

2. This formulation permits a direct comparison of the regulatory abilities of a number of different species from different habitats, in quantitative terms.

3. Even the most stenohaline species discussed possess some ability to regulate body volume and water content, both to a greater extent than expected in a theoretical "simple osmometer."

4. Species which are hyperosmotic regulators when adapted to low salinities have a much greater control over water content and body volume than do species which are osmoconformers.

5. For the regulation of total water content, the worms seem to be monitoring the osmotic (or ionic) characteristics of the internal body fluids, and not those of the external medium.

6. Results of a comparative analysis of water-content regulation in seven species of worms are consistent with information on all other aspects of their osmotic behavior, including (a) tolerance of low salinities in the laboratory; (b) ability to regulate body volume after salinity transfers; (c) lower critical salinity for maintenance of hyperosmotic body fluids; (d) whether the species is a hyperosmotic regulator in low salinities; (e) extent of penetration into estuaries; and (f) relative distribution within estuaries compared with other species.

7. Use of this formulation does not require any assumptions about the relative importance of the several mechanisms involved in regulation of water content and body volume.

ACKNOWLEDGMENTS

I would like to express my profound thanks to Professor John Shaw of the University of Newcastle-upon-Tyne, whose "doodling" one evening generated the basic idea discussed in this paper, and to my wife, Dr. Alice Shoemaker Oglesby, for her valiant attempts to develop logical thinking and clarity in this presentation. My thanks go to Dr. Denis G. Baskin for helpful comments, and to Mr. Jeffrey A. Stephenson and Mr. E. Wayne Hogue for their assistance in writing the computer program. Portions of the research discussed were supported by National Science Foundation research grant GB-442 by a North Atlantic Treaty Organization postdoctoral fellowship to the University of Newcastle-upon-Tyne, and by a research grant from Pomona College.

LITERATURE CITED

Adolph, E. F. 1936. Differential permeability to water and osmotic exchanges in the marine worm *Phascolosoma*. J. Cell. Comp. Physiol. 9:117-35.

Bair, T. D. and M. Peters. 1971. Osmoregulation in the trematode *Haematoloechus medioplexus* (Looss). Comp. Biochem. Physiol. 39A:165-71.

Beadle, L. C. 1937. Adaptation to changes of salinity in the polychaetes. I. Control of body volume and of body fluid concentration in *Nereis diversicolor*. J. Exp. Biol. 14:56-70.

Bogucki, M. 1954. Adaptacja *Nereis diversicolor* (O. F. M.) do rozcieńczonej wody morskiej i wody słodkiej. Polskie Arch. Hydrobiol. 2:237-51.

_____. 1963. The influence of salinity on the maturation of gametes of *Nereis diversicolor* O. F. Müller. Polskie Arch. Hydrobiol. 11:343-47.

_____ and A. Wojtczak. 1964. Content of body water in *Nereis diversicolor* O. F. M. in various medium concentrations. Polskie Arch. Hydrobiol. 12:125-43.

Croll, N. A. and D. R. Viglierchio. 1969. Osmoregulation and the uptake of ions in a marine nematode. Proc. Helminth. Soc. Washington 36:1-9.

Dice, J. F. Jr. 1969. Osmoregulation and salinity tolerance in the polychaete annelid *Cirriformia spirabrancha* (Moore, 1904). Comp. Biochem. Physiol. 28:1331-43.

Florey, E. 1966. An Introduction to General and Comparative Animal Physiology. Philadelphia: W. B. Saunders Co.

Gross, W. J. 1954. Osmotic responses in the sipunculid *Dendrostom zostericolum*. J. Exp. Biol. 31:402-23.

Hoar, W. S. 1966. General and Comparative Physiology. Englewood Cliffs, N. J.: Prentice-Hall.

Hogue, E. W. and L. C. Oglesby. 1972. Further observations on sal balance in the sipunculid worm *Themiste dyscritum*. Comp. Biochem. Physiol. 42A:915-26.

Ivleva, I. V. and M. I. Popenkina. 1966. Vliyanie obschego soderzhaniya solei v srede na teplovuyu ustoichivost' myshechn tkani polikhet. In: Fiziologiya Morskikh Zhivotnykh, pp. 156 75. Moscow: Akademia Nauk S.S.S.R.

Jørgensen, C. B. and R. P. Dales. 1957. The regulation of volume and osmotic regulation in some nereid polychaetes. Physiol. Comp. Oecol. 4:357-74.

Kamemoto, F. I. and E. J. Larson. 1964. Chloride concentrations in the coelomic and nephridial fluids of the sipunculid *Dendrostomum signifer*. Comp. Biochem. Physiol. 13:477-80.

_____ and J. Y. Nitta. 1964. Volume regulation in *Sipunculus*. Amer. Zool. 4, Abstract 48.

Koller, G. 1939. Über die Nephridien von *Physcosoma japonicum*. Verhandl. Deutsch. Zool. Gesell. 41:440-47.

Krishnamoorthi, B. 1962. Salinity tolerance and volume regulation in four species of polychaetes. Proc. Indian Acad. Sci. 55 (6, Sec. B):363-71.

Krishnan, G. 1952. On the nephridia of Nereidae in relation to habitat. Proc. Nat. Inst. Sci. India 18:241-55.

Muus, B. J. 1967. The fauna of Danish estuaries and lagoons. Distribution and ecology of dominating species in the shallow reaches of the mesohaline zone. Medd. Danmarks Fiskeri-og Havunders. (New series) 5:1-316.

Oglesby, L. C. 1965a. Steady-state parameters of water and cloride regulation in estuarine nereid polychaetes. Comp. Biochem. Physiol. 14:621-40.

_____. 1965b. Water and chloride fluxes in estuarine nereid polychaetes. Comp. Biochem. Physiol. 16:437-55.

_____. 1968a. Responses of an estuarine population of the polychaete *Nereis limnicola* to osmotic stress. Biol. Bull. 134:118-38.

_____. 1968b. Some osmotic responses of the sipunculid worm *Themiste dyscritum*. Comp. Biochem. Physiol. 26:155-77.

_____. 1969a. Inorganic components and metabolism; ionic and osmotic regulation: Annelida, Sipuncula, and Echiura. In: Chemical Zoology, Vol. 4 (M. Florkin and B. T. Scheer, eds.), pp. 211-310. New York: Academic Press.

_____. 1969b. Salinity-stress and desiccation in intertidal worms. Amer. Zool. 9:319-31.

_____. 1970. Studies on the salt and water balance of *Nereis diversicolor*. I. Steady-state parameters. Comp. Biochem. Physiol. 36:449-66.

_____. 1972. Studies on the salt and water balance of *Nereis diversicolor*. II. Components of total sodium efflux. Comp. Biochem. Physiol. 41A:765-90.

_____. 1973. Salt and water balance in lugworms (Polychaeta: Arenicolidae), with particular reference to *Abarenicola pacifica* in Coos Bay, Oregon. Biol. Bull. 145:180-99.

Peebles, F. and D. L. Fox. 1933. The structure, functions, and general reactions of the marine sipunculid worm *Dendrostoma zostericola*. Univ. Calif. Scripps Inst. Oceanogr. Bull. Tech. Ser. 3(9):201-24.

Smith, R. I. 1950. Embryonic development in the viviparous nereid polychaete, *Neanthes lighti* Hartman. J. Morph. 87:417-65.

_____. 1955. Comparison of the level of chloride regulation by *Nereis diversicolor* in different parts of its geographical range. Biol. Bull. 109:453-74.

_____. 1957. A note on the tolerance of low salinities by nereid polychaetes and its relation to temperature and reproductive habit. Ann. Biol. 33:93–107.

_____. 1963a. On the occurrence of *Nereis (Neanthes) succinea* at the Kristeneberg Zoological Station, Sweden, and its recent northward spread. Arkiv Zool. 15:437–41.

_____. 1963b. A comparison of salt loss rate in three species of brackish-water nereid polychaetes. Biol. Bull. 125:332–43.

_____. 1964a. D_2O uptake rate in two brackish-water nereid polychaetes. Biol. Bull. 126:142–49.

_____. 1964b. On the early development of *Nereis diversicolor* in different salinities. J. Morph. 114:437–63.

_____. 1970a. Chloride regulation at low salinities by *Nereis diversicolor* (Annelida, Polychaeta). II. Water fluxes and apparent permeability to water. J. Exp. Biol. 53:93–100.

_____. 1970b. Hypo-osmotic urine in *Nereis diversicolor*. J. Exp. Biol. 53:101–8.

Wesenberg-Lund, E. 1958. Lesser Antillean polychaetes, chiefly from brackish water, with a survey and a bibliography of fresh and brackish-water polychaetes. Studies Fauna Curaçao Other Carib. Islands 8(30):1–41.

Halophytes: Adaptive mechanisms

W. H. Queen

Relatively few vascular plant species inhabit the more saline regions of estuaries. Nonetheless, these few species are exceedingly important to the estuarine systems with which they are associated. Their function in marsh formation, shoreline erosion control, and their essentiality as a component of the habitat of many estuarine animals have been well documented (Chapman, 1960; Lauff, 1967; Waisel, 1972; Reimold and Queen, 1974). The value of these plants as a food source for estuarine animals has also been examined (Odum and de la Cruz, 1963). Recently, a role for these species in the maintenance of estuarine water quality was postulated (Grant and Patrick, 1970). The great importance of these habitat roles fully justifies the attention they have received. In addition to the studies of habitat functions, investigations have also been directed during the past several years to defining characteristics of estuarine plants which permit them to thrive in an environment which is hostile to most other higher plant species. Mechanisms for coping with a high salt concentration, the most serious stress, have been emphasized in these studies. These mechanisms and the work that has been done to elucidate them will be reviewed in this paper.

Salinity is such a significant environmental factor that classification schemes have been devised for dividing plant species into various groups on the basis of their ability to tolerate salt. Plants unable to withstand high salinities are termed glycophytes; those tolerant of high salinities (generally, soil salt concentrations of 0.5% and higher) are referred to as halophytes. Waisel (1972) has made a detailed classification of halophytes. Considerable differences in growth form are found among estuarine halophytes. Submerged rooted aquatics are algal-like, mangroves are small trees, and many of the salt marsh species are herbaceous. Many typical halophyte features (e.g., succulence) are also characteristic of other plant groups which are not salt-tolerant. Most plant genera having halophytic estuarine species also have nonestuarine species which are not particularly salt-resistant. Tolerance of high salt concentrations has resulted from a number of specific adaptations which have occurred in several different plant taxa. Some of the mechanisms used by halophytes for coping with high salt concentrations are also found in glycophytes. In these instances the differences are more quantitative than qualitative. Hence, a sharp separation cannot always be made between the two groups.

THE SALT PROBLEM

Major obstacles related to the presence of a high salt concentration must be overcome if plant growth is to occur in an estuarine environment. Foremost among these are: (1) acquisition of water from an external solution with a high osmotic pressure; (2) maintenance of internal ionic balance within narrow limits; and (3) absorption of sufficient essential nutrients from a medium with an ionic mix unfavorable to higher plants.

If the internal osmotic pressure is lower than the external pressure, water will flow out of the plant to the external medium. Desiccation and death will eventually follow if the gradient is not reversed. One method by which plants respond to this problem is by absorbing additional salt to increase the internal osmotic pressure above that of the external medium (Bernstein, 1961). Acquisition of water will then continue, even from a highly saline environment.

While absorption of salt from the external medium sufficient to increase the osmotic pressure within the plant above that of the external medium does permit a continuation of water absorption, it does not resolve all problems related to high salinity. In fact, an increased rate of salt absorption creates additional problems. Initially, increased absorption can lead to internal ionic imbalance. Many of the better understood adaptive mechanisms of halophytes function to reduce these imbalances. Even if the necessary ionic balances are maintained, the problem of an exceedingly high internal salt content remains. The toxic effect of such concentrations is well known. Undoubtedly, cellular mechanisms of halophytes are better adapted to function under such conditions than those of glycophytes, but our knowledge of these adaptations is exceedingly limited.

A third major obstacle relating to plant growth in a highly saline environment concerns the acquisition of essential nutrients. Generally, concentrations of these essential nutrients are higher in

seawater than in agricultural soils. The problem relates to an abundance of other ions, not present in crop soils, which compete with
the essential nutrients for the active transport sites of the root
membranes. If adequate quantities of nutrients are to be absorbed,
active transport mechanisms concerned with their uptake must be able
to discriminate to a very high degree between the essential nutrients
and their nonessential competitors. Recent studies indicate that
halophytes have a much greater capability in this regard than glycophytes.

REGULATION OF SALT CONTENT

Halophytes have a variety of mechanisms for regulating their
internal ionic balance and concentration. Salt glands are the best
known. These structures, which were first studied in the mid-
nineteenth century (Waisel, 1972), are highly selective salt excre-
tors. Frequently, the exudate from salt glands of estuarine species
consists almost exclusively of sodium and chloride ions. These ions
are not required by higher plants, except in very small quantities by
a few species. Their occurrence in estuarine plants in high concentrations is related to the necessity these plants have for maintaining an internal osmotic pressure greater than that of the external
medium in order to acquire water. As stated in the previous section,
absorption of additional salt from the external medium is one method
by which plant species achieve a higher internal osmotic pressure.
However, absorption proceeds in a much less selective manner during
periods of rapid salt uptake. Nonessential ions, such as sodium and
chloride, are absorbed in large quantities. Prolonged absorption of
unneeded ions, in the absence of structures like salt glands, would
result in the development of cellular ionic imbalances which would
be highly toxic. By selectively absorbing and excreting undesirable
ions from the remainder of the plant, salt glands assist in preventing the development of toxic ionic imbalances in halophytes. Salt
glands do not, however, prevent the development of high internal ion
concentrations in halophytes growing in a highly saline medium.
Additional structures, such as salt hairs and salt bladders, which
have functions similar to those of salt glands, are also found in
estuarine halophytes.

Other methods for reducing undesirable internal salt concentrations are possessed by estuarine halophytes. These include the
following: (1) the shedding of salt packed leaves, (2) the translocation of salt back to the root system followed by its excretion to
the soil, and (3) the development of succulent leaves. With the
development of succulence, cell volume is increased and ionic concentration is reduced by dilution. Even though some of these mechanisms are able to reduce the internal salt concentration, as well as
prevent the development of toxic ionic imbalances, only very small
reductions can occur because of the necessity these plants have for

maintaining an internal osmotic pressure sufficiently high to acqui
water.

Even though the mechanisms of "salt removal" are more thorough
understood, mechanisms for regulating salt absorption may prove to
more significant in controlling the internal ionic balance of estua
rine halophytes. The fact that cellular membranes have control ove
types and quantities of ions absorbed from the external medium has
been known for half a century (Steward and Sutcliffe, 1959). However, this control over salt absorption is not absolute. Potentiall
toxic ions, if present in the external medium, will be absorbed.
Also, control is much less effective under highly saline conditions
ions are somewhat indiscriminately absorbed in order to achieve an
internal osmotic pressure sufficiently high to permit the acquisition of water. While active transport and mineral absorption under
saline conditions have been studied for years, data indicating that
significant differences exist between the active transport mechanism
of salt-tolerant and salt-sensitive species have been obtained only
during the past decade. The first major breakthrough occurred with
the discovery that ions of many elements are absorbed by a dual
"pathway" (Epstein, Rains, and Elzam, 1963). Two separate mechanism
1 and 2, have been found for a great number of ions in a variety of
plant tissues (Epstein, 1969). Mechanism 1 is best characterized
by its ability to absorb ions of essential nutrient elements from ar
external medium with an exceedingly low concentration of these ions
and to build up cellular concentrations which greatly exceed those
of the external environment. The rate of absorption for mechanism 1
ceases to increase when the external concentration of the essential
ion increases from extremely low to moderately low or high levels.
Epstein (1969) has reported increased potassium absorption by
mechanism 1 of tall wheatgrass up to concentrations of approximately
0.5 mM. Potassium absorption by mechanism 1 is not increased when
the external concentration is increased above 0.5 mM even though the
rate of absorption does increase with an increase in the external
concentration of potassium. However, mechanism 2 has been shown to
be responsible for this increased absorption. Thus, mechanism 2,
which is inactive at exceedingly low concentrations, can be characte
ized as being more responsible for the absorption of essential ions
at moderate to high external concentrations.

Rains and Epstein (1967) reported that *Avicennia marina*, a mangrove, has absorptive mechanisms for potassium that differ from thos
of glycophytes in two important respects. First, mechanism 1 of
Avicennia was found to have a much weaker affinity for potassium tha
mechanism 1 of crop plants. Also, mechanism 2 of *Avicennia* was foun
to be much less sensitive to the presence of competing ions, such as
sodium, than mechanism 2 of nonhalophytes. These differences have
major adaptive significance. Crop plants need an absorptive mechanism (1) with a strong affinity for ions of essential elements like
potassium because these elements often occur in low concentrations i
agricultural soils. The inhibitory effects of high concentrations o
competing ions on mechanism 2 are not particularly important because

high concentrations of competing ions like sodium are rarely found in agricultural soils. With halophytes, the situation is reversed. A mechanism 1 with a strong affinity for essential ions is not required because ions of elements that must be built up to high cellular concentrations are almost always found in seawater at moderate to high concentrations. However, a mechanism 2 that is resistant to the presence of competing ions is highly important to halophytes because of the the presence of high concentrations of these competing ions in seawater.

CELLULAR ADAPTATIONS

Mechanisms discussed in the previous section for regulating salt absorption and for removing excessive amounts of undesirable ions are certainly significant in the adaptation of estuarine halophytes to a highly saline environment. However, these mechanisms are only a partial solution to the problem. They assist only in maintaining a tolerable ionic balance within the plant. Because the internal ionic concentration must be maintained at a level higher than that of the external soil solution in order for the plant to acquire water, the problem of a high cellular salt content remains. Two important theories have been advanced to account for the greater tolerance of higher cellular salt concentrations by halophytes than by glycophytes. One of these is that salt-sensitive enzymes of halophytes are not inhibited as much by high cellular salt concentrations as those of glycophytes. A second theory is that critical cellular mechanisms in halophytes are protected by subcellular compartmentalization to a greater degree than in glycophytes.

There have been relatively few studies of the effects of salt on the activity of halophyte enzymes. Moreover, results from these investigations have been contradictory. Data both supporting and refuting the idea that salt-sensitive enzymes of halophytes are more tolerant than those of glycophytes have been obtained. Many of the more interesting investigations have involved malate dehydrogenase (MDH). Lee (1971) reported that _in vitro_ MDH activity increased with increases in the sodium chloride concentration of the assay medium up to 0.16 M for *Spartina alterniflora* and 0.20 M for *Salicornia virginica*. Further increases in concentration were inhibitory. Similar results were obtained with ammonium chloride and potassium chloride. Both *Spartina alterniflora* and *Salicornia virginica* are important estuarine species, particularly along the Atlantic and Gulf coasts of the United States. Yopp (1974) has reported comparable results for *Salicornia pacifica*, a west coast species. Maximal MDH activity was found at a sodium chloride concentration of 0.20 M. These concentrations for maximal MDH activity are considerably higher than those for MDH of glycophytes. MDH activity has been found to be maximal at sodium chloride concentrations of 0.02 M for peas(Weimberg, 1967), 0.04 M for spinach (Hiatt and Evans, 1960) and 0.06 M for corn

(Lee, 1971). While the salt concentration for maximal MDH activity
of halophytes is quite high in relation to glycophytes, it is low in
relation to that of halophytic bacteria. Maximal MDH activity has
been reported to occur at a sodium chloride concentration of 1.0 M
for *Halobacterium solinarium* (Holms and Halvorson, 1965). These
studies indicate that cellular tolerance of high salt concentrations
can be attributed, at least in part, to enzyme adaptation.

Before enzyme adaptation can be fully accepted as an important
factor in halophyte tolerance of high cellular salt concentrations,
additional data will be required. If confirming data are not forth-
coming, alternative theories of cellular salt tolerance must be
pursued. Among these, subcellular compartmentalization of ions is
most frequently mentioned (Epstein, 1969; Greenway and Osmond, 1970.
Presently, data supporting compartmentalization are no stronger than
those supporting enzyme adaptation.

LITERATURE CITED

Bernstein, L. 1961. Osmotic adjustment of plants to saline media.
    I. Steady state. Amer. J. Bot. 50:360-70.
Chapman, V. J. 1960. Salt Marshes and Salt Deserts of the World.
    London: Leonard Hill Books Ltd.
Epstein, E. 1969. Mineral metabolism in halophytes. In: Ecologi-
    cal Aspects of the Mineral Nutrition of Plants (I. H. Rorison,
    ed.), pp. 345-55. Oxford: Blackwell.
_____, D. W. Rains, and O. E. Elzam. 1963. Resolution of dual
    mechanisms of potassium absorption by barley roots. Proc. Natr
    Acad. Sci. U. S. 49: 684-92.
Grant, R. R. and R. Patrick. 1970. *Tinicum* marsh as a water puri-
    fier. In: Two Studies of a *Tinicum* Marsh, pp. 105-23. Washing-
    ton, D. C: The Conservation Foundation.
Greenway, H. and C. B. Osmond. 1970. Ion relations, growth and
    metabolism of *Atriplex* at high external electrolyte concentra-
    tions. In: The Biology of *Atriplex* (R. Jones, ed.), pp. 49-56.
    Canberra, Australia: Div. Plant Industry C.S.I.R.O.
Hiatt, A. T. and H. J. Evans. 1960. Influence of salts on activity
    of malic dehydrogenase from spinach leaves. Plant Physiol.
    35:662-72.
Holms, P. K. and H. Halvorson. 1965. Purification of a salt-requir-
    ing enzyme from an obligately halophilic bacterium. J. Bacteri
    90:312-16.
Lauff, G. A. (Ed.). 1967. Estuaries, Publ. No. 83. Washington,
    D. C.: Amer. Assoc. Adv. Sci.
Lee, J. 1971. Studies of malate dehydrogenase from selected salt
    marsh plants. Master's thesis, University of South Carolina.
Odum, E. G. and A. A. de la Cruz. 1963. Detritus as a major com-
    ponent of ecosystems. AIBS Bulletin 13:39-40.

Rains, D. W. and E. Epstein. 1967. Preferential absorption of potassium by leaf tissue of the mangrove *Avicennia marina*: An aspect of halophytic competence in coping with salt. Aust. J. Biol. Sci. 20:847-57.

Reimold, R. J. and W. H. Queen (Eds.). 1974. Ecology of Halophytes. New York: Academic Press.

Steward, F. C. and J. F. Sutcliffe. 1959. Plants in relation to inorganic salts. In: Plant Physiology, A Treatise (F. C. Stewart, ed.), Vol. 2, pp. 253-478. New York: Academic Press.

Waisel, Y. 1972. Biology of Halophytes. New York: Academic Press.

Weimberg, R. 1967. Effect of sodium chloride on the activity of a soluble malate dehydrogenase from pea seeds. J. Biol. Chem. 242:3000-6.

Yopp, J. 1974. Effects of low water potential on the activity of mitochondrial, chloroplast and supernatant malic dehydrogenase from the halophyte, *Salicornia pacifica*. Trans. Ill. Acad. Sci. 66: 1.

Aspects of individual adaptation to salinity in marine invertebrates

H. Theede

It is well known that organisms living in brackish water as well as in estuaries and the intertidal region may be exposed to severe regular and irregular fluctuations of salinity and other ambient factors. The mechanisms allowing survival under such environmental conditions involve different types of responses, which may be classified under the terms of reactions, regulations, and adaptations. The literature on this topic has recently been extensively reviewed (Kinne, 1971; Schlieper, 1971; Lange, 1972; and Vernberg and Vernberg, 1972). This report will therefore be restricted to only those aspects about which some work has been done in our laboratory.

Important reactions in this context, according to Kinne (1964, 1967), comprise avoidance or escape movements as well as responses which lead to a temporary reduction of contact with the adverse environment (e.g., closure of shells, retreat into holes and burrows, withdrawal of sensitive body parts, and production of mucus). These reactions will allow tolerance to more extreme salinity fluctuations than does the full active state. However, in habitats with strong salinity fluctuations, the intermediate periods with "normal salinities" will have to be sufficiently long for food intake and other obligative activities.

Regulatory devices have been repeatedly summarized under aspects of ionic, volume, and osmoregulation (Schlieper, 1958, 1971; Potts and Parry, 1964; Potts, 1968; Kinne, 1964, 1971; Lange, 1972). The degree of euryhalinity of poikilosmotic marine animals is

closely connected with regulations and tolerance on the tissue or
cellular level, whereas in homoiosmotic species, the efficiency of
regulatory capacity on the organismic level is decisive.

Nongenetic adaptation (acclimatization, acclimation) to
environmental factors involves adjustments, which may be recognized
on all levels of organismic functions and structures, and which
finally are advantageous in an ecological sense (Kinne, 1964, 1967)
Adaptation to salinity may comprise changes of the ranges of resistance and tolerance. Within these ranges, different physiological
functions, such as activity rates, regulatory capacities (ionic,
volume, and osmoregulation), permeability, metabolic rate and
efficiency, growth, and others may be modified. Regulations and
acclimations may be differentiated by the fact that regulations do
not change their preexisting regulatory organs, whereas acclimations
may influence all response mechanisms, including different regulatory devices. Besides, the time course of the development of
acclimation may be divided into the following three phases: the
immediate responses, the period of stabilization, and the new steady
state (Kinne, 1964, 1971).

Poikilosmotic Animals

In several cases it has been found that specimens of the same
euryhaline species from localities with different habitat salinities
have different ranges of tolerance to salinity. Also, maximum
activity rates are only maintained within an optimum range of
salinity which may depend on the salinity of adaptation. Thus,
common mussels *(Mytilus edulis)* from the German North Sea (with
about 30 o/oo S) and the Baltic (with about 15 o/oo S) have quite
different optima for their filtering activity. The filtration
rates are strongly reduced after direct transfer into sub- or
supraoptimal salinities. After a period of acclimation to the
changed osmotic conditions the filtering activities increase. But
even after 7 to 10 days of adaptation, the two populations still
prove to be different in their dependence of filtration activity on
salinity (Theede, 1963). Pierce (1971) found that the activity of
specimens of the genus *Modiolus*, as measured by valve movements, is
reduced when the salinity of the water is strongly changed.
According to Bayne (1973), the rate of oxygen consumption declines
in the mussel *(Mytilus edulis)* when the salinity of the water is
reduced. Both valve movements and oxygen consumption appear to
return to near "normal" after a period of time dependent on the
magnitude of the salinity step and the speed of the salinity change.

It has been speculated whether differences in salinity ranges
of tolerance and activity in different populations are mainly due
to individual adaptation or whether they are a product of selection
and thus represent genetically fixed physiological races. An
answer to this question can be obtained from long-term acclimation
experiments in several cases. If the differences of activities and
of resistance ranges do not disappear in these experiments, this

may, of course, indicate nongenetic transmission or "dauermodifica-
tion" (Remane, 1971). Then breeding experiments are required.

Acclimation experiments have been carried out with the small
snail *Theodoxus fluviatilis* (Neumann, 1961). In specimens from
fresh water and brackish water (5 to 9 o/oo S), the upper salinity
limits of this species differ considerably. This holds true also
for representatives of these populations after preacclimation to
the same salinity (e.g., 0.2 or 6.2 o/oo S, respectively) up to
20 days. However, the differences of the salinity limits disappear
by stepwise acclimation if the animals are exposed to a step of
2 o/oo S every 4 days. Khlebovich and Kondratenkov (1973) have
demonstrated that the resistance range of *Hydrobia ulvae* from the
White Sea (habitat salinity 20.2 o/oo) could be enlarged from 14 to
34 o/oo S up to 6 to 76 o/oo S by stepwise acclimation (Fig. 1).

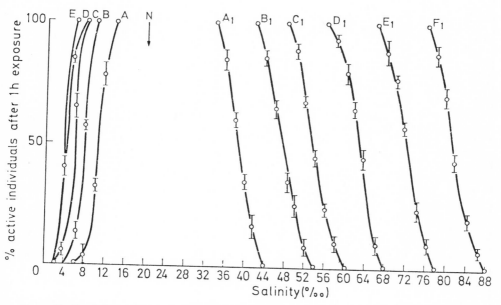

Fig. 1. *Hydrobia ulvae*, from the White Sea. Increase of the
tolerance range by stepwise acclimation to different
levels of salinity. After Khlebovitch and Kondratenkov,
1973.

The exposure time to the different levels of salinity was 10 to
15 days. A total adaptation time of more than two months was
required. The final tolerance limits attained after this stepwise
acclimation obviously represent the genotypically fixed salinity
range for this species.

The time required to attain the final level (steady state) of
individual adaptation to a changed salinity depends on the species,
on the salinity of preacclimation, and on the magnitude of the
salinity step. Moreover, it obviously depends on the type of
performance or resistance which is taken as a measure for the

adaptation or its action. This has been documented for different types of marine animals, particularly for those having shells, such as bivalves, snails, and barnacles. In such forms, osmotic and ionic equilibrium between the external and internal medium will be accomplished much quicker within the range of full mechanical activity. Outside of this range the osmotic balance may be delayed by protective responses, such as the closure of shells.

After a change of salinity, osmotic equilibrium is attained within a few hours in active *Hydrobia ulvae*, whereas it may require several days in specimens with closed shells after exposure to a greater salinity step. In addition, the behavior of the animals and the development of salinity adaptation after a change of osmotic conditions will depend on other simultaneously acting environmental factors, such as temperature (Fig. 2).

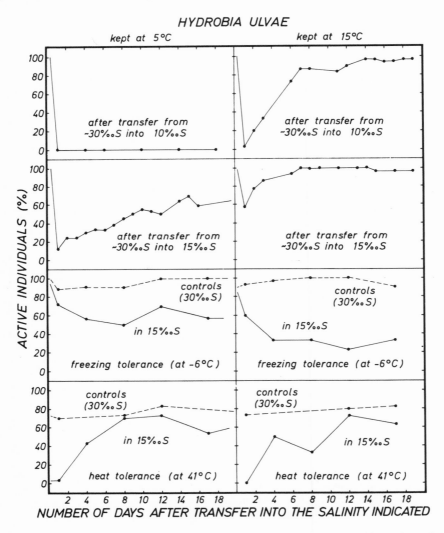

Fig. 2. *Hydrobia ulvae*. Changes of activity rate and tolerances during the course of acclimation to salinity (Theede and Saffé, unpublished).

If specimens from the North Sea (30 o/oo S) are transferred into brackish water of 10 o/oo S, they remain inactive and close for some weeks at a temperature of 5°C. At 15°C,however, only a temporary reduction of activity is observed, and the number of active animals increases within about 7 days. Also after transfer into water of 15 o/oo S, great differences in the speed of acclimation are noticed at different temperatures. When individuals are exposed to an additional environmental stress, such as freezing (13 min at -6°C)or heat (20 min at 41°C)after different times of salinity acclimation, changes of tolerances are found. There is a surprising difference in the reaction to freezing and to heating during salinity acclimation. The sensitivity to freezing is slowly reduced during several days of acclimation, whereas maximum sensitivity to heat is observed shortly after transfer into low salinity water.

Similar results have been obtained with *Littorina saxatilis* after transfer from normal seawater into brackish water. In addition, a decrease of resistance to hydrostatic pressure in the course of about two weeks was detected. In the reverse case of transferring animals of this species from brackish water (15 o/oo S) into normal seawater (30 o/oo S), a development of higher resistances to different abiotic factors (e.g., heat, freezing) is observed. But the acquisition of a high freezing tolerance requires more time than the changes of the other tolerances. This points to a special mechanism operative in the establishment of a high freezing resistance which has ecological importance.

The fact that low-salinity adaptation in many marine animals living in brackish waters is correlated to a reduction of resistance to different abiotic factors seems to be an interesting aspect for the ecologist as well as for the physiologist (Table 1). Several littoral marine species which have a high resistance to heat, freezing, and dehydration when they live in normal seawater, lose these abilities in strongly diluted brackish water (Theede and Lassig, 1967). The marine bivalves from the Gulf of Finland, living there at salinities around 6 o/oo, have almost completely lost their ability to survive freezing (at -10°C),whereas populations of these species from water with higher salinity are able to survive partial freezing of short duration, such as during low tide on the dry mud flats of the North Sea. In the inner Baltic these species will only survive cold in winter if the individuals remain covered by water. The loss of resistance, especially of freezing resistance, seems to be a contributory factor for the "upper submergence" of several species.

Similar effects of salinity adaptation on the resistance pattern as in whole animals are also observed on the cellular level. Gill tissue pieces from bivalve populations of the same species from the North Sea, the western and the eastern Baltic display quite different salinity ranges. The broadest range is found in specimens from normal seawater. It gradually gets narrower in habitats of lower salinity, and at the same time the tolerance range shifts to lower salinities. Comparing specimens from marine and brackish habitats, changes of the lower salinity limits are found to be obviously smaller than changes of the upper salinity limits. One should keep in mind, however, that small salinity changes at the lower end of this scale tend to have more pronounced ecological consequences. This tendency is obvious in all species tested and probably represents a general phenomenon (Theede and Lassig, 1967).

TABLE 1
Comparison of resistances of isolated gill tissues of specimens of *Mytilus edulis* from the Baltic and the North Sea to different abiotic factors.

Factor	Additional specifications[1]	Locality of the tested specimens	
		North Sea (30 o/oo)	Baltic Sea (15 o/oo)
Salinities[2]	24 hr at 10°C	65 o/oo	45 to 40 o/oo
Heat[2]	36°C	90 to 100 min	40 to 50 min
Freezing[2]	-10°C	70 to 80 min	7 to 8 min
Hydrostatic pressure[3]	5°C, 24 hr	600 atm	500 atm
O_2-deficiency + H_2S[4]	0.15 ml O_2/1 +50 mg $N_2S \cdot 9H_2O$, 10°C	120 to 144 hr	96 to 120 hr

[1] Complete cessation of ciliary activity was found after tissue exposure to the following conditions for the periods of time indicated.
[2] Theede and Lassig, 1967.
[3] Ponat, 1967.
[4] Theede, et al., 1969.

The results of reciprocal adaptation of mussels to normal seawater from the North Sea and to brackish water from the Baltic (Fig. 3) indicate that the differences of resistance ranges found in the different populations are of a nongenetic nature. After transfer of mussels from brackish water into normal seawater, an increase of the upper cellular salinity limits is found as well as of the freezing resistance (Theede, 1965; Fig. 4). The acclimation times for acquisition of the final resistance levels are definitely longer than the time required for osmotic adjustment to the changed salinity. It should be emphasized that the slow process of adaptation is by no means completed after osmotic balance between the internal and the external medium has been attained.

If the acclimation to high salinities is combined with low temperatures, the freezing resistance is greater than when acclimated to higher temperatures. This is true for whole animals (*Enchytraeus*) (Kähler, 1970), as well as for isolated tissues (*Mytilus*) (Theede, 1969b, 1972). The degree of this effect is genetically fixed and correlated to the ecological behavior and the geographical distribution of the species. The small increase of freezing resistance caused by a decrease of temperature in strongly diluted brackish water is without ecological significance. In every case, low temperatures combined with high salinities are necessary for the evolution of a high freezing resistance (Fig. 5). One should therefore consider this effect as a consequence of combined temperature-salinity adaptation.

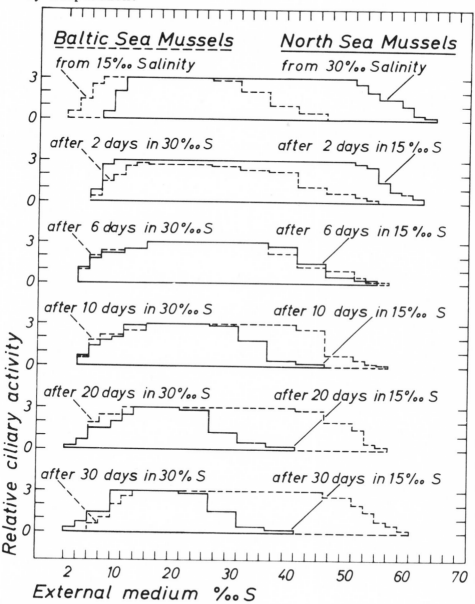

Fig. 3. *Mytilus edulis*. Change of the individual cellular salinity ranges during cross acclimation of specimens from the North Sea (30 o/oo S) and the Baltic (15 o/oo S). After Theede, 1965.

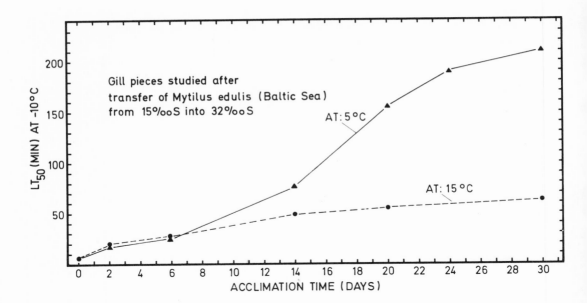

Fig. 4. *Mytilus edulis.* Changes of cellular freezing resistance
during acclimation of specimens of the Baltic (from
15 o/oo S) to 32 o/oo S at two different temperatures.
After Theede, 1972.

Fig. 5. *Mytilus edulis.*
Cellular freezing
resistance during
acclimation of
specimens of the
Baltic (from
15 o/oo S) to
32 o/oo S at two
different
temperatures.
After Theede, 1972.

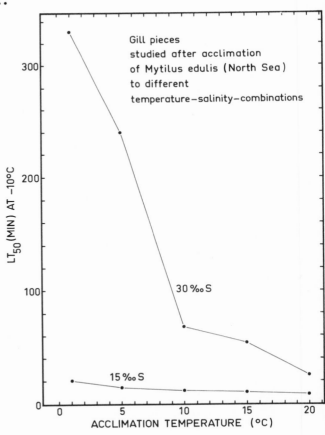

A fundamental process which accounts for salinity resistance and the degree of euryhalinity is the faculty to keep the cell volume within more or less narrow limits by isosmotic regulation (Lange, 1970). At different external salinities, the intracellular isosmotic regulation is correlated with marked changes in the content of amino acids (Lange, 1972) and changes of protein degradation and synthesis (Florkin and Schoffeniels, 1965, 1969). Berger and Kharazova (1971) studied the rate of uptake of marked amino acids by different tissues of *Littorina littorea* and *L. obtusata* after exposure of the animals to reduced salinities. Some hours after transfer, the protein synthesis was temporarily reduced and within about 8 days adjusted to about the original level. Also changes of the intensity of absorption of neutral red in these species indicate corresponding changes of cellular activity. The oxygen consumption of mussels and periwinkles after transfer from 25 to 14 o/oo S attains a steady state within 10 to 15 days (Berger, Lukanin, and Lapshyn, 1970).

Homoiosmotic Animals

Contrary to the particular significance of cellular regulations and tolerances for the degree of euryhalinity in poikilosmotic species, the survivability of homoiosmotic invertebrates at fluctuating salinities mainly depends on their capacity of osmotic and ionic regulation on the level of the body fluids. Salinity adaptations may affect the regulatory functions and organs. It was observed (Theede, 1969a) that specimens of *Carcinus maenas* grown at low salinities in the western Baltic display a more effective osmotic and sodium regulation after transfer into diluted brackish water than animals from North Sea water with higher salinity (Fig. 6). Acclimation experiments proved that the capacity of osmoregulation of North Sea specimens (from Helgoland) at low salinities of 5 and 10 o/oo S can be improved by a slow acclimation to brackish water of 15 o/oo S. Changes of the osmoregulatory capacity induced by individual acclimation to salinity were also observed in *Callinectes sapidus* (Anderson and Prosser, 1953).

Tissues and cells of homoiosmotic animals often have a smaller range of tolerance to salinity changes and a smaller, genetically fixed capacity of adaptation than those of corresponding euryhaline poikilosmotic species, but the regulatory response seems to be basically the same. In crustaceans osmoregulation by ions and low molecular organic compounds is relatively quick, with the main part occurring within about 20 to 30 hours or a few days (Siebers, 1970). According to Spaargaren (1971), the electrolyte content of the cells of *Crangon crangon* is held on an almost constant level and proves to be relatively insensitive to temperature changes in a certain range. At low salinities, the osmotic equilibration of the intracellular fluids with the osmotic concentration of the blood is connected with an efflux of nonelectrolytes into the blood. At high salinities, the intracellular content of nonelectrolytes increases.

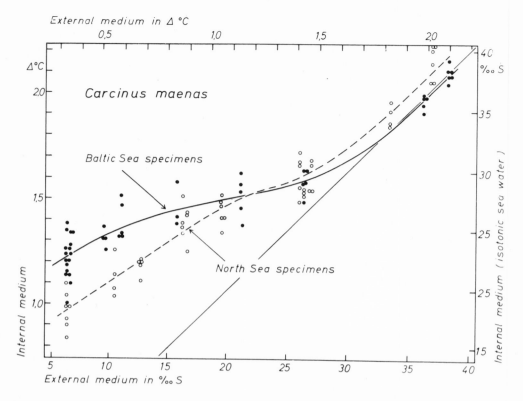

Fig. 6. *Carcinus maenas*. Osmoregulation capacities of North Sea
and Baltic specimens in different salinities of the
external medium. After Theede, 1969a.

The temperature-dependent shift of nonelectrolytes requires more
time than the change of electrolyte content of the blood caused by
salinity changes. In *Crangon crangon* the combination of low temperature with low salinities results in a minimum intracellular
concentration on nonelectrolytes. High temperatures together with
high salinities lead to a maximum. At extreme conditions of this
kind, the osmotic concentration of the intracellular fluids cannot
maintain isotonicity with respect to the blood without change of
electrolyte content. Experiments indicate an increased mortality
rate under such conditions. Animals in their natural habitat try
to avoid adverse combinations of environmental factors by selective
behavior and migrations.

In many species, changes of oxygen consumption have been
observed during the course of acclimation to salinity (Kinne, 1971;
Schlieper, 1971). Also, recent papers confirm specific relations
between metabolic rates and external salinity (Bulnheim, 1972;
Dorgelo, 1973). Formerly these changes were explained by correlating
metabolic rate of whole animals or tissues during osmotic stress
with expenditure of energy for osmotic and ionic regulation or by
correlating metabolic changes with the degree of hydration
(Schlieper, 1958). But these explanations did not seem to be quite

satisfactory (Potts and Parry, 1964). However, permanent and transitory changes of metabolic rates have often not been differentiated clearly. Short-term metabolic changes may be interpreted in terms of immediate responses due to osmotic stress. Gross (1957) and Newell (1970) point to a possible role of locomotor activity. Gilles (1973) proposes that the variations in oxygen consumption of *Callinectes sapidus*, the blue crab, could be related, at least partly, to changes in the amino acid catabolism which occur during the process of cellular osmotic adjustment, and which seem to be strictly connected with the isosmotic regulation of cell volume.

These processes may lead to long-term metabolic changes. Another recently described effect of salinity concerns the diffusion rate of oxygen (Lange, Staaland, and Mostad, 1972). The diffusion rate of oxygen varies proportionally to the solubility of oxygen in water; it decreases with increasing salinity. Also, it is thought, to influence the respiratory rate of different types of aquatic invertebrates.

In this report I would not like to speculate more about the mechanisms which may account for the reported effects. For a better basis to an understanding of the changes of reactions and regulations taking place in connection with salinity adaptation, much further biochemical work on the dependence of enzymatic reactions on internal and external factors and on possible hormonal control is required. Newell (1970) has pointed out how difficult it is to bring the situation found on the cellular and subcellular level in vitro into a useful context with the reactions observed in whole animals.

LITERATURE CITED

Anderson, D. and C. L. Prosser. 1953. Osmoregulating capacity in populations occurring in different salinities. Biol. Bull. Woods Hole 105:369.

Bayne, B. 1973. The responses of three species of bivalve mollusc to declining oxygen tension at reduced salinity. Comp. Biochem. Physiol. 45A:793-806.

Berger, V. J. and A. D. Kharasova. 1971. The investigation of substantial changes and of protein synthesis during the adaptation to lowered salinities of the environment in some White Sea snails. Russ. Cytologia 13:1299-1303.

_____, V. V. Lukanin, and V. V. Lapshyn. 1970. Oxygen consumption of some littoral White Sea molluscs in the process of acclimation to salinity changes. Russ. Ecologia Sverdlovsk 1:68-72.

Bulnheim, H. P. 1972. Vergleichende Untersuchungen zur Atmungsphysiologie euryhaliner Gammariden unter besonderer Berücksichtigung der Salzgehaltsanpassung. Helgoländer wiss. Meeresunters. 23:485-534.

Dorgelo, J. 1973. Comparative ecophysiology of Gammarids (Crustacea: Amphipoda) from marine, brackish and fresh-water habitats exposed to the influence of salinity-temperature combinations. III. Oxygen uptake. Netherl. J. Sea Res. 7: 253–66.

Florkin, M. and E. Schoffeniels. 1965. Euryhalinity and the concept of physiological radiation. In: Studies in Comparative Biochemistry (K. A. Munday, ed.), pp. 6–40. London: Pergamon Press.

_____ and _____. 1969. Molecular Approaches to Ecology. New York and London: Academic Press.

Gilles, R. 1973. Oxygen consumption as related to the amino-acid metabolism during osmoregulation in the blue crab *Callinectes sapidus*. Netherl. J. Sea Res. 7:280–89.

Gross, W. J. 1957. An analysis of response to osmotic stress in selected decapod crustacea. Biol. Bull. Woods Hole 112:43–62.

Kähler, H. H. 1970. Über den Einfluß der Adaptationstemperatur und des Salzgehaltes auf die Hitze- und Gefrierresistenz von *Enchytraeus albidus* (Oligochaeta). Mar. Biol. 5:315–24.

Khlebovitch, V. V. and A. P. Kondratenkov. 1973. Stepwise acclimation – A method for estimating the potential euryhalinity of the gastropod *Hydrobia ulvae*. Mar. Biol. 18:6–8.

Kinne, O. 1964. Non-genetic adaptation to temperature and salinity. Helgoländer wiss. Meeresunters. 9:433–58.

_____. 1966. Physiological aspects of animal life in estuaries with special reference to salinity. Netherl. J. Sea Res. 3:222–44.

_____. 1967. Physiology of estuarine organisms with special reference to salinity and temperature: General aspects. In: Estuaries (G. H. Lauff, ed.), pp. 525–40. Washington, D. C.: American Association for the Advancement of Science.

_____. 1971. Salinity: animals-invertebrates. In: Marine Ecology (O. Kinne, ed.), pp. 821–995. London: Wiley Interscience.

Lange, R. 1970. Isosmotic intracellular regulation and euryhalinity in marine bivalves. J. Exp. Mar. Biol. Ecol. 5:170–79.

_____. 1972. Some recent work on osmotic, ionic and volume regulation in marine animals. Oceanog. Mar. Biol. Ann. Rev. 10:97–136.

_____, H. Staaland, and A. Mostad. 1972. The effect of salinity and temperature on solubility of oxygen and respiratory rate in oxygen-dependent marine invertebrates. J. Exp. Mar. Biol. Ecol. 9:217–29.

Neumann, D. 1960. Osmotische Resistenz und Osmoregulation der Flußdeckelschnecke *Theodoxus fluviatilis* L. Biol. Zbl. 79: 585–605.

Newell, R. 1970. Biology of Intertidal Animals. London: Logos Press.

Pierce, S. K. 1971. Volume regulation and valve movements by marine mussels. Comp. Biochem. Physiol. 39A:103–17.

Ponat, A. 1967. Untersuchungen zur zellulären Druckresistenz verschiedener Evertebraten der Nord- und Ostsee. Kieler Meeresforsch. 23:21-47.

Potts, W. T. W. 1968. Osmotic and ionic regulation. Ann. Rev. Physiol. 30:73-104.

_____ and G. Parry. 1964. Osmotic and Ionic Regulation in Animals. Oxford: Pergamon Press.

Remane, A. 1971. Ecology of brackish water. In: Biology of Brackish Water (A. Remane and C. Schlieper, eds.), pp. 1-210. New York: Wiley Interscience.

Schlieper, C. 1958. Physiologie des Brackwassers. In: Die Biologie des Brackwassers (A. Remane and C. Schlieper, eds.), pp. 219-330. Stuttgart: Schweizerbart'sche Verlagsbuch-handlung.

_____. 1971. Physiology of brackish water. In: Biology of Brackish Water (A. Remane and C. Schlieper, eds.), pp. 211-350. New York: Wiley Interscience.

Siebers, D. 1970. Mechanismen der intrazellulären isosmotischen Regulation der Aminosäurekonzentration bei dem Flußkrebs Oronectes limosus. Ph.D. dissertation, Free University of Berlin.

Smith, R. I. 1970. The apparent water-permeability of Carcinus maenas (Crustacea, Brachyura, Portunidae) as a function of salinity. Biol. Bull. 139:351-62.

Spaargaren, D. H. 1971. Aspects of osmotic regulation in the shrimps Crangon crangon and Crangon allmanni. Netherl. J. Sea Res. 5:275-335.

Theede, H. 1963. Experimentelle Untersuchungen über die Filtrationsleistung der Miesmuschel Mytilus edulis L. Kieler Meeresforsch. 19:20-41.

_____. 1965. Vergleichende experimentelle Untersuchungen über die zelluläre Gefrierresistenz mariner Muscheln. Kieler Meeresforsch. 21:153-66.

_____. 1969a. Einige neue Aspekte bei der Osmoregulation von Carcinus maenas. Mar. Biol. 2:114-20.

_____. 1969b. Experimentelle Untersuchungen über physiologische Unterschiede bei Evertebraten und Fischen aus Meer- und Brackwasser. Limnologica (Berlin) 7:119-28.

_____. 1972. Vergleichende ökologisch-physiologische Untersuchungen zur zellulären Kälteresistenz mariner Evertebraten. Mar. Biol. 15:160-91.

_____ and J. Lassig. 1967. Comparative studies on cellular resistance of bivalves from marine and brackish waters. Helgoländer wiss. Meeresunters. 16:119-29.

_____, A. Ponat, K. Hiroki, and C. Schlieper. 1969. Studies on the resistance of marine bottom invertebrates to oxygen-deficiency and hydrogen sulphide. Mar. Biol. 2:325-37.

Vernberg, W. B. and F. J. Vernberg. 1972. Environmental Physiology of Marine Animals. New York and Berlin: Springer-Verlag.

Cellular adjustments to salinity variations in an estuarine barnacle, *Balanus improvisus*

H. J. Fyhn and J. D. Costlow

The estuary is characterized by large variations in environmental parameters and in order to be successfully adapted to this environment the inhabitants must tolerate these variations. The seawater salinity is one of the variables and poses osmotic problems for the animals. Acorn barnacles make up a quantitatively important component of intertidal and estuarine communities and are euryhaline forms. Some species like *Balanus amphitrite* and *B. improvisus* have been reported to occur even in fresh water (Shatoury, 1958; Zullo, Beach, and Carlton, 1972). This euryhalinity is remarkable, since recent studies have shown that cirripeds are osmoconformers (Foster, 1970; Fyhn, Petersen, and Johansen, 1972). The mechanism allowing cirripeds to exhibit such a high euryhalinity is therefore located at the cellular level. In the present paper some data are given on the cellular response to variations in seawater osmolality of the estuarine barnacle, *B. improvisus*.

MATERIALS AND METHODS

Specimens of *B. improvisus* Darwin were collected in September 1972 at Cherry Branch, Neuse River, North Carolina, where they were attached to submerged roots of dead trees in water of 25°C and 12 ppt. Pieces of roots, each having 100 to 150 unharmed animals, were kept in aerated aquaria (volume 15L) at 25°C ± 1°C. The animals

were exposed to slowly changing seawater osmolalities by dripping
distilled water or seawater into the aquaria at a rate which halved
or doubled the osmolality in 48 hrs. The animals were maintained
at their test osmolalities for 5 to 10 weeks before experimentation.
With this procedure the animals behaved normally throughout the
acclimation period as judged by their cirri beating, feeding, and
molting activity. The animals were fed newly hatched *Artemia* nauplii
(Metaframe, San Francisco Bay Brand) every second day and the water
was changed weekly. No food was given during the last 1½ to 2 days
before sampling. All animals used were in stage C (interecdysis)
of the molt cycle (Davis, Fyhn, and Fyhn, 1973).

Thorax musculature was used as the test tissue. Dissected free
from cirri, oral cone, and hind gut and lightly blotted with filter
paper, the excised thorax gives an almost pure muscle sample of 2 to
5 mg wet weight in adult *B. improvisus*. The relative water content
of the thorax musculature was calculated from wet weight and dry
weight determinations of single thoraxes using a Sartorius micro-
balance (Model 1802) with an accuracy of 10^{-3} mg. The tissue was
dried to constant weight at 105°C. Tissue homogenization and protein precipitation were carried out in ice-cold 80% ethanol. Three
to four thoraxes were pooled to give a total wet weight of about
10 mg. The dissection, weighing, and freezing (-20°C) of each
thorax took 5 to 6 minutes. The thoraxes were homogenized four times
and the precipitate was spun down at 18,000 RPM for 15 min at 4°C in
a Sorwall RC2-B centrifuge. The precipitate was washed once. The
supernatant was evaporated to dryness at 70 to 75°C and the deposit
redissolved in 250 to 600 µl citrate buffer (0.2 N) of pH 2.2 with
0.5 mM norleucine for analysis of ninhydrin-positive substances
(NPS) and amino acids. The NPS values were determined according to
the photometric method of Moore and Stein (1948). Triplicate samples
(10 µl) were read in a Beckman DB-G spectrophotometer at 570 nm with
the sample diluent as the blank. Norleucine standards were run with
each analysis. The amino acid analyses were carried out on a Beckman
amino acid analyzer (Model 116) on single samples (200 µl) and the
added norleucine functioned as an internal standard.

RESULTS

Figure 1 shows the relative water content of the thorax musculature of animals maintained for 5 weeks at their respective test
osmolalities. There are only small changes in the relative water
content (less than 5 mg water per 100 mg of tissue) in spite of a
forty times increase in the seawater osmolality. The stippled line
shows the expected changes in the relative water content of the tissue if the muscle fibers behave as ideal osmometers. This line is
calculated by the following formula (Fyhn, Petersen, and Johansen,
1972) based on a muscle tissue with a relative water content of 83.9%
at an original extracellular osmolality of 250 m-osmoles:

$$RW_2 = \frac{100 \text{ osm}_1}{\text{osm}_1 + [(100/RW_1)-1]\text{osm}_2}$$

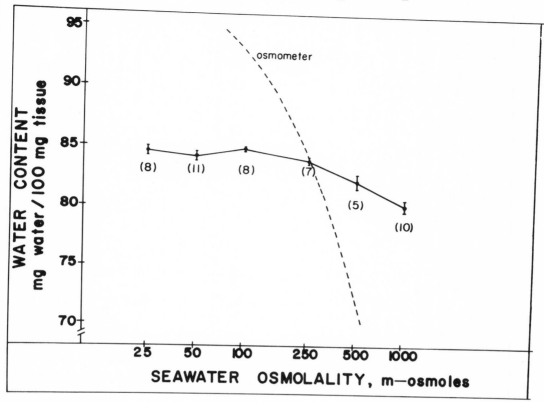

Fig. 1. Relative water content of the thoracic musculature of *Balanus improvisus* acclimated for 5 wks to the various seawater osmolalities. Mean values with standard error are given, with the number of animals in parentheses. The stippled curve describes the behavior of an ideal osmometer as calculated by the formula in the text.

RW_1 and RW_2 are the relative water contents of the tissue at the original (osm_1) and at the changed (osm_2) extracellular osmolality, respectively. As the extracellular osmolality is decreased, a passive influx of water would swell an osmometer and result in the steeply increasing relative water content (Fig. 1). The actual measurements show that the muscle fibers are not behaving as osmometers but are actively regulating the cellular hydration.

Figure 2 shows the concentration of free NPS of the thorax musculature of animals maintained for 8 to 10 weeks at their respective test osmolalities. The NPS analyses were carried out on pooled samples of 3 to 4 thoraxes and each point represents, therefore, the mean of 3 to 4 animals. The data show that the NPS concentration is

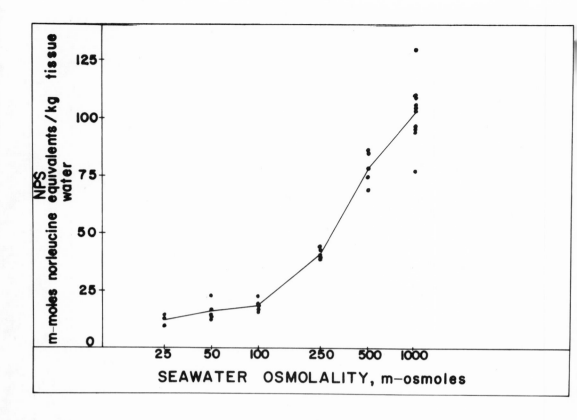

Fig. 2. Concentration of ninhydrin-positive substances (NPS) of
the thorax musculature of *Balanus improvisus* acclimated for
8 to 10 weeks to the various seawater osmolalities. Each
point represents measurements on a pooled sample of 3 to 4
thoraxes.

rapidly increasing with increasing seawater osmolality above 100 m-
osmoles. Below 100 m-osmoles, however, there is almost no change
in the NPS concentration.
     Figure 3 and Tables 1 and 2 show the concentrations of individ-
ual neutral and acidic amino acids of the thorax musculature of
animals maintained for 8 to 10 weeks at their respective test osmola
lities. The analyses were carried out on pooled samples of 3 to 4
thoraxes and mean values of 2 to 4 analyses are plotted in Figure 3.
Individual data of the analyses are given in Tables 1 and 2. Figure
3 shows that the concentration of the four dominating amino acids
increases with increasing seawater osmolality above m-osmoles and ha
a stable level below this osmolality. Alanine, glycine, and taurine
are part of the NPS and added together, they make up 60% of the NPS
at 1,000 m-osmoles. Proline, however, is not revealed by the genera
ninhydrin procedure and its presence is additional to the measured
NPS. Above 500 m-osmoles proline is the dominant amino acid of the
muscle tissue and at 1,000 m-osmoles it comprises more than all the
other amino acids together. Below a seawater osmolality of

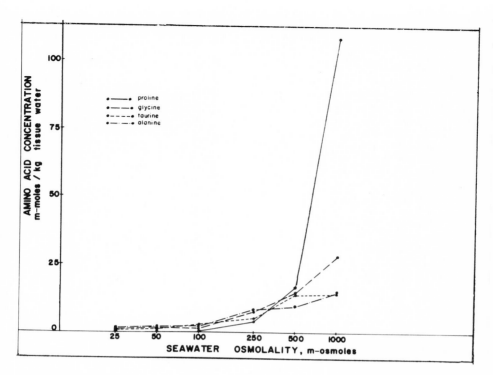

Fig. 3. Concentration of amino acids of the thorax musculature of *Balanus improvisus* acclimated for 8 to 10 weeks to the various seawater osmolalities. The analyses are carried out on pooled samples of 3 to 4 thoraxes and mean values of 2 to 4 analyses are given.

100 m-osmoles, proline is reduced to undetectable values while alanine, glycine, and taurine stabilize in a concentration of 2 to 3 m-molal.

Other neutral and acidic amino acids are present in the muscle tissue in concentrations lower than 5 m-molal (Tables 1 and 2). The concentration of glutamic acid, serine, and to some degree valine increases with seawater osmolality (Table 1). Eight amino acids (Table 2) are present in mean concentrations of 1 m-molal or lower and do not show a clear correlation with the seawater osmolality.

The total concentration of α-amino compounds (the sum of proline and NPS) is shown in Table 3. The mean concentration of the α-amino compounds increases 15 to 20 times from the lowest to the highest seawater osmolality. Relative to the seawater osmolality, these compounds make up a stable proportion of about 0.20 above 100 m-osmoles while the proportion increases below this osmolality.

TABLE 1
Major neutral and acidic amino acids of the thorax musculature of
*Balanus improvisus**

Amino Acid	Seawater Osmolality (m-osmoles)					
	25	50	100	250	500	1000
Alanine	0.92	1.53	1.82	7.88	8.99	13.59
	1.60	2.39	1.95	8.02	10.00	14.30
			2.58	9.00	10.63	15.84
						16.00
Mean	1.26	1.96	2.12	8.30	9.87	14.93
Glycine	0.73	1.22	1.25	7.06	12.98	23.08
	1.02	1.96	1.38	7.71	13.39	26.99
			1.67	8.64	17.22	30.94
						32.80
Mean	0.88	1.59	1.43	7.80	14.53	28.45
Glutamic acid	1.14	1.54	1.41	3.22	4.61	4.75
	1.36	2.09	2.00	3.47	4.88	5.28
			2.12	3.59	5.33	5.72
						6.43
Mean	1.25	1.21	1.84	3.46	4.94	5.55
Proline	$-^{\dagger}$	–	–	3.18	14.66	96.57
	–	–	–	3.60	17.06	106.33
			0.43	5.04	19.20	110.16
						120.84
Mean	–	–	0.14	3.94	16.97	108.48
Serine	0.58	0.70	0.51	0.52	2.74	1.24
	0.67	0.77	0.56	0.87	3.89	3.32
			0.62	1.03	4.78	6.02
Mean	0.63	0.74	0.56	0.81	3.80	3.53
Taurine	0.57	1.17	1.95	4.48	12.82	12.62
	0.64	1.35	2.32	5.60	13.83	13.84
			3.00	5.67	16.36	15.97
						16.39
Mean	0.61	1.26	2.42	5.25	14.34	14.71
Valine	0.53	0.64	0.65	1.36	1.52	1.67
	0.70	0.69	0.67	1.47	1.53	2.16
			0.75		1.62	2.36
						2.53
Mean	0.62	0.67	0.69	1.42	1.56	2.18

*Acclimated for 8 to 10 weeks to the various seawater osmolalities.
The analyses are carried out on pooled samples of 3 to 4 thoraxes
and the amino acids are expressed in m-moles per kg tissue water.
†- Undetectable amounts.

TABLE 2
Minor neutral and acidic amino acids of the thorax musculature of
*Balanus improvisus**

Amino Acid	Seawater Osmolality (m-osmoles)					
	25	50	100	250	500	1000
Aspartic acid	-†	-	-	0.22	-	-
	-	-	-	0.34	0.11	0.32
			0.05	0.37	0.13	0.36
Mean	-	-	0.02	0.31	0.08	0.23
Cystine	0.39	0.69	0.35	0.53	0.47	0.67
	0.53	0.70	0.41	0.75	0.63	1.05
			0.53		0.69	1.23
						1.70
Mean	0.46	0.70	0.43	0.64	0.60	1.16
Isoleucine	-	-	0.09	0.25	0.21	0.21
	0.09	0.08	0.10	0.28	0.29	0.52
			0.25		0.37	0.63
						0.74
Mean	0.05	0.04	0.15	0.27	0.29	0.53
Leucine	-	-	0.15	0.39	0.50	0.56
	0.08	0.16	0.21	0.47	0.64	0.58
			0.25		1.03	0.63
						0.68
Mean	0.04	0.08	0.20	0.43	0.72	0.61
Methionine	-	-	0.06	0.25	0.57	0.22
		0.13	0.09	0.28	0.64	0.23
			0.17		0.80	0.45
						0.71
Mean	-	0.07	0.11	0.27	0.67	0.40
Phenylalanine	-	-	-	-	0.28	0.20
	-	-	-	0.25	0.46	0.27
			0.20		0.60	0.28
						0.54
Mean	-	-	0.07	0.13	0.45	0.32
Threonine	0.23	0.11	-	0.19	0.43	0.24
	0.50	0.22	0.04	0.20	0.57	0.68
			0.08	0.23	0.84	0.97
Mean	0.37	0.17	0.04	0.21	0.61	0.63
Tyrosine	-	-	-	-	0.55	0.06
	-	-	-	0.08	0.74	0.23
			0.18		0.77	0.62
						0.90
Mean	-	-	0.06	0.04	0.69	0.60

*Acclimated for 8 to 10 weeks to the various seawater osmolalities.
The analyses are carried out on pooled samples of 3 to 4 thoraxes
and the amino acids are expressed in m-moles per kg tissue water.
†- Undectable amounts.

TABLE 3
Total concentration of proline plus ninhydrin-positive substances (NPS) of the thorax musculature of *Balanus improvisus* at and relative to various seawater osmolalities[*]

Seawater (m-osmoles)	Proline + NPS (m-molal)	Proline + NPS / Seawater
1000	211.1	0.21
500	95.8	0.19
250	45.2	0.18
100	18.9	0.19
50	16.2	0.32
25	12.4	0.50

[*]The animals are acclimated for 8 to 10 weeks. The analyses are carried out on pooled samples of 3 to 4 thoraxes and mean values are given.

DISCUSSION

The evasive behavior of closing the opercular valves for long periods of time is typical of barnacles encountering adverse environmental conditions (Barnes, Finlayson, and Piatigorsky, 1963; Foster and Nott, 1969). When exposed to changes in salinity, the effect of this reaction is to retard the osmotic equilibration between hemolymph and seawater. Thus, barnacles of the genera *Balanus* and *Elminius* needed 5 to 8 days to equilibrate osmotically when the animals were transferred directly from 100% to 50% seawater (Foster, 1970). An acclimation time of several weeks, therefore, seems necessary when studying the osmotic response of barnacles. In the present study the animals were exposed to slowly changing seawater osmolalities before experimentation. The data presented, therefore, describe the steady state condition of the animals at the various seawater osmolalities.

Crustaceans have low levels of free amino acids in the hemolymph, generally 2 to 5 m-molal in total concentration (Florkin and Schoffeniels, 1969; Gerard and Gilles, 1972). This applies to the gooseneck barnacle *Pollicipes polymerus*, in which NPS concentrations of about 1 m-molal were found irrespective of seawater osmolality (Fyhn, Petersen, and Johansen, 1972). The α-amino compounds of the thorax musculature of *B. improvisus* are therefore assumed to be intracellularly located and changes in the concentration of these compounds result in correspondingly large changes in the intracellular osmolality. Passively, as shown by the osmometer curve (Fig. 1), a decrease in the extracellular osmolality would swell the cells and thus increase the relative water content of the tissue. The thorax musculature, however, has an almost unchanged water content, which

indicates that the muscle fibers actively regulate the cellular
hydration. This means a regulation of cell volume since water is
the major component of the cells. The adjustment of the intracellular α-amino compounds may be partly responsible for this cell volume
regulation. A decrease in the amount of intracellular osmotic-
active substances will counteract the osmotic effect of decreased
extracellular osmolality by reducing the water influx to the cell
and thereby will stabilize the cell volume (Lange and Fugelli, 1965).
This homeostatic process has been termed the isosmotic intracellular
regulation (Jeuniaux, Bricteux-Grégoire, and Florkin, 1961) and has
been found to function in many invertebrate groups (Schoffeniels and
Gilles, 1970). The present findings show also that *B. improvisus*
depends on the isosmotic intracellular regulation for its euryhalinity. However, in *B. improvisus*, substances other than α-amino compounds are also involved in the cell volume regulation since these
compounds make up only 20% of the intracellular osmolality at seawater osmolalities above 100 m-osmoles. Comparable values were
found in the gooseneck barnacle *P. polymerus* where the NPS accounted
for 20 to 25% of the intracellular osmolality of animals in seawater
between 500 and 900 m-osmoles (Fyhn, Petersen, and Johansen, 1972).

  No data can yet be given to explain the constancy in cellular
hydration at seawater osmolalities below 100 m-osmoles. If the
osmoconforming behavior is retained by the animals at these low
salinities, the constancy in the concentration of the α-amino compounds would result in an increased proportion of these compounds to
the intracellular osmolality (Table 3). The use of intracellular
osmotic effectors to regulate cellular hydration then must have been
shifted from α-amino compounds to other organic or inorganic compounds. A reason for this could be to maintain critical levels of
amino acids for protein metabolism. If, however, the hemolymph of
B. improvisus does not conform osmotically with the seawater at the
lowest osmolalities, the constancy in cellular hydration could
reflect a constancy in the extracellular osmolality. Some data favor-
ing such a possibility are in fact given by Foster (1970) in a study
of *B. improvisus*, although he interpreted his data differently. From
determinations of hemolymph osmolality of animals maintained for 1
to 2 days in salinities between 34 and 1.7 ppt (about 1,000 to 50
m-osmoles), Foster concluded that *B. improvisus* is an osmoconformer.
His data, however show a deviation from the osmoconforming behavior
of animals in seawater of salinities between 3.0 and 1.7 ppt. This
deviation, Foster concluded, would disappear if the animals were
given ample acclimation time. The deviation from the osmoconforming
behavior started at a seawater osmolality of about 100 m-osmoles and
our data of the NPS and amino acid concentrations likewise have the
breaking point of the curves at an osmolality of about 100 m-osmoles.
The data of NPS and amino acids do show the steady state condition
and the constancy in their concentrations below a seawater osmolality
of 100 m-osmoles suggests that the hemolymph osmolality may also
have a stable level below this osmolality. The findings of constancy
in cellular hydration as well as in NPS and amino acid concentrations

at the lowest seawater osmolalities may reflect a constancy in hemolymph osmolality. To learn whether the deviation from the osmoconforming behavior in B. *improvisus* is of temporary or permanent chara
ter, studies will be carried out on long-term acclimated animals.

SUMMARY

These data show that the euryhalinity of B. *improvisus* is
dependent upon a regulation of the cell volume. Amino acids and
other α-amino compounds participate in this regulation and especiall
proline, but also alanine, glycine, and taurine are of importance ir
adjusting the intracellular osmolality. Below a seawater osmolality
of about 100 m-osmoles the importance of the intracellular osmotic
effectors may have shifted from α-amino compounds to other substances, or the hemolymph osmolality may not conform osmotically wit
the seawater for unknown reasons.

ACKNOWLEDGMENTS

This research was supported in part by contract No. NR 104-194
from the Office of Naval Research. H. J. Fyhn acknowledges support
from The American-Scandinavian Foundation and from the Norwegian
Research Council for Science and the Humanities. The authors wish
to thank Mr. C. William Davis for pointing out the locality for
B. *improvisus* and for identifying the species. They are further
indebted to Dr. J. Bolling Sullivan for his generosity in providing
the use of some of the equipment needed during this study.

LITERATURE CITED

Barnes, H., D. M. Finlayson, and J. Piatigorsky. 1963. The effect
    of desiccation and anaerobic conditions on the behaviour, sur-
    vival, and general metabolism of three common cirripeds. J.
    Anim. Ecol. 32:233-52.
Davis, C. W., U. E. H. Fyhn, and H. J. Fyhn. 1973. The intermolt
    cycle of cirripeds: Criteria for its stages and its duration ir
    Balanus amphitrite. Biol. Bull. 145:
Florkin, M. and E. Schoffeniels. 1969. Molecular Approaches to
    Ecology. New York: Academic Press.
Foster, B. A. 1970. Responses and acclimation to salinity in the
    adults of some balanomorph barnacles. Phil. Trans. Roy. Soc.
    London, B 256:377-400.
_____ and J. A. Nott. 1969. Sensory structures in the opercula of
    the barnacle *Elminius modestus*. Mar. Biol. 4:340-44.

Fyhn, H. J., J. A. Petersen, and K. Johansen. 1972. Eco-physiological studies of an intertidal crustacean, *Pollicipes polymerus* (Cirripedia, Lepadomorpha). I. Tolerance to body temperature change, desiccation and osmotic stress. J. Exp. Biol. 57:83-102.

Gerard, J. F. and R. Gilles. 1972. The free amino-acid pool in *Callinectes sapidus* (Rathbun) tissues and its role in the osmotic intracellular regulation. J. Exp. Mar. Biol. Ecol. 10: 125-36.

Jeuniaux, Ch., S. Bricteus-Grégoire, and M. Florkin. 1961. Contribution des acides aminés libres à la regulation osmotic intracellulaire chez deux crustacés euryhalins, *Leander serratus* F. et *Leander squilla* L. Cah. Biol. Mar. 2:373-80.

Lange, R. and K. Fugelli. 1965. The osmotic adjustment in the euryhaline teleosts, the flounder, *Pleuronectes flesus* L. and the three-spined stickleback, *Gasterosteus aculeatus* L. Comp. Biochem. Physiol. 15:283-92.

Moore, S. and W. H. Stein. 1948. Photometric ninhydrin method for use in the chromatography of amino acids. J. Biol. Chem. 176: 367-88.

Schoffeniels, E. and R. Gilles. 1970. Osmoregulation in aquatic arthropods. In: Chemical Zoology (M. Florkin and B. T. Scheer, eds.), vol. V, part A, pp. 255-86. New York: Academic Press.

Shatoury, H. H. 1958. A freshwater mutant of *Balanus amphitrite*. Nature 181:790-91.

Zullo, V. A., D. B. Beach, and J. T. Carlton. 1972. New barnacle records (Cirripedia, Thoracica). Proc. Calif. Acad. Sci. 39: 65-74.

The uptake and incorporation of glycine by the gills of
Rangia cuneata (Mollusca: Bivalvia)
in response to variations in salinity and sodium

J. W. Anderson

The bivalve *Rangia cuneata* (Gray) is a prominent member of estuarine ecosystems along the northern border of the Gulf of Mexico and has recently become abundant in the estuaries of the Atlantic Coast of the United States from Virginia to Florida. This clam is found in regions of the estuary which exhibit fluctuations between 0 and 25 parts per thousand salinity (o/oo S) (Woodburn, 1962; Chanley, 1965; Hopkins, 1970) and has been reported to be capable of living in fresh water (below 0.3 o/oo S) for a period of at least seven months (Hopkins and Andrews, 1970).

Bedford and Anderson (1972a) have shown that *Rangia* is perhaps the only estuarine or marine bivalve studied thus far that is capable of significant hyperosmotic blood regulation at salinities below 10 o/oo. A differential of 55 to 65 milliosmoles/l above the environment was maintained at media concentrations between 16 and 100 milliosmoles/l (approximately 0 to 5 o/oo S). More recently, the relationship between salinity and the uptake and incorporation of glycine by whole clams has been reported (Anderson and Bedford, 1973). While uptake of glycine from dilute solution in the ambient medium was found to decrease with decreasing salinity, the percent incorporation of accumulated glycine increased. Evidence suggested that at salinities below 5 o/oo, the osmotically active glycine was rapidly converted to protein. While the respiration of *Rangia* acclimated to various salinities was not significantly different (Bedford and Anderson, 1972b), the total uptake and utilization of accumulated glycine by the clams increased with increased salinity, between 1 and 10 o/oo S.

Although previous research had answered several questions regarding the effects of salinity on the physiology of this extremely successful estuarine bivalve, the following questions remained:

1. Since isolated gill tissue was shown to accumulate glycine actively (Anderson and Bedford, 1973), are the cells of the gill also capable of incorporation and utilization?
2. Are uptake and accumulation of glycine dependent upon the total osmotic concentration of the medium or upon its ionic composition?
3. What are the primary constituents of the blood, when it is hyperosmotic to the environment?

This study was undertaken to resolve these questions in an attempt to understand more completely the regulatory mechanisms utilized by *Rangia* which allow it to flourish in an environment subject to numerous natural and man-made stresses.

MATERIALS AND METHODS

Clams were collected from the upper reaches of the Galveston Bay and San Antonio Bay estuaries of Texas. They were transported in ice chests, without water, to the laboratory at College Station. Animals were maintained in aquaria for at least 2 days at the collection salinity before acclimation to experimental salinities, which unless noted, was greater than 7 days. Maintenance and experimental temperatures ranged between 21 and 22°C, as controlled by laboratory air conditioning, but variations during a given experiment were less than 1°C.

All water used in maintenance and experimental procedures consisted of a synthetic seawater mixture (Instant Ocean, Eastlake, Ohio) diluted with distilled water to the required salinity. Dilutions of this mixture, which will be referred to as seawater, were made from full strength medium (34 o/oo S) by calculations, and the final solution checked with a refractometer (American Optical Instruments). In experiments regarding low sodium solutions, the synthetic seawater was prepared from a formula listed in Cavanaugh (1964), and sodium chloride was replaced by choline chloride. Afte a stock solution of this low sodium solution was prepared (equivale to sea water at 34 o/oo S in total osmolarity), dilutions were made on a volumetric basis, and the chlorinity tested with a chloridomet (Buchler-Cotlove). By replacing nearly all of the sodium with choline, the sodium content of the stock solution and dilutions was reduced to a level such that these preparations will be referred to as "sodium-free" (Na-free), but this assumption was not confirmed by analyses.

In previous studies, the relationship between salinity and the percentage water and percentage ash in whole clams and gill tissue was determined. Regardless of the experimental treatment, all uptake data were converted to counts per minute of C^{14}-radioactivity per milligram ash-free dry weight (cpm/mg). In instances when it was more beneficial to examine the percentage of the total radio-activity which was incorporated into alcohol-insoluble (80% ethanol) material, the data were first converted to cpm/mg ash-free dry weight, then expressed as a percentage. After an exposure to uniformly labeled C^{14}-glycine, the gills were rinsed three times for 10 min each in identical solutions without glycine, blotted, weighed, and placed into 10 ml of 80% ethanol for 48 hrs. The ethanol extracts were counted and the gills blotted and placed into scintillation vials with 1 ml of Protosol (New England Nuclear). The solubilized gills were counted 24 hrs later after adding 10 ml of Aquasol (New England Nuclear) to the vials and shaking. By use of a standard curve relating wet weight to ash-free dry weight, the C^{14}-radioactivity of the ethanol-soluble (alcohol) and insoluble fractions was calculated. The number of gills used for each specific data point varied from 4 to 8, and the standard deviation from the mean was calculated in each instance. Sample counting of water, alcohol extracts, and solubilized gills was performed on a Beckman LS-200 liquid scintillation counter using an Aquasol cock-tail. While data are available which would allow for very minor corrections in quenching and for the conversion of C^{14}-radioactivity to micromoles of glycine, these steps were not taken. The purpose of this paper was to compare the levels of uptake by gills in solutions of different salinity and ionic composition. Therefore, in all experiments the concentration of C^{12}- (3-4 × 10^{-6} M) and C^{14}- (3-4 × 10^{-7} M) glycine was held constant and the treatment of gills was precisely the same. In this way, the only variable being studied was the salinity or composition of the acclimation and exposure media.

To determine the chloride concentration and level of total ninhydrin-positive substance (NPS) in the blood of *Rangia*, a specific experiment was conducted. Clams were transferred to salinities of 0, 2, 4, 6, 8, 10, and 15 o/oo at a rate of 2 o/oo per day. They were maintained at these salinities for a period of 2 wks before experimentation. From six *Rangia* at each salinity, a total of approximately 250µl of blood was withdrawn from the hearts with a Pasteur pipette and rubber bulb. Aliquots of the pooled blood from each salinity were measured for chloride (chloridometer) and total NPS (Clark, 1964). The chloride concentrations of stand-ards and samples of the acclimation media were also determined. Standard solutions of glycine were analyzed simultaneously and used to calculate the concentration of NPS in aliquots of the blood.

RESULTS

Studies utilizing whole *Rangia* had shown that the organisms maintained at low salinities (0 to 5 o/oo) rapidly removed the accumulated glycine from the free pool and incorporated it into protein and other high molecular weight compounds (Anderson and Bedford, 1973). In the present investigation, isolated gills from salinity-acclimated clams were tested for their ability to accomplish this same conversion, and even more dramatic effects were demonstrated (Fig. 1).

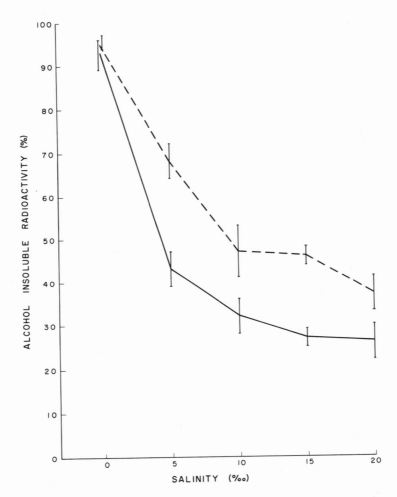

Fig. 1. The effect of salinity on the percent of the total accumulated C^{14}-glycine which was incorporated into alcohol-insoluble material by *Rangia* gills. The solid line represents gills extracted immediately after a 1-hr exposure, while the others were postincubated for 5 hrs (dashed line). Each point is the mean of 6 gills, with the standard deviations (S.D.) shown as vertical bars.

In this experiment a large number of *Rangia* were held at each acclimation salinity for 2 wks and then the uptake and incorporation of whole animals were compared to those of isolated gills from other animals in the same aquaria. Figure 1 illustrates the relationship between salinity and the percentage of the total activity in the alcohol-insoluble fraction of the gill tissue after a 1-hr exposure to C^{14}-glycine, and after a 5-hr postincubation period in the respective salinity without glycine. Incorporation after 1 hr was approximately equal at 10, 15, and 20 o/oo S, while there was a significantly higher percent incorporation at 5 o/oo S. The percentage values increased from approximately 30% at 10, 15, and 20 o/oo S, to 43% at 5 o/oo, and to a remarkable level of 93% incorporation at 2 o/oo. After the 5-hr postincubation period, the second group of gills showed increased incorporation at all salinities tested. At salinities of 10, 15, and 20 o/oo, the increase was on the order of 10 to 15%, while gills acclimated and tested at 5 o/oo S showed an increase of 25% (43 to 68% alcohol-insoluble radioactivity). An interesting aspect of these results was that almost no increase (2%) occurred in the incorporation of glycine by gills acclimated and tested at 2 o/oo S.

Whole clams from the same acclimation aquaria were also examined for their ability to incorporate the accumulated glycine. Due to the individual variability of organisms and the resulting large standard deviations, it is difficult to discuss the significance of differences between the animals tested at each salinity. The trend toward increased incorporation at lower salinities was in agreement with the results of earlier studies on whole *Rangia* (Anderson and Bedford, 1973). Percentage of incorporation after 1 hr varied from a mean low of 21% at 15 and 20 o/oo S, to a mean high of 43% at 2 o/oo S. The increase in mean incorporation after 5 hrs at each salinity was between 10 and 20%, but was not clearly related to salinity.

Utilizing the gill tissue of *Rangia*, studies were conducted to determine the effect of the ionic composition of dilute seawater on the uptake and incorporation of glycine. Since it has been shown that estuarine waters at salinities between 0 and 5 o/oo often contain a relatively higher proportion of calcium and lower proportion of sodium (Khlebovich, 1968), the effect of these two ions was examined. Clams were acclimated in salinities of 0, 2, 4, 6, 8, and 10 o/oo for 1 wk. After the salinity acclimation, calcium chloride solution was added to the 0, 2, 4, 6, and 8 o/oo solutions such that the final calcium content of each salinity was equivalent to that of 10 o/oo seawater (approximately 120 parts per million). In two different experiments, the glycine uptake of gills was tested after 3-hr or 72-hr incubation (whole *Rangia*) at the above salinities, with the supranormal calcium content. The alcohol-soluble and insoluble C^{14}-radioactivity of these gills were compared to those values obtained for gills acclimated and exposed in an identical manner, but in water containing calcium concentrations normally present in the diluted water. Neither 3-hr nor 72-hr

incubation in the high calcium water significantly altered the
general pattern of salinity-dependent accumulation and incorporati
of glycine.

Utilizing artificial seawater with sodium chloride replaced b
choline chloride (Na-free), the effect of sodium was studied.   Cla
were acclimated for approximately 10 days at salinities of 2, 4, 6
8, and 10 o/oo seawater. As usual, the gills of each group were
carefully dissected, but in this case gills from one side were
placed in Na-free water at the same osmotic concentration as the
other side which went into the acclimation salinity of seawater.
Since our recent unpublished data have indicated that the individu
variation between gills was decreased by comparing matched gills,
this technique should provide the best comparative results. The
two groups of gills from each salinity were preincubated for 3 hrs
in their respective solutions, before exposure to the C^{14}-labeled
glycine for 1 hr. The resulting alcohol extracts of the gills
exposed in seawater illustrated the usual increase with increasing
salinity, while those tested in Na-free water exhibited C^{14}-activi
only slightly above background, with one exception. Only the gill
from animals which had been acclimated at 2 o/oo S showed a signif
cant amount of glycine uptake, which was approximately equal to
that of the 2 o/oo S gills tested in the presence of sodium. The
analysis of the alcohol-insoluble radioactivity demonstrated that
there was no incorporation by gills in Na-free water except those
acclimated to 2 o/oo S. Figure 2 illustrates the relationship
between salinity and percent incorporation of glycine by matched
gills tested in dilutions of seawater or Na-free seawater.

Fig. 2. The effect of
   acclimation salinity on the
   percent incorporation of
   accumulated glycine by
   gills preincubated and
   exposed in either diluted
   seawater (solid line), or
   dilutions of Na-free
   seawater (dashed line). At
   each salinity, gills tested
   in the Na-free water were
   compared to gills from the
   opposite side of the same
   animal exposed in dilute
   seawater. Each point is
   the mean of 4 gills
   (vertical bars = S.D.).

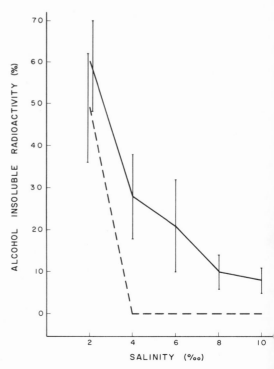

While the pattern in seawater of decreasing percent incorporation with increasing salinity was consistent with other data, the effect of Na-free water was quite interesting. Incorporation was completely inhibited by the lack of sodium at salinities of 4, 6, 8, and 10 o/oo, while the percent incorporation of gills at 2 o/oo S in Na-free water was not significantly different from that in 2 o/oo seawater.

Since the 3-hr preincubation in Na-free water appeared sufficient to affect a change in uptake response by the gills, this time interval was used in subsequent studies. To test the effect of acclimation salinity on the uptake of gills in Na-free water and seawater, a specific experiment was conducted. Clams had been acclimated to 2, 4, 6, 8, and 10 o/oo S for 2 wks. The gills were dissected from each group and one side placed in Na-free water, while the other side was placed in seawater of the equivalent osmotic concentration. In addition, the solutions in which each group was preincubated for 3 hrs were at a concentration 2 o/oo less than their acclimation salinity. This procedure should provide data regarding the effect of both sodium and acclimation. The results of this experiment are presented in Figure 3, where the log of the total C^{14}-radioactivity is plotted against salinity for both gills exposed to seawater and Na-free water. Those gills exposed to C^{14}-glycine in seawater at 2, 4, 6, and 8 o/oo S, after acclimation in seawater 2 o/oo greater in salinity, showed approximately equal amounts of C^{14}-activity.

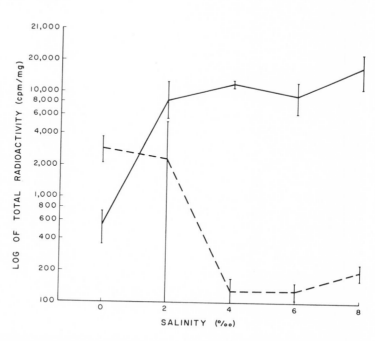

Fig. 3. The effect of salinity on the log of the total C^{14}-activity associated with *Rangia* gills. All gills were preincubated for 3 hrs before exposure in solutions which were reduced 2 o/oo below the acclimation salinity. The solid line represents uptake in dilutions of seawater, while the dashed line illustrates uptake in Na-free water of various osmotic concentrations. Four matched sets of gills were used at each salinity to determine the mean (vertical bars = S.D.).

Uptake and incorporation at 0 o/oo S seawater were suppressed by about one order of magnitude. While the total uptake by gills at 0 o/oo S was quite low, the percent incorporation of accumulated glycine at 0 and 2 o/oo S was approximately equal and 25% greater than incorporation at the higher salinities. An interesting response was shown by the gills exposed in Na-free seawater, which was also 2 o/oo lower in osmoconcentration than the acclimation salinity. Both the alcohol-soluble and insoluble radioactivities were only slightly above background for gills tested at 4, 6, and 8 o/oo S in Na-free water. However, those gills preincubated for 3 hrs and exposed to C^{14}-glycine in 2 and 0 o/oo S Na-free water demonstrated very significant amounts of radioactivity in both the soluble and insoluble fractions. The extremely large variations shown by individual gills tested at 2 o/oo S will be discussed later.

Since the lack of sodium nearly or completely inhibited uptake and incorporation of glycine, at least in solutions with osmotic concentrations equivalent to 4 to 10 o/oo S, it was felt that uptake should be examined at a constant sodium concentration. In this experiment, animals had been acclimated for approximately 2 wks at salinities between 0 and 15 o/oo. The dissected gills were placed in media which contained sodium equivalent to 6 o/oo S seawater, but varied in total osmotic concentration from 6 to 15 o/oo S. Therefore, gills which had been acclimated to 0, 2, and 4 o/oo S were subjected to a salinity increase, as well as an increase in sodium. Gills from *Rangia* acclimated to salinities of 6, 8, 10, and 15 o/oo were subjected to the usual total osmotic concentration (by additions of Na-free water), but only the amount of sodium equivalent to 6 o/oo seawater. The preincubation in these media was again 3 hrs, before a 1-hr exposure to the C^{14}-labeled glycine in fresh solutions of the same characteristics. To obtain maximum data from these experimental animals, three additional parameters were measured.

After exposure to glycine as described above, half of the gills (6) from each test medium were rinsed as usual, but then placed in Gilson respirometer flasks and the respiratory rate was measured for 150 min at 20°C. During respiratory measurements, the flasks were shaken at a rate of 124 cycles per minute and CO_2 was absorbed in KOH. These gills will be referred to as the 4-hr post-incubation group, since this period of time elapsed between the placement of these gills into alcohol and the extraction of other gills immediately after rinsing.

At the time that clams were dissected for the removal of gills, the blood was withdrawn from the ventricle as described on page 3. The pooled blood from each group of six animals (approximately 250μl) was measured for chloride and total NPS.

The resulting total C^{14}-activity of the gills (log scale) initially and after the 4-hr postincubation is plotted against the acclimation salinity in Figure 4. It should be noted that regardless of the salinity to which clams were acclimated, the radioactivity of gills preincubated for 3 hrs and exposed for 1 hr was approximately the same. There was a general overlap of the vertical bars, indicating standard deviations from the mean. Since the sodium concentration of each solution was equal, the values obtained for gills placed in alcohol immediately after rinsing indicate that uptake and incorporation are sodium-dependent. The levels of total radioactivity in gills which were used for respiratory measurements and later (4 hrs) placed in alcohol are more difficult to explain.

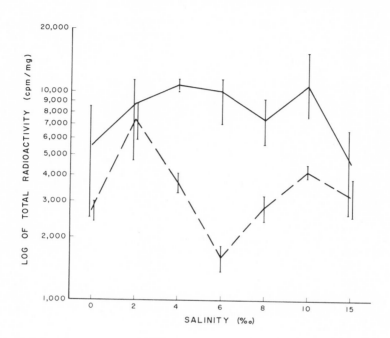

Fig. 4. The effect of salinity acclimation on the log of total uptake by gills preincubated and exposed in water with a sodium content equivalent to 6 o/oo S seawater. At each salinity, 6 gills were extracted immediately after exposure (solid line), while 6 others were postincubated for 4 hrs (dashed line) as their respiratory rate was measured (vertical bars = S.D.).

The significantly low total C^{14}-activity in the gills from clams acclimated to 6 o/oo S was due almost entirely to decreases in the alcohol-soluble fraction over the 4-hr period. The C^{14}-O_2 collected during the respiratory measurements showed that the gills at 6 o/oo S also metabolized the greatest amount of C^{14}-glycine. The C^{14}-O_2 values for 6 o/oo S were 2 to 17 times greater than those of gills at other salinities. While the gills from animals which had been acclimated to 6 o/oo S for 2 wks showed the greatest C^{14}-O_2 production and the largest decrease in total activity, they also exhibited the lowest respiratory rate (Table 1). These data show that the only significantly different gill respiratory rates were exhibited by gills taken from *Rangia* acclimated to either 4 or

6 o/oo S. It would appear that either an increase or a decrease in sodium concentration, over the acclimation concentration, caused a significant increase in respiration. It should be noted that gills were allowed to adjust to the new sodium concentrations for approximately 5.5 hrs before respiratory measurements were taken.

TABLE 1
Respiratory rate of *Rangia* gill tissue in media of equal sodium content, but differing total osmotic concentration. Each value represents the mean respiration of six gills, with standard deviations shown

Acclimation salinity (o/oo) (2 weeks)	0	2	4	6	8	10	15
Test salinity (o/oo) (total osmotic equivalent)	6	6	6	6	8	10	15
Sodium concentration (salinity equivalent, o/oo)	6	6	6	6	6	6	6
Mean respiration (6 gills) (ml O_2/g/hr)	2.79	3.03	2.39	2.30	2.91	3.13	3.16
Standard deviations	0.06	0.26	0.12	0.14	0.27	0.14	0.35

Figure 5 shows the effects of sodium concentration on the percentage incorporation of C^{14}-glycine into the alcohol-insoluble fraction.

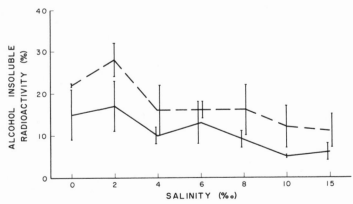

Fig. 5. The effect of a constant sodium concentration on the incorporation of gills previously acclimated to various salinities. The solid line indicates the percent incorporation of gills from Fig. 4, which were extracted after exposure, and the dashed line represents the incorporation exhibited by gills after 4 hrs postincubation (vertical bars = S.D.).

The results indicate that relatively little C^{14}-radioactivity from glycine was converted to alcohol-insoluble compounds, even after a 4-hr postincubation period. At both the initial and 4-hr periods, the mean value for percent incorporation at each salinity was close to or within the standard deviations of all other values at the same time interval. There was a general trend toward higher values of incorporation after 4 hrs, but in several cases, these differences were not significant. The lack of significant differences in incorporation at both time intervals suggests that the transfer of glycine from the alcohol-soluble pool into larger molecules is also dependent upon sodium concentration.

The final aspect of this study was to determine the concentrations of chloride and NPS in the blood of clams acclimated to various salinities. The findings from these analyses are listed in Table 2.

TABLE 2
Relationship between acclimation salinity and the concentrations of chloride and ninhydrin-positive substance (NPS) in the blood of *Rangia*

Approximate salinity (o/oo)	Medium Chloride (mEq/1)	Blood Chloride (mEq/1)[*]	Blood NPS (millimoles/1)[*]
0.5	5	20	1.3
2	50	39	2.1
5	96	67	1.8
8	154	134	2.5
10	195	166	4.5
11	237	203	4.2
17	366	242	5.5

[*]The blood of 6 clams/salinity was pooled, and 4 aliquots were used in both determinations.

At the lowest salinity tested (approximately 0.5 o/oo), the chloride concentration of the blood was slightly higher than that of the medium, while all other values for the blood were slightly lower than the media. Over the range of salinities tested (0.5 to 17 o/oo), the concentration of NPS only varied from 1.3 to 5.5 mM/1. It is possible that the relatively small changes which occurred in the concentration of NPS may, in part, be responsible for the maintenance of a hyperosmotic state by the blood at salinities below 10 o/oo S.

DISCUSSION

In an earlier study on the relationship between salinity and the uptake of glycine by *Rangia*, Anderson and Bedford (1973) showed that the total amount of uptake and incorporation (alcohol-soluble and insoluble fractions) decreased as salinity decreased below 6 o/oo. In one phase of this earlier work, the alcohol-soluble C^{14}-radioactivity in gill tissue was also shown to decrease sharply at lower salinities. The results of the present research (Fig. 1) are consistent with earlier findings, but more clearly represent the fate of accumulated glycine. It is evident that within the 1-hr exposure to C^{14}-labeled glycine, the gill tissue which had been acclimated and exposed in salinities of 2 and 5 o/oo had already incorporated large amounts of the glycine-derived radio-activity into larger and less soluble compounds. The fact that little alcohol-soluble radioactivity was exhibited by gill tissue tested at 2 o/oo was a result of the extremely rapid incorporation of nearly all of the accumulated radioactivity (93%). While a post-incubation period of 5 hrs resulted in increased incorporation of the accumulated glycine at all salinities, only a slight increase was shown by gills at 2 o/oo S, where little soluble activity remained after the 1-hr exposure. The present findings, combined with those of earlier work, have shown that the rate of incorporation of accumulated glycine by whole *Rangia* was related to salinity, but the rate and extent of this conversion were small compared to the same processes in isolated gill tissue.

Though it is difficult to assess the significance of a process from the response of isolated tissue, it is quite possible that the activity of intact gills would be similar. Gills of bivalves, as well as of crustaceans, are generally the most active tissues of the body. In bivalves, the cells of these structures are responsible for water transport, respiratory exchange, food sorting, and ionic and osmotic regulation. This latter function is still open to question since, in general, marine and estuarine bivalves are considered to be osmoconformers (Robertson, 1964; Pierce, 1970; Schoffeniels and Gilles, 1972). While the mechanism of regulation is still not clear, *Rangia* has been shown to regulate its blood above that of the environment at salinities below 10 o/oo (Bedford and Anderson, 1972b). Molluscs have been shown to release amino acids intact from their free amino acid pools to the environment when subjected to a decrease in salinity (Lynch and Wood, 1966; Virkar and Webb, 1970). In most molluscan species examined, taurine was high in concentration at higher salinities and was released to the medium to a greater proportion than other amino acids when the the animals were exposed to a decrease in salinity (Bricteux-Gregoire et al., 1964a, b; Pierce, 1971; Pierce and Greenberg, 1972) It is unusual, but perhaps consistent with other findings, that *Rangia*, which can tolerate and osmoregulate in fresh water, has been shown to form taurine from precursors, but then release it to the medium (Allen and Awapara, 1960). These authors compare the lack of taurine retention by *Rangia* to the absence of taurine in all freshwater animals. Allen (1961) studied the changes which took place in the concentrations of alanine, glycine, glutamic acid, and aspartic acid in the tissue of *Rangia* as salinity varied. The concentrations of these four amino acids increased linearly from a total of 18 μMoles at 3 o/oo S to 325 μMoles per gram dry weight at 17 o/oo S. These data will be discussed later in light of the NPS found in the blood of *Rangia*, but they serve now to indicate that *Rangia*, as other marine bivalves tested, decreased the level of free amino acids in the pool in response to a decrease in salinity.

The point in question is, does *Rangia* merely release amino acids from the pool to the environment, or might the decrease be at least partially due to increased levels of incorporation into larger and osmotically inactive molecules? Studies by several investigators on crustaceans have shown that the free amino acid content varies in relation to salinity (Schoffeniels and Gilles, 1970; Vincent-Marique and Gilles, 1970), and that enzymes active in amino acid metabolism are modified by changes in salinity (Schoffeniels, 1966; Florkin and Schoffeniels, 1969). It is believed that during hypo-osmotic stress, crustaceans decrease the rate of amino acid formation and increase the rate of amino acid catabolism, with the subsequent increase in ammonia production and excretion. Apparently, the aspect of increased incorporation of free amino acids at low salinity has not been investigated in crustaceans, or at least, has not been reported. Since the crustaceans tested have been shown incapable of glycine uptake (Anderson and Stephens, 1969), the labeled amino acids would have to be introduced in a different manner. The only report of increased incorporation of amino acids into protein by molluscs in response to a decrease in salinity was by Bedford (1971). She demonstrated an increase in the C^{14}-alanine-derived radioactivity in the protein of foot tissue from the gastropod, *Melanopsis trifasciata*. This increase in the protein fraction was at the expense of the amino acid radioactivity, as shown by the increased ratio of protein to amino acid radioactivity.

From the results of our research, exemplified by Figure 1, it is evident that in response to decreased salinity, the cells of *Rangia* gill tissue rapidly convert large amounts of exogenously supplied glycine into less soluble and less osmotically active compounds. From studies on whole clams it is quite probable that the majority of these compounds are proteins (Anderson and Bedford, 1973). If the gill cells are capable of converting and, therefore, retaining the exogenous glycine, it would seem quite likely that during periods of dilution, major fractions of endogenous amino acids are incorporated into proteins intracellularly. In doing so, the organism is able to conserve chemical energy, and at the same time, reduce the osmotic concentration of the cell in response to the environment. It is not suggested that all of the free amino acids are retained, since prior studies have demonstrated the release of accumulated glycine by clams in response to a salinity shock (Anderson and Bedford, 1973).

In an attempt to determine which of the inorganic constituents of seawater may be responsible for the regulation of uptake and incorporation, both calcium and sodium were tested. Numerous studies (reviewed by Kinne, 1971) have demonstrated the relationship between calcium and the permeability of cell membranes to inorganics, and others have reported on the proportion of calcium in estuarine waters. Since the proportion of calcium in dilute estuarine water is greater than would be expected if natural or synthetic seawater were diluted with distilled water, earlier findings based on such dilutions were in doubt. However, these experiments have shown that the normal patterns of glycine accumulation and incorporation, relative to salinity, are unaffected by increased levels of calcium concentration. Even when an amount of calcium equivalent to the concentration in 10 o/oo S seawater (about 120 ppm) was added to 2 o/oo S seawater, the uptake and incorporation of glycine were not significantly different from that in normal 2 o/oo S water. The lack of effect was consistent over short and long acclimations (72 hrs) to high calcium concentrations.

Research involving the effect of sodium on glycine accumulation and incorporation has been much more productive. The findings illustrated in Figures 2 through 5 clearly demonstrate the inhibitory effect of low sodium concentrations on the uptake and incorporation of glycine. The data presented in Figures 2 and 3 show that gills from *Rangia* acclimated at salinities between 4 and 10 o/oo and tested in sodium-free water of the same osmolarity exhibit little or no uptake or incorporation of glycine. The glycine-derived radioactivity in the alcohol-soluble and insoluble fractions of gills exposed at 2 and 0 o/oo S is a bit more difficult to explain. To attempt an explanation of these surprising phenomena, some review of sodium-linked membrane transport is required.

In his review of literature regarding the movement of sugars or amino acids across cell membranes, Stein (1967) examined and presented evidence for cotransport systems which demonstrated sodium-dependent sugar and amino acid movement. The studies reviewed utilized the intestinal wall of toads, frogs, rats, hamsters, and rabbits, as well as Erlich ascites tumor cells. By using lithium or choline chloride to replace sodium, and such inhibitors as phlorizin, ouabain, and strophanthidin, considerable data have been compiled which indicate sodium-dependent metabolite transport. Extensive studies by Vidaver (1964a, b, c) on "ghosts" of pigeon red cells have clearly shown the dependence on sodium for glycine transport across the cell membrane. While many models for membrane transport of sugars and amino acids have been proposed, one of the most attractive was suggested by Curran (1968). Curran's work with rabbit ileum showed that not only did increased sodium in the bathing medium stimulate the uptake of alanine, but also sodium influx was stimulated by increased alanine concentrations. The observed rates of alanine uptake relative to sodium internal and external concentrations correlated closely to his proposed model and resulting formula. It was demonstrated that as long as external sodium concentration remained high, alanine influx occurred, regardless of decreased intracellular sodium. An important aspect of his proposed transport model was that intracellular sodium had to be low relative to the bathing solution, and that this differential was maintained by an energy-requiring sodium pump at the serosal side of the cell. Using this model as a guide, it is possible to explain the lack of glycine uptake in sodium-free water, which was demonstrated in most of our work with *Rangia* gill tissue. The presence of uptake exhibited by gills acclimated or exposed in salinities of 0 or 2 o/oo must be explained in terms of sodium fluxes. When the cells of gills which were acclimated to salinities greater than 2 o/oo were placed in sodium-free water, a rapid efflux of sodium was likely. However, it is quite possible that cells could not withstand a loss of sodium below a concentration approximately equal to the content in 2 o/oo S seawater. If it is assumed that no major efflux is occurring into the sodium-free water when cells had already acclimated to 2 o/oo S, then it is possible to interpret the observed results. The efflux of sodium would not allow the membrane transport system to accumulate glycine, but for 0 and 2 o/oo S acclimated gills, which restricted this efflux, the intracellular sodium might participate in glycine transport.

The results shown in Figure 3 may be used to discuss this hypothesis further. It will be recalled that 3 hrs before the determination of uptake, the gills were transferred to solutions which were at an osmotic concentration 2 o/oo S less than the acclimation media. In those transferred to sodium-free water at an osmotic concentration of 2 o/oo (from 4 o/oo), a tremendous variation in glycine uptake was exhibited by the individual gills. It may be that during the 3-hr preincubation and the 1-hr exposure to C^{14}-glycine, the cells were in the process of reducing their osmotic content, including sodium efflux. If some cells had reached a new equilibrium relative to 2 o/oo seawater and others were still in the process, then the variations in glycine uptake would represent this lack of uniformity. If, as suggested above, we assume that even in 0 o/oo water the cells maintain an internal concentration of sodium equivalent to 2 o/oo seawater, then the significant uptake and small variation demonstrated by gills in 0 o/oo sodium-free water would be explained.

The sodium dependence of uptake by the clam gills is illustrated in Figures 4 and 5. Even after a relatively short preincubation to the same sodium concentration (equivalent to 6 o/oo S seawater), gills from *Rangia* acclimated to a wide range of salinities (0 to 15 o/oo S) exhibited total glycine-derived radioactivity of approximately the same magnitude. The total radioactivity was quite similar for gills previously acclimated to 2 through 10 o/oo S seawater, while the uptake by gills subjected to the largest change in sodium (0 and 15 o/oo S) was somewhat suppressed. The dramatic decrease in total C^{14}-activity after the postincubation period, exhibited by gills previously acclimated to 6 o/oo S, indicates that the enzymes active in glycine metabolism were also affected by sodium and total osmotic alterations. It is apparent that the decrease was a result of metabolic degradation since the percentage incorporation did not significantly increase and the amount of collected C^{14}-O_2 was significantly greater at this salinity. Further evidence of the dependence on sodium for the conversion of glycine into the alcohol-insoluble fraction is the uniform level of incorporation exhibited by gills acclimated to different salinities, but preincubated and exposed at the same level of sodium (Fig. 5).

Finally, the possibility that chloride or NPS may be responsibl for the hyperosmotic state of the blood at salinities below 10 o/oo was investigated. Levels of chloride in the blood were slightly less than those of the ambient media at all test salinities except the lowest. In tap water (5 mEq chloride/1), the blood of *Rangia* was about 15 mEq/1 more concentrated in chloride. This differential is probably not of any significance and chloride is obviously not being regulated in dilute solutions of seawater. The determinations of NPS were somewhat more interesting, in that the concentration of this material only varied from 5.5 to 1.3 mM/1 as salinity decreased from approximately 17 to 0.5 o/oo S. It is not possible from these data, or other data available, to determine the significance of these levels of NPS in the blood. If the data from Allen (1961) on four amino acids are converted by use of additional data on percent water, it is possible to obtain estimates of the molar concentration of these combined amino acids in the tissue water of *Rangia*. Such calculations lead to the assumption that the tissue fluid of *Rangia* varies from a concentration of 108 mM/1 at 17 o/oo to 4 mM/1 at 3 o/oo S. These are only approximations, but it would appear that the content of amino acids in the blood at low salinity is not significantly greater than that of the tissue fluid. Subsequent studies are planned which will more closely examine the regulation of amino acids in the tissues and blood of *Rangia*. Additional information will be obtained regarding the concentrations of other inorganic ions in the blood at various salinities.

Two aspects of amino acid uptake investigations have not been discussed thus far. Since this paper has dealt primarily with isolated gill tissue and the relative uptake of glycine in response to salinity and sodium, the aspect of nutrition has not been considered. While it is obvious that the gill is a very active site of glycine accumulation, any discussion of uptake as a supple-mental source of nutrition should be related to the whole animal. For an excellent review of this subject, the reader is referred to Stephens (1972). Stephens also discussed the leakage of accumulated amino acids from the tissue to the environment. It may be suggested that differences in the amounts of glycine accumulated by euryhaline invertebrates at various salinities are due to the effects of the media on the leakage rate. In a specific experiment designed to answer this question, *Rangia* gills were first exposed to C^{14}-glycine at 2, 5, 10, and 15 o/oo S for 1 hr. Half of the gills from each

salinity were immediately placed in ethanol, while the remaining
gills from each salinity were placed in a flowing system, containing
water of the suitable salinity plus 3×10^{-6} Molar C^{12}-glycine.
During 3 hrs of constant flow at a rate of 7 ml/min, duplicate
samples from each effluent were collected. By comparing acidified
to nontreated effluent samples, it was concluded that all radioactivity obtained was in the form of C^{14}-O_2. Changes in the alcohol-
insoluble C^{14}-activity of gills before and after the 3 hrs gave
further evidence of the increased rate of incorporation at low
salinity.

SUMMARY

Earlier studies on whole *Rangia* had shown that while the
incorporation of glycine, which had been accumulated from the
ambient medium, was stimulated by decreased salinity, the total
amount of glycine-derived radioactivity was reduced. It seemed,
from these results and from the low amounts of alcohol-soluble
glycine associated with gill tissue exposed at low salinity (below
8 o/oo S), that uptake was suppressed by low salinities. For whole
animals this is apparently the case, but in this study the total
uptake of glycine by isolated gill tissue was not reduced by
salinities as low as 2 o/oo. Instead, it has been shown that within
a 1-hr exposure to C^{14}-glycine, over 90% of the accumulated glycine
has been converted to large insoluble and osmotically inactive
compounds. The rate and extent of transfer into the alcohol-
insoluble fraction were a function of salinity. It would appear
that the cells of bivalve gills are not only the major site of
uptake, but also possess the capacity to regulate rapidly the
metabolic pathways receiving the accumulated glycine. The fact
that whole *Rangia* exhibit reduced total uptake at low salinity may
be due either to the inability of such regulation by other tissues,
to a much reduced rate of regulation, or to the lack of sufficient
transport of glycine from the gills to other parts of the body.

It has been demonstrated that the concentration of sodium and
not calcium is critical in the regulation of glycine uptake.
Accumulation and incorporation by the gills were generally inhibited
in sodium-free water and these same functions were quite uniform
when osmotic concentration was varied, but sodium content held
constant. The variations from this general response, as well as
the basis for such findings, are probably linked to a sodium-
dependent, cotransport system. In low sodium solutions the efflux
of sodium from the cells may inhibit uptake. When intracellular
sodium is already minimal (0 to 2 o/oo), efflux may be slowed and
limited sodium-linked transport of glycine made possible. Experiments utilizing labeled sodium and labeled glycine are planned.

The concentrations of chloride and NPS in the blood of *Rangia*
decreased as the salinity of the medium decreased. While chloride

was apparently not being regulated, the relatively small change in NPS (5 to 1 mM/l) over the salinity range of 0 to 17 o/oo may indicate some contribution of this substance to the hyperosmotic state of the blood at low salinities.

ACKNOWLEDGMENTS

The author wishes to gratefully acknowledge the competent and diligent aid of Mrs. Christine McCarthy and Mr. Thomas Dillon in the performance of these experiments. Financial support for a portion of this study was provided by a contract from the American Petroleum Institute, in an attempt to establish baseline data for use in evaluating the effect of oil hydrocarbons on marine organisms.

LITERATURE CITED

Allen, K. 1961. The effect of salinity on the amino acid concentration in *Rangia cuneata* (Pelecypoda). Biol. Bull. 121:419-24.

_____ and J. Awapara. 1960. Metabolism of sulfur amino acids in *Mytilus edulis* and *Rangia cuneata*. Biol. Bull. 118:173-82.

Anderson, J. W. and W. B. Bedford. 1973. The physiological response of the estuarine clam, *Rangia cuneata* (Gray), to salinity. II. Uptake of glycine. Biol. Bull. 114:229-247.

_____ and G. C. Stephens. 1969. Uptake of organic material by aquatic invertebrates. VI. Role of epiflora in apparent uptake of glycine by marine crustaceans. Mar. Biol. 4:243-49.

Bedford, J. J. 1971. Osmoregulation in *Melanopsis trifasciata*. IV. The possible control of intracellular isosmotic regulation. Comp. Biochem. Physiol. 40A:1015-28.

Bedford, W. B. and J. W. Anderson. 1972a. The physiological response of the estuarine clam, *Rangia cuneata* (Gray) to salinity. I. Osmoregulation. Physiol. Zool. 45:255-60.

_____ and _____. 1972b. Adaptive mechanisms of the estuarine bivalve, *Rangia cuneata* to a salinity stressed environment. Amer. Zool. 12:528.

Bricteux-Grégoire, S., G. Duchâteau-Bosson, C. Jeuniax, and M. Florkin. 1964a. Constituants osmotiquement actifs des muscles adducteurs d'*Ostrea edulis* adaptée à l'eau de mer ou à l'eau saumatre. Arch. Int. Physiol. Biochim. 72:267-75.

_____. 1964b. Constituants osmotiquement actifs des muscles adducteurs de *Gryphaea angulata* adaptée à l'eau saumatre.

Cavanaugh, G. M., ed. 1964. Marine Biological Laboratory, Formulae and Methods. Woods Hole, Mass.: Marine Biological Laboratory.

Chanley, P. 1965. Larval development of the brackish water mactrid clam, *Rangia cuneata*. Chesapeake Sci. 6:209-13.

Clark, M. C. 1964. Biochemical studies on the coelomic fluid of
    Nephtys hombergi (Polychaeta: Nephtyidae), with observations
    on changes during different physiological states. Biol. Bull.
    127:63-84.
Curran, P. F. 1968. Coupling between transport processes in the
    intestine. The Physiologist 11:3-23.
Florkin, M. and E. Schoffeniels. 1969. Molecular Approaches to
    Ecology. New York: Academic Press.
Hopkins, S. H. 1970. Studies on the brackish water clams of the
    genus *Rangia* in Texas. Nat. Shellfish. Ass., Proc. 60:5-6.
_____ and J. D. Andrews. 1970. *Rangia cuneata* on the East Coast:
    thousand mile range extension or resurgence? Science 167:686.
Khlebovich, V. V. 1968. Some peculiar features of the hydrochemical
    regime and the fauna of mesohaline waters. Mar. Biol. 2:47-49.
Kinne, O. 1971. Marine Ecology, Vol. 1, Part 2. New York and
    London: Wiley-Interscience.
Lynch, M. P. and L. Wood. 1966. Effect of environmental salinity
    on free amino acids of *Crassostrea virginica* Gmelin. Comp.
    Biochem. Physiol. 19:783-90.
Pierce, S. K., Jr. and M. J. Greenberg. 1970. The water balance
    of *Modiolus* (Mollusca: Bivalvia: mytilidae): Osmotic concen-
    trations in changing salinities. Comp. Biochem. Physiol. 36:
    521-33.
_____. 1971. A source of solute for volume regulation in marine
    mussels. Comp. Biochem. Physiol. 38A:619-35.
_____. 1972. The nature of cellular volume regulation in marine
    bivalves. J. Exp. Biol. 57:681-92.
Robertson, J. D. 1964. Osmotic and ionic regulation. In:
    Physiology of Mollusca, Vol. 1 (K. M. Wilbur and C. M. Yonge,
    eds.), pp. 283-311. New York: Academic Press.
Schoffeniels, E. 1966. The activity of the L-glutamic acid
    dehydrogenase. Archs. Int. Physiol. Biochim. 74:665-76.
_____ and R. Gilles. 1970. Osmoregulation in aquatic arthropods.
    In: Chemical Zoology, Vol. 5 (M. Florkin and B. Scheer, eds.),
    pp. 255-86. New York and London: Academic Press.
_____. 1972. Ionoregulation and osmoregulation in Mollusca. In:
    Chemical Zoology, Vol. 7 (M. Florkin and B. Scheer, eds.),
    pp. 393-420. New York and London: Academic Press.
Stein, W. D. 1967. The Movement of Molecules Across Cell Membranes.
    New York: Academic Press.
Stephens, G. C. 1972. Amino acid accumulation and assimilation in
    marine organisms. In: Symposium on Nitrogen Metabolism and
    Environment (J. W. Campbell and L. Goldstein, eds.), pp. 155-84.
    New York and London: Academic Press.
Vidaver, G. A. 1964a. Transport of glycine by pigeon red cells.
    Biochem. 3:662-67.
_____. 1964b. Glycine transport by hemolyzed and restored pigeon
    red cells. Biochem. 3:795-99.
_____. 1964c. Some tests of the hypothesis that the sodium-ion
    gradient furnishes the energy for glycine-active transport by
    pigeon red cells. Biochem. 3:803-808.

Vincent-Marique, C. and R. Gilles. 1970. Modification of the amino
    acid pool in blood and muscle of *Eriocheir sinensis* during
    osmotic stress. Comp. Biochem. Physiol. 35:479-85.

Virkar, R. A. and K. L. Webb. 1970. Free amino acid composition
    of the softshell clam, *Mya arenaria*, in relation to salinity
    of the medium. Comp. Biochem. Physiol. 32:775-83.

Woodburn, D. D. 1962. Clams and oysters in Charlotte County and
    vicinity. Florida State Board Conserv. FSBC Marine Laboratory,
    1962--No. 12.

Reproduction in bivalve molluscs under environmental stress

B. Bayne

The seasonal nature of gametogenesis in bivalve molluscs is well known. This gametogenic cycle involves the initiation of gametogenesis; the growth of the spermatocytes and oocytes, including the processes of vitellogenesis in the female; the physiological "ripening" of the full-grown gametes; and their release from the adult (spawning). Intimately linked with this cycle may be a cycle of the synthesis and storage of reserves of carbohydrate and lipid, followed by their utilization. These nutrient reserves may be concentrated either in follicle cells in the gonad, as in *Mya arenaria* (Coe and Turner, 1938), *Paphia staminea* (Quayle, 1943), and *Macoma balthica* (Caddy, 1967), or in the connective tissue of the gonad, as in *Crassostrea virginica* (Loosanoff, 1965), *Ostrea edulis* (Loosanoff, 1963), and *Mytilus edulis* (Daniel, 1922). The detailed sequence of the different stages of the gametogenic cycle varies between species. In many species gametogenesis is initiated immediately or very shortly after the end of spawning, and may then continue throughout the winter. This is true for *Mercenaria mercenaria* (Loosanoff, 1937), *Venus striatula* (Ansell, 1961), and *Mulinia lateralis* (Calabrese, 1969). Gametogenesis may cease during the winter, however, as in *Macoma balthica* (Caddy, 1967). In some other species, the initiation of gametogenesis may be delayed for some time after spawning. During this "resting" period the animal may lay down nutrient reserves in the follicle cells or connective tissue of the gonad. Such a period of synthesis of reserves occurs in such species as *Mya arenaria* (Shaw, 1962), *Mytilus edulis*

(Daniel, 1921, 1922), and *Ostrea edulis* (Walne, 1970), but is lacking in *Cardium edule* (Boyden, 1971).

Many of the studies quoted above have been concerned primarily with describing the gametogenic cycle. However, the influence of environmental variables such as temperature and food on gametogenesis are of considerable importance and have been studied recently in *Mercenaria mercenaria* (Ansell, Loosmore, and Lander, 1964), *Aequipecten irradians* (Sastry, 1963, 1966, 1970), *Ostrea edulis* (Gabbott and Walker, 1972; Helm, Holland, and Stephenson, 1973), and *Mytilus edulis* (Bayne and Thompson, 1970; Gabbott and Bayne, 1973). In this paper, some of these studies, and the results of some of our recent experiments with *Mytilus edulis*, are reviewed.

ENVIRONMENTAL EFFECTS ON GAMETOGENESIS

Studies over the years have implicated temperature as the major environmental variable which "triggers" the initiation of gametogenesis and determines the time-course of growth of the gametes and spawning. However, the nutritional conditions have also been shown to be important. Recent studies by Sastry (1966, 1970) on *Aequipecten irradians* and Gimazane (1972) on *Cerastoderma (Cardium) edule* indicate how temperature and ration interact. In both species, gametogenesis can be initiated in the resting stage by an increase in temperature, but only if sufficient nutritional reserves within the animal or food in the environment is present. Once triggered, gametogenesis proceeds at a rate dependent in part on the ambient temperature. Reduction in temperature may retard gametogenesis, but will not invoke resorption of gametes unless food becomes scarce. It has proved possible, therefore, to hold animals at a particular stage of the gametogenic cycle by reducing the water temperature and providing food (Loosanoff and Davis, 1951; Bayne, 1965; Gruffydd and Beaumont, 1972). Gimazane (1972) has further shown that the delay between administering a temperature stimulus and the initiation of gametogenesis may vary according to what point in the resting stage the stimulus is applied. In addition, whereas starvation early in the resting stage prevents the initiation of gametogenesis, starvation halfway through the resting stage has no effect.

Inherent in these and other discussions of gametogenesis is the concept of a neurosecretory cycle which controls the general gametogenic rhythm (Lubet, 1955) while temperature and ration act as "triggers" and synchronizers. Two rather different phenomena can be recognized. First, the production of gametes represents an expenditure of energy by the animal, for which a supply of nutrient must be present, either in the form of abundant food in the environment, or as nutrient reserves in the animal. In addition, environmental variables, the chief of which is temperature, are relied upon as synchronizers of the basic seasonal rhythm of gametogenesis.

TABLE 1

Production of ova by *Mytilus edulis* as a proportion of total production at 15°C in the laboratory

Experiment	Assimilated ration (calories/day;Ab)[a]	Metabolic rate (calories/day;R)[b]	Scope for Growth (calories/day;Ab-R)	Net energy balance over 30 days (P+G)	Calories produced as spawned ova (G)[c]	$\frac{G}{(P+G)} \cdot 100$
1.	77.1	48.1	29.0	888	135	15.2
2.	84.2	38.3	45.9	1374	199	14.5

[a]Ration was determined as filtration (=feeding) rate multiplied by the calorific value of the number of cells available per animal (Widdows and Bayne, 1971); to derive assimilated rations, the value for ration was corrected by the assimilation efficiency (from Thompson and Bayne, 1972).

[b]The metabolic rate was measured as the rate of oxygen consumption (Bayne, 1973) and converted to calories by the standard oxy-calorific coefficient (4.8 cal/ml oxygen).

[c]After 30 days the animals were induced to spawn, and the weight of spawned ova converted to calories on the assumption that the ova contained 5.5 cal/mg.

These two aspects are illustrated by data on the production of gametes by *Mytilus edulis* in the United Kingdom. Table 1 lists the results of one experiment in which animals were brought into the laboratory at an early stage of gametogenesis (i.e., early in the cytoplasmic growth phase of the oocytes) and held under conditions of adequate ration (i.e., a mean of 2.2% of the body weight per day as cells of the flagellate *Tetraselmis suecica*) at a temperature of 15C (5C above field ambient). At intervals over a period of one month the "scope for growth" of these animals was measured (Widdows and Bayne, 1971; Bayne, Thompson, and Widdows, 1973) by determining the rate of oxygen consumption and the feeding rate (and correcting the latter for the efficiency of assimilation of the food), and then calculating the calories remaining for growth and the production of gametes by difference. After one month the animals were induced to spawn in individual dishes, and the ova collected separately, washed, and dried. The weight of eggs was converted to a figure in calories by assuming a calorific content of 5.5 cals mg^{-1}, and the production of ova (Pr) then was calculated as a percentage of the total production (Pg + Pr), which was the sum over days of the scope for growth. The Pr was approximately 15% of the total production. Ansell, Loosmore, and Lander (1964) estimated gamete production by *Mercenaria mercenaria*, as based on field data, to be 26.2% of the total production. Hughes (1970) used a similar procedure to estimate the Pr of a mud-flat population of *Scrobicularia plana*; values ranged from 24 to 52% of the total production, with a population mean of 48%. Trevallion (1971) has published values that range from 9 to 62% for a sand beach population of *Tellina tenuis*. Values of Pr derived by induced spawning of *Mytilus* in the laboratory may be slight underestimates due to the incomplete release of gametes. Nevertheless, the figures taken together show that gametogenesis represents a very significant proportion of total energy expenditure by bivalve molluscs.

Data are presented in Figure 1A and B on the gametogenic cycle in *Mytilus* over 3 yrs, based on the recognition by histological examination of the following five stages. Stage 0 is the "resting spent" stage (Chipperfield, 1953) which follows spawning, when the gonad is depleted of gametes and the connective tissue in the mantle is devoid of stored food reserves. Stage I is the stage of synthesi of glycogen reserves in the mantle, preceding the initiation of gametogenesis (Williams, 1969; Gabbott and Bayne, 1973). During Stage II, gametogenesis is renewed and spermatogonia and oogonia are distinguishable in the follicles in the mantle (Chipperfield, 1953; Lubet, 1955); this is the stage at which vitellogenesis occurs in the female. At Stage III the gametes are fully grown and physiologically ripe, in the sense that the germinal vesicle of the ovum breaks down if the ova are released into seawater, and the sperm become active. Stage IV is the stage of spawning, when the gametes are released from the body. The percentage abundances of these stages in a population of *Mytilus* have been plotted against "day-degrees" in Figure 1A, where "day-degrees" is the product of time in days and the mean sea temperature, summed for all the periods of observation, zero being taken as the first date on which all individuals in a sample of 20 were at Stage 0. In Figure 1B the measured sea temperatures for the 3 yrs are plotted.

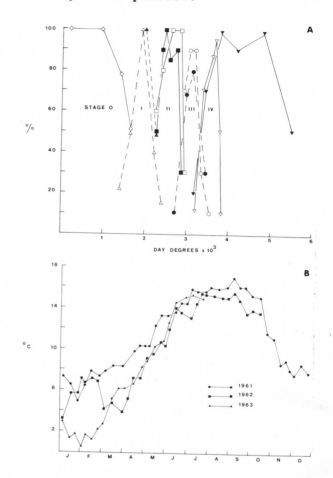

Fig. 1A. The seasonal gametogenic cycle in *Mytilus edulis*, represented as the percentage occurrence of five stages and plotted against "day-degrees" (see text). The data for 1961/1962 are represented by empty symbols. Notice the similarity between the two seasonal cycles, with the exception of the duration of Stage IV. B. Mean water temperatures taken daily at a site close to the population sampled for Fig. 1A. Notice the differences in the first 4 mo over the 3 yrs.

Two points emerge. First, in spite of considerable differences in the thermal regime of late winter of 1961/1962 and 1962/1963, the timing and duration of Stages I, II, and III, and the onset of Stage IV were very similar in terms of day-degrees though not in terms of calendar dates. This finding is suggestive of accurate synchronization by environmental temperature with time. Second, in 1963, spawning (Stage IV) lasted considerably longer, in terms of day-degrees, than in 1962, in spite of rather similar absolute temperature regimes. Apparently, temperature did not provide the control of the duration of the spawning period. Perhaps nutritional conditions earlier in the cycle were more significant determinants of the length of Stage IV. Loosanoff (1965), when assessing thirteen years of observations on *Crassostrea virginica*, suggested a correlation between the thickness of the gonad (and therefore, by implication the fecundity) and the abundance of food in the environment during the preceding autumn.

GAMETOGENESIS IN ANIMALS UNDER STRESS

Variations in ration, and deviations of temperature from norma
seasonal values have recently been recognized as imposing a stress
on *Mytilus* when reproduced in the laboratory (Bayne and Thompson,
1970; Bayne, 1973; Gabbott and Bayne, 1973). In this context,
stress is taken to mean a disturbance of a physiological steady-
state condition. In response to an environmental change the animal
may alter the steady-state of particular physiological rate functions, so that the degree of stress experienced by the animal may
be quantified as the difference between two steady-state values,
before and following the environmental change. In the context of
the gametogenic cycle, there is clearly a spectrum of response
ranging from a failure to initiate gametogenesis to the resorption
of full-grown gametes. But what of the less extreme responses?
How much is known of the subtle effects of slight environmental
changes on gametogenesis?

In a series of experiments with *Mytilus edulis* over 3 yrs,
animals were exposed in the laboratory to a temperature from 2 to
10°C higher than the seasonal field ambient temperature, at a ration
that was less than the maintenance requirement (Bayne and Thompson,
1970; Gabbott and Bayne, 1973). Under these conditions, there was
often a decline in body weight caused by utilization of energy
reserves in the tissues. This was accompanied by a decline in the
rate of oxygen consumption from routine to standard values, and a
disturbance of the normal catabolic balance between protein,
carbohydrate, and lipid (Bayne, 1973).

Figure 2 shows the seasonal cycle of gametogenesis in *Mytilus*
in terms of an index of the developmental stage of the gametes, and
experimental values for such a gonad index in animals exposed to
15°C and a low ration in the laboratory. The normal cycle is one of
gradual growth and maturation of gametes from October to May,
followed by spawning in the summer, and a period of quiescence from
August through September. In the autumn and the winter (experiments
1 and 2), increased temperature in the laboratory induced more
rapid development of the gametes than in the field. In the spring

Fig. 2. Changes in an index
   of the stage of gamete
   development in *Mytilus
   edulis* with season (Δ)
   and during temperature
   and nutritive stress in
   the laboratory (●). The
   horizontal bars represent
   the duration of four
   experiments. Redrawn
   from Gabbott and Bayne
   (1973).

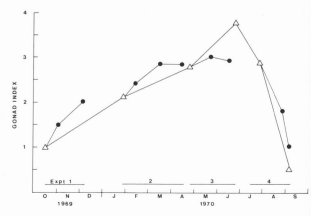

(experiment 3), the animals maintained a high gonad index in the laboratory. In the summer (experiment 4), whereas animals in the field spawned, with the consequent decline in the gonad index, experimental animals did not spawn, but they resorbed the full-grown gametes, also resulting in a decline in the gonad index. These results confirmed earlier suggestions that *Mytilus* can continue gametogenesis during temperature and nutritive stress, as long as the gametes are not physiologically "ripe." After the complete maturation of the gametes (experiment 4), however, stress results in the resorption of the sperm and ova, and the recession of the gonad.

This apparent ability to "buffer" events in the gonad from the implications of the general stress response is best illustrated by the results of an experiment carried out from January to March. During the summer months *Mytilus* accumulates a considerable store of glycogen in the digestive gland and the mantle tissues (Fig. 3).

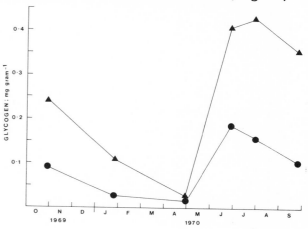

Fig. 3. Seasonal changes in the glycogen content, as mg/g dry tissue weight, of *Mytilus edulis*. ▲, mantle tissue; ●, remaining "nonmantle" tissues.

During the winter this store is gradually utilized, and by January the store of glycogen has been depleted markedly. When placed under a temperature and nutritive stress at this time, the animal is forced into a condition of negative scope for growth, due in part to a high metabolic rate and the lack of sufficient food. In order to balance its own energy budget, energy reserves from the body are rapidly utilized, especially the protein (Table 2). Glycogen is also utilized, but to a lesser extent, and it appears that the glycogen store in the female mantle is conserved to some extent for use in the synthesis of yolk. This in turn is an energy-demanding process and leads to a high rate of oxygen consumption. The general requirement for "energy of maintenance" is met at the same time by catabolism of protein, lipid, and some glycogen. An end-result of these processes is the continuation of gametogenesis in spite of a very marked stress condition.

The evidence suggests, therefore, that when a temperature increase is not so great as to bring about complete cessation of gametogenesis and subsequent resorption of gametes, and when a certain minimum amount of food and nutrient reserves are present, the animal can continue gametogenesis and vitellogenesis (Ansell, Loosmore, and Lander, 1964; Loosanoff, 1965; Sastry, 1966). Two problems are posed, however. First, to what extent are the gametes which are produced under these stress conditions normal? Also, what is known about the mechanism underlying the apparent capacity to "buffer" the gonad from the somatic response to stress?

TABLE 2
Losses of protein, lipid and carbohydrate, as calories, in *Mytilus edulis* during stress in the laboratory from January to March

Tissue	Energy losses in calories				Percentage of total energy loss
	Protein	Lipid	Carbohydrate	Total	
Mantle	571	147	79	797	33.8
"Nonmantle"[a]	1,202	206	156	1,564	66.2
Total	1,773	353	235	2,361	
Percentage of total energy loss	75.1	15.0	9.9		

[a]"Nonmantle" tissues include all the body tissues except mantle (from Gabbott and Bayne, 1973).

DEVELOPMENT OF LARVAE FROM ADULTS UNDER STRESS

Mytilus were held under conditions of temperature and nutritive stress in the laboratory for 6 to 10 wks, and they were then induced to spawn. The success of fertilization and embryogenesis, and the subsequent rate of growth of the larvae, were then assessed in comparison with control animals not subjected to stress in the laboratory. The results clearly indicated a higher failure rate in larvae from stressed adults, especially during the period of intense morphogenesis during cleavage, gastrulation, and development to the first shelled larva (Bayne, 1972). In a similar study on the oyster *Ostrea edulis*, Helm, Holland, and Stephenson (1973) reared the larvae from parents that were held under two different conditions of ration in the laboratory. They measured the "vigor" of the larvae as their rate of growth, and the yield of metamorphosed spat at the end of the larval period. The authors also attempted to equate larval viability with the gross biochemical composition of the larvae immediately after release from the parent. The results showed that larvae liberated from adults which were held at low ration (unsupplemented natural seawater) had a slower rate of growth than larvae from adults at high ration (seawater with flagellates added); and the rate of growth of the larvae over 96 hrs was itself predictive of spat yield. Further, these researchers found that larval vigor was significantly correlated with the total neutral lipid content (Fig. 4). These studies indicate, therefore, that whereas animals under sublethal temperature and nutritive stress can and do continue to produce gametes, these gametes in turn develop into embryos and larvae that are less viable than embryos produced by adults not under stress.

Fig. 4. The relationship between the proportion of neutral lipid in newly liberated *Ostrea edulis* larvae, and their growth in the 96-hr period following liberation. o, adults with unsupplemented ration; ●, adults with a supplemented ration. From Helm, Holland, and Stephenson, 1973.

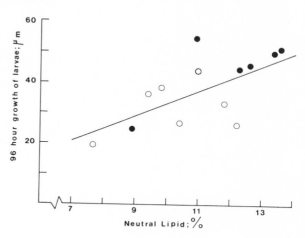

PASSAGE OF NUTRIENTS FROM THE DIGESTIVE GLAND TO THE OVA

In a search for possible mechanisms underlying the relationships between ration, gametogenesis, and viability of the larvae, the "first step" must be to examine the conditions under which nutrients, derived from the food, are utilized in the growth of the gametes. Bivalves in general seem able to build up reserves of glycogen in the digestive gland, and some species can do so also in the gonad. Sastry and Blake (1971) provided some evidence that material in the digestive gland of *Aequipecten* could be mobilized to the gonad during growth of the oocytes, and they suggested that the rate of mobilization was dependent upon both temperature and the stage of gametogenesis.

In *Mytilus* also there is evidence of transfer of nutrients from the digestive gland to the mantle tissues. In the laboratory experiments, a ^{14}C-label was introduced into the animals via the food cells *(Tetraselmis)*. Figure 5A and B illustrates the decline of ^{14}C-labeled acid-soluble substances (e.g., small molecular weight sugars, amino acids, and fatty acids) and lipid in the digestive gland, and the coincident increase of these labeled substrates in the mantle. The time-course of mobilization of material from the digestive gland and incorporation into the female mantle varies with the season. For example (Fig. 6A and B), transfer of material in animals that have been fed a low ration is complete within 48 hrs in September, but continues over a period of at least 35 days in the winter (January). This seasonal difference is probably due to variations in the levels of reserves already present in the mantle. In the autumn, when reserve levels are high (Gabbott and Bayne, 1973), nutrients from the food are largely stored in the digestive gland, and the ration at which the animals are kept makes little difference to the general distribution of the ^{14}C-label. In January, however, as already discussed, nutrient reserves in the mantle are partially depleted, and these are replenished when possible from the food intake. At this time, animals that are fed at a high ration reach a steady-state distribution of the ^{14}C-label within 10 days, but in starved animals, with a continuing necessity to utilize nutrient reserves, the label is more slowly but more consistently lost from the digestive gland over the 35 days of the experiment.

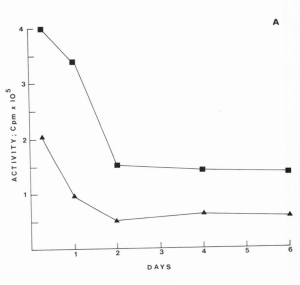

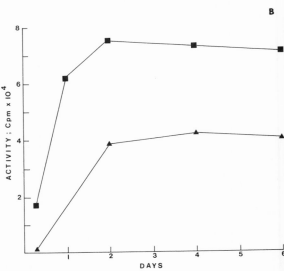

Fig. 5A. The decline in ^{14}C-labeled "acid-soluble" (■) and lipid (▲) fractions from the digestive gland of *Mytilus edulis* at 15°C in the autumn. The label was administered in the food. B. The increase in the ^{14}C-labeled "acid-soluble" (■) and the lipid (▲) fractions in the mantle tissue of *Mytilus edulis* at 15°C in the autumn. The label was administered in the food. Data for 5A and B from Widdows (1972).

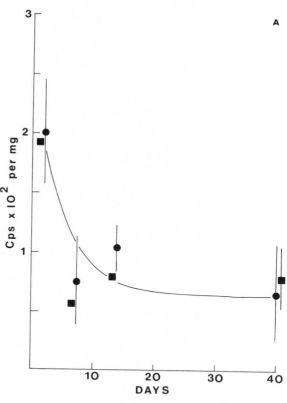

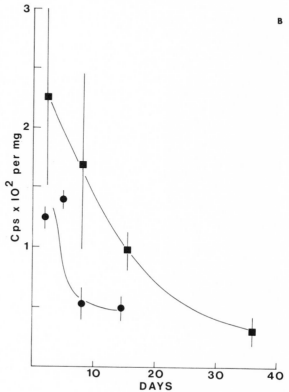

Fig. 6A. The distribution of ^{14}C, which was administered in the food, in the digestive gland of *Mytilus edulis* during September; ●, fed animals; ■, starved animals. Mean values plus the total range of counts. B. The distribution of ^{14}C, which was administered in the food, in the digestive gland of *Mytilus edulis* during January; ●, fed animals; ■, starved animals. Mean values plus the total range of counts. Data for 6A and B from Thompson (1972).

Recently Vassallo (1973) has identified the transfer of lipid from the digestive gland to the gonad in the scallop, *Chlamys hericia*. This avenue of transfer seems established, therefore (Sastry and Blake, 1971; Thompson, 1972). How is this material utilized within the mantle?

Experiments were run in which *Mytilus* were pulse-fed a single meal of ^{14}C-labeled food cells after an initial period of acclimation in the laboratory, and maintained for 4 weeks at (1) field-ambient temperature and high ration (the control group), (2) ambient temperature +9°C with high ration (the "high ration" group), and (3) ambient temperature +9°C with a low ration (the "low ration" group). All the animals were then given a uniform spawning stimulus and the spawned ova were collected and freeze-dried. The lipids were extracted from the dried material, the lipid fractions then separated by thin-layer chromatography, and the thin-layer plates scanned for the presence of radioactivity. The results are summarized in Table 3 and Figure 7.

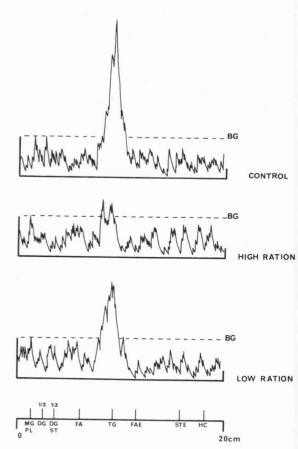

Fig. 7. The scan of a thin-layer chromatography plate to identify the distribution of ^{14}C in the lipid fractions extracted from spawned ova (see text). BG, maximum background count rate; PL, phospholipids; MG, monoglycerides; 1:2 DG, 1:2 diglycerides; 1:3 DG, 1:3 diglycerides; ST, sterols; FA, fatty acids; TG, triglycerides; FAE, fatty acid esters; STE, sterolesters; HC, hydrocarbons. Control animals held at field-ambient temperature and high ration. High ration animals held at ambient temperature +9°C and high ration. Low ration animals held at ambient +9°C and low ration. Notice the concentration of radioactivity in the triglyceride fraction. From Bayne and Gabbott, unpublished data.

TABLE 3
Distribution of radioactivity in ova which were spawned from adult *Mytilus edulis*[a]

Condition	Mean "scope for growth"[b] (cal/day)	Total production as calories over 34 days	Calories per female as spawned ova	Radioactivity; cpm mg^{-1} [c]		
				Total	Lipid	Neutral lipid
Low ration	- 42.8	- 1452	80	1380	312 (22%)	177 (58%)
High ration	+ 29.8	+ 1012	151	340	86 (25%)	54 (63%)
Control	+ 16.1	+ 550	107	1453	407 (28%)	266 (65%)

[a]Held under three conditions of temperature and ration (see text) in the laboratory.
[b]For determination of the "scope for growth," see legend to Table 2.
[c]Radioactivity counts in the total lipids extracted from the ova are presented as percentages of the total counts, and in the neutral lipid fraction as percentages of the counts in lipid (Bayne and Gabbott, unpublished data).

The low-ration animals had a markedly negative scope for growth during the experiment, and they produced an average of 80 calories per animal as spawned ova. The high-ration group had a positive scope for growth, and they spawned an average of 151 calories per animal as ova. The control group had received a lower ration than the high-ration animals, and although their scope for growth was positive during the experiment, the absolute value was lower than in high-ration animals. These animals produced an average of 107 calories per animal as spawned ova. These data indicate, therefore, that fecundity (measured as calories of spawned ova per female) was reduced in animals kept at low ration during the "conditioning" period of 4 wks.

No significance will be attached at this time to the difference in the total amounts of radioactivity in the ova in the three different conditions of the experiment. However, in all three treatments, an average of 25% of the total radioactivity was in the lipid fraction, and of this lipid activity, an average of 62% was found in the neutral lipid, or triglyceride, fraction. This high percentage in the triglycerides is of interest because of the correlation between neutral lipid and larval vigor reported by Helm, Holland, and Stephenson (1973) for *Ostrea* larvae. The data also confirm that considerable nutrient imbalance is not a barrier to the mobilization of ingested food material for the synthesis of lipids in the ova.

CONCLUSION

The ambient temperature and the food level serve to synchronize different stages in the gametogenic cycle of bivalves. There are limits to change in either of these variables beyond which gametogenesis fails or does not occur. Within these limits, however, the animals show a considerable tolerance, even to conditions under which they are stressed to the point of having to utilize their own tissue reserves to meet their basal requirements for energy. Under such conditions of stress, growth and maturation of the gametes proceed in spite of the significant energy drain that this imposes. However, in such a stress regime, fecundity is reduced and the vitality of the gametes and the vigor of the larvae that are produced from them are impaired to some degree; there is evidence that this may be at least partially attributable to reduced synthesis of neutral lipid in the developing ova.

Within this scheme of events, the interrelationships between the food, the sites of energy reserve in the animal, and the utilization of materials in gametogenesis are very complex. Figure 8 shows a simple schematic model of some of these events in *Mytilus edulis* during the early winter months, which is the main period of gamete growth.

Fig. 8A. A schematic diagram
of the net distribution
of calories over 30 days,
together with possible
"flow paths," in *Mytilus
edulis* held at 15°C, in
the winter, at high ration.
From the digestive gland
(DG) and mantle (Man),
150 calories are dis-
tributed to the ova;
1,728 calories are lost
as the equivalent of
oxygen consumed (R) and
173 as amino-nitrogen (U).
The net maintenance re-
quirement is 2,385 calo-
ries when calculated on
the basis of an assimila-
tion efficiency of 86%.
R.S.T = somatic tissues
other than the digestive
gland. B. A schematic
diagram of the net dis-
tribution of calories
over 30 days, together
with possible "flow
paths," in *Mytilus edulis*
held at 15°C, in the
winter, at zero ration.
From the digestive gland
(DG) and mantle (Man),
80 calories are distrib-
uted to the ova; 726 calories are lost as the
equivalent of oxygen
consumed (R) and 290 as
amino-nitrogen (U).

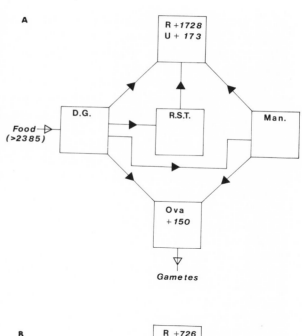

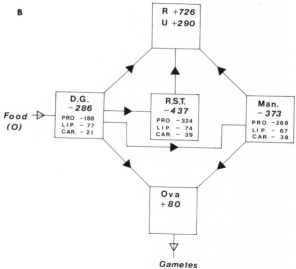

These requirements for energy are met by utilizing 286 calories
from the digestive gland (distributed to other tissues, and the
ova), 373 from the mantle, and 437 from somatic tissues other
than the digestive gland (R.S.T). Approximate losses due to
protein (PRO), lipid (LIP), and carbohydrate (CAR) are indicated
for each tissue "reservoir."

The figure describes the events in two female mussels, of 1 g initial
dry tissue weight, during 30 days at 15°C. One animal has been held
at a ration greater than the maintenance requirement (Fig. 8A) and
the other animal has been held at zero ration (Fig. 8B). The numbers
represent total calories which, over the 30-day period, are either

lost in metabolism, accumulated as ova, or lost as depleted body reserves during starvation. In the animal which is fed at high ration, the metabolic rate, estimated as the rate of oxygen consumption, is high (0.5 ml O_2 hr^{-1}) and the loss of amino-N, as calories, amounts to 10% of the calorific loss determined as oxygen uptake (Bayne, 1973). The fecundity of the animal is high, at 150 calories as spawned ova. The estimated maintenance requirement in terms of assimilated calories is 2,051 calories in 30 days. If an assimilation efficiency of 0.86 is assumed (Thompson and Bayne, 1972), food intake must be approximately 80 calories per day (or 14.3 mg) for the animal to gain weight. In other words, the maintenance requirement is approximately 1.5% of the dry body weight per day.

In the animal at zero ration, metabolic rate is reduced (from 0.5 to 0.25 ml O_2 hr^{-1}), but the rate of loss of amino-N is increased to 40% of the calorific equivalent of the rate of oxygen consumption. Fecundity is reduced to 80 calories per animal as spawned ova. Nutrient reserves in the body must be utilized to meet these demands for energy. Of the three sites of reserves that have been considered, the mantle loses material to the extent of 34% of the total energy loss; the digestive gland, 26%; and the remaining somatic tissue, 40% (Thompson, 1972; Gabbott and Bayne, 1973). Values are also shown for the separate contribution of protein, lipid, and carbohydrate, from each site of reserves to the total energy utilization. On the average, protein accounts for 70%, lipid for 22%, and carbohydrate for 8% of the total loss of reserves, although there are significant differences between tissues. However, in spite of the considerable loss of reserves, gametogenesis continues although with a reduced fecundity.

ACKNOWLEDGMENTS

I am extremely grateful to Ray Thompson, John Widdows, and Pete Gabbott for allowing me to quote from their unpublished data. My thanks also to Dave Holland and Mike Helm for permission to use their data on *Ostrea* larvae. Some of the research discussed was funded by N.E.R.C.

LITERATURE CITED

Ansell, A. D. 1961. Reproduction, growth and mortality of *Venus striatula* (da Costa) in Kames Bay, Millport. J. Mar. Biol. Ass. U. K. 41:191–215.

_____, F. A. Loosmore, and K. F. Lander. 1964. Studies on the hard-shell clam, *Venus mercenaria*, in British waters. II. Seasonal cycle in condition and biochemical composition. J. Appl. Ecol. 1:83–95.

Bayne, B. L. 1965. Behaviour and ecology of young of *Mytilus edulis* L. Ph.D. dissertation, University of Wales.

_____. 1972. Some effects of stress in the adult on the larval development of *Mytilus edulis*. Nature 237:459.

_____. 1973. Physiological changes in *Mytilus edulis* L. induced by temperature and nutritive stress. J. Mar. Biol. Ass. U. K. 53:39-58.

_____ and R. J. Thompson. 1970. Some physiological consequences of keeping *Mytilus edulis* in the laboratory. Helgolander. Wiss. Meeres 20:526-52.

_____, _____, and J. Widdows. 1973. Some effects of temperature and food on the rate of oxygen consumption by *Mytilus edulis* L. In: Effects of Temperature on Heterothermic Organisms (W. Weiser, ed.). Berlin: Springer-Verlag. In press.

Boyden, C. R. 1971. A comparative study of the reproductive cycles of the cockles *Cerastoderma edule* and *C. glaucum*. J. Mar. Biol. Ass. U. K. 51:605-22.

Caddy, J. F. 1967. Maturation of gametes and spawning in *Macoma balthica* (L). Can. J. Zool. 45:955-65.

Calabrese, A. 1969. Reproductive cycle of the coot clam, *Mulinia lateralis* (Say), in Long Island Sound. Veliger 12:265-69.

Chipperfield, P. N. J. 1953. Observations on the breeding and settlement of *Mytilus edulis* (L.) in British waters. J. Mar. Biol. Ass. U. K. 32:449-76.

Coe, W. R. and H. J. Turner. 1938. Development of the gonads and gametes in the soft-shell clam *(Mya arenaria)*. J. Morphol. 62:91-111.

Daniel, R. J. 1921. Seasonal changes in the chemical composition of the mussel *(Mytilus edulis)*. Rep. Lancs. Sea-Fish. Labs. 30:205-21.

_____. 1922. Seasonal changes in the chemical composition of the mussel *(Mytilus edulis)*. Rep. Lancs. Sea-Fish. Labs. 31:27-50.

Gabbott, P. A. and B. L. Bayne. 1973. Biochemical effects of temperature and nutritive stress on *Mytilus edulis* L. J. Mar. Biol. Ass. U. K. 53:269-86.

_____ and A. J. M. Walker. 1972. Changes in the condition index and biochemical content of adult oysters (*Ostrea edulis* L.) maintained under hatchery conditions. J. Cons. Perm. Int. Explor., Mer. 34:99-106.

Gimazane, J. P. 1972. Etude expérimentale de l'action de quelques facteurs externes sur la reprise de l'activité génitale de la Coque, *Cerastoderma edule* L., Mollusque bivalve. C. r. Séanc. Soc. Biol. 166:587-89.

Gruffydd, L. D. and A. R. Beaumont. 1972. A method for rearing *Pecten maximum* larvae in the laboratory. Mar. Biol. 15:350-55.

Helm, M. M., D. L. Holland, and R. R. Stephenson. 1973. The effect of supplementary algal feeding of a hatchery breeding stock of *Ostrea edulis* L. on larval vigour. J. Mar. Biol. Ass. U. K. 53:

_____. In press.

Hughes, R. N. 1970. An energy budget for a tidal flat population of the bivalve *Scrobicularia plana* (Da Costa). J. Anim. Ecol. 39:333-56.

Loosanoff, V. L. 1937. Seasonal gonadial changes of adult clams *Venus mercenaria* (L). Biol. Bull. Mar. Biol. Lab., Woods Hole 72:406-16.

_____. 1963. Gametogenesis and spawning of the European oyster *O. edulis* in waters off Maine. Biol. Bull. Mar. Biol. Lab., Woods Hole 122:86-94.

_____. 1965. Gonad development and discharge of spawn in oysters of Long Island Sound. Biol. Bull. Mar. Biol. Lab., Woods Hole 129:546-61.

_____ and H. C. Davis. 1951. Delaying of spawning of lamellibranch by low temperature. J. Mar. Res. 10:197-202.

Lubet, P. 1955. Cycle neurosécrétoire chez *Chlamys varia* L. et *Mytilus edulis* L. (Mollusques lamellibranches). C. r. hebd. Séanc. Acad. Sic., Paris 241:119-21.

Quayle, D. B. 1943. Sex, gonad development and seasonal gonad changes in *Paphia staminea* Conrad. J. Fish. Res. Bd. Can. 6: 140-51.

Sastry, A. N. 1963. Reproduction of the bay scallop, *Aequipecten irradians* Lamarck. Influence of temperature on maturation and spawning. Biol. Bull. Mar. Biol. Lab., Woods Hole 125:146-53

_____. 1966. Temperature effects in reproduction of the bay scallop, *Aequipecten irradians* Lamarck. Biol. Bull. Mar. Biol. Lab., Woods Hole 130:118-34.

_____. 1970. Environmental regulation of oocyte growth in the bay scallop *Aequipecten irradians* Lamarck. Experientia 26:1371-7

_____ and N. J. Blake. 1971. Regulation of gonad development in the bay scallop *Aequipecten irradians* Lamarck. Biol. Bull. Mar. Biol. Lab., Woods Hole 140:274-83.

Shaw, W. N. 1962. Seasonal gonadal changes in female soft-shell clams, *Mya arenaria*, in the Tred Avon River, Maryland. Proc. Natl. Shellfish Ass. 53:121-32.

Thompson, R. J. 1972. Feeding and metabolism in the mussel *Mytilus edulis* L. Ph.D. dissertation, University of Leicester

_____ and B. L. Bayne. 1972. Active metabolism associated with feeding in the mussel *Mytilus edulis* L. J. Exp. Mar. Biol. Ecol. 9:111-24.

Trevallion, A. 1971. Studies on *Tellina tenuis* Da Costa. III. Aspects of general biology and energy flow. J. Exp. Mar. Biol Ecol. 7:95-122.

Vassallo, M. T. 1973. Lipid storage and transfer in the scallop *Chlamys hericia* Gould. Comp. Biochem. Physiol. 44A:1169-75.

Walne, P. 1970. The seasonal variation of meat and glycogen content of seven populations of oysters *Ostrea edulis* L. and a review of the literature. Fish. Invest., Ser. II 26:1-35.

Widdows, J. 1972. Thermal acclimation by *Mytilus edulis* L. Ph.D dissertation, University of Leicester.

_____ and B. L. Bayne. 1971. Temperature acclimation of *Mytilus edulis* with reference to its energy budget. J. Mar. Biol. Ass. U. K. 51:827–43.

Williams, C. S. 1969. The effect of *Mytilicola intestinalis* on the biochemical composition of mussels. J. Mar. Biol. Ass. U. K. 49:161–73.

Physiology and ecology of reproduction in marine invertebrates

A. N. Sastry

Reproduction in sexually reproducing marine invertebrates is a cyclical physiological process. Populations of marine invertebrates reproduce annually, semiannually, or continuously within the year. The patterns of reproductive periodicity show a relationship to the climatic conditions to which they are exposed. In most species reproduction occurs on a predictable basis which may be either for a short period or for several months in the year.

A number of marine invertebrates occurring in the temperate climatic zones exhibit an annual reproductive cycle (Giese, 1959). In a seasonally changing environment, the timing of the breeding period, when the environmental conditions are optimal for development and growth of the offsprings, is adaptive. A precise timing of the breeding period requires that a population initiate and regulate the course of gonad development to maturation in coordination with the changes in the external environment. The environment-organism interactions in the determination of the course of reproductive cycle and the variations among populations occurring in different climatic zones have not been examined for many species. In this paper an attempt will be made to review briefly some aspects of environment-organism interactions in the determination of the course of an annual reproductive cycle with emphasis on the experimental results obtained with the bay scallop, *Aequipecten irradians* Lamarck.

REPRODUCTIVE CYCLE

To determine the periods of gonad development and breeding in marine invertebrates during the course of the reproductive cycle a number of methods have been used (Giese, 1959). One useful qualitative and quantitative method for defining the course of the reproductive cycle is the determination of the gonad index (gonad weight/body weight × 100) combined with the observations on changes in gametogenesis samples collected at intervals during the year. Correlated with the stage of gametogenesis, the mean monthly gonad index values and oocyte diameter provide information on the average stage of gonad development. From such observations, it is possible to delineate the events and to determine the duration of these events during the course of the annual reproductive cycle. The reproductive cycle thus determined includes a series of events, vegetative stage, activation, growth and gametogenesis, maturation, spawning, and resting stage. The course of these events could be synchronous or asynchronous in members of a population.

The population of *A. irradians* at Beaufort, North Carolina, exhibits an annual reproductive cycle (Fig. 1). The gonad index values increase rapidly after activation of the gonads and reach an annual maximum before spawning (Fig. 1A). The annual gonad peak is followed by a decrease in the gonad index values during the breeding period. After the breeding period, the gonad index values remain at a sustained level during the resting stage.

Primary germ cells and oogonia are the characteristic stages of gametogenesis during the vegetative stage (Fig. 1B). After activation of the gonads, the oocytes enter the growth phase and develop to maturity by the time the gonad index reaches the annual maximum. The gonad growth and oocyte growth are apparently the result of one and the same phenomenon. Regression of the gonads occurs with spawning during the breeding period.

The population of *A. irradians* has a synchronized annual reproductive cycle. In each individual of the population, however, oocytes of different sizes occur during the period of gonad development and spawning. The easily recognizable sequence of events in the synchronized annual reproductive cycle of this species makes it well suited for experimental studies to determine the influence of different environmental factors on the annual pattern.

ENVIRONMENTAL FACTORS INFLUENCING THE REPRODUCTIVE CYCLE

Factors Correlating with the Reproductive Cycle

Temperature, salinity, day length, and food abundance influence the reproductive cycle (Giese, 1959). The sequential events in the reproductive cycle of a population of *A. irradians* at Beaufort,

North Carolina, closely correlate with the seasonal changes in
temperature, day length, and food abundance (Figs. 1 and 2). In
early spring when the temperature and the day length are increasing
and when the food abundance is still low, the primary germ cells
and oogonia are developed in this population. The gonad growth and
oocyte growth begin when the temperature is about 20°C, the day
length is maximum, and food is abundant in May through June. Gonad
growth to the annual maximum occurs during the summer when food is
abundant, but the rapid increase in gonad index occurs in early and
late summer. In midsummer, the gonad index remains at a steady
level, suggesting that the amplitude of the gonad index is affected
by the temperature level and food concentrations in the environment.
Maturation, occurring in the middle of September, is followed by
the breeding period when temperature, food abundance, and day length
are decreasing. The breeding period ends by the middle of November
when temperature is about 17 to 18°C. Later, the population remains
in the resting stage during the winter when temperature, food
abundance, and day length are near the annual minimum. The close
relationship between the sequence of reproductive events and

Fig. 1. Annual reproduc-
tive cycle of *A. ir-
radians* from Beaufort,
North Carolina.
A. Changes in monthly
mean gonad index
values, showing the
annual gonad cycle.
B. Changes in monthly
mean oocyte diameter
of samples, showing
the gametogenic cycle.

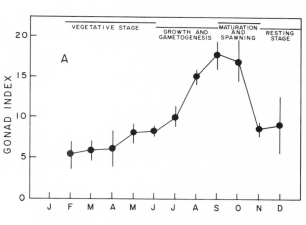

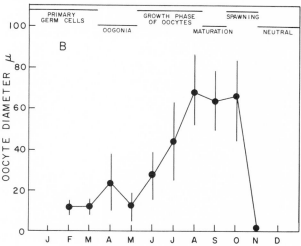

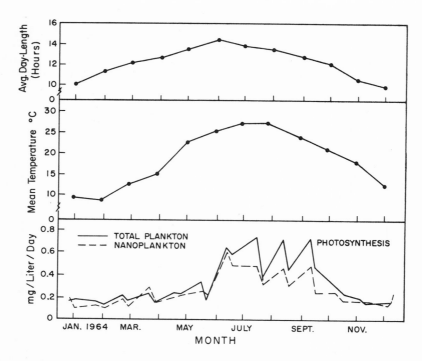

Fig. 2. Environmental factors correlating with the annual reproductive cycle of A. *irradians* from Beaufort, North Carolina. (Phytoplankton data from Williams and Murdoch, 1966.)

seasonal changes in the environment suggests a dynamic interaction between the organism and the environment which might be regulating the pattern of the annual reproductive cycle.

Temperature

Among the various factors influencing the reproductive cycle, the seasonal change in temperature has been considered as an important factor in the determination of an annual pattern (Orton, 1920; Runnström, 1927; Gunter, 1957; Giese, 1959; Kinne, 1970). Temperature was also found to influence reproduction in some species in which gonad development has been stimulated in the resting stage animals (Sastry, 1968). In others undergoing gametogenesis, the development of gametes to maturation has been advanced or delayed by exposing the animals to appropriate temperatures in the presence of food (Loosanoff and Davis, 1950; Crisp, 1957; Sastry, 1963; Bayne, 1965). Although these studies suggest temperature influence on gonad development, the nature of temperature interaction with the sequence of events in the reproductive cycle has remained unclear. Sastry (1966a) showed that the sequential events in the reproductive cycle of A. *irradians* are affected differently by temperature. The vegetative stage scallops exposed to a series of temperatures failed to initiate gonad growth and oocyte growth in the absence of food. After exposure to these temperatures, the primary germ cells and oogonia were resorbed and gonad and digestive gland indices decreased (Table 1). Similar results were obtained when the resting stage scallops were exposed to different temperatures without food. In contrast to these stages, the scallops with minimum gonad reserves and oocytes in the growth phase developed to maturity between 20 and 30°C in the absence of food supply. The oocytes of the scallops held at 10 and 15°C, however, cytolyzed during vitellogenesis and resorbed. Apparently, the scallops in the vegetative and resting stages require both suitable temperatures and adequate food supply to initiate the gonad growth. However, the food supply does not seem critical after the gonads have accumulated certain minimum reserves. After this stage, the development of gonads to maturation occurs at a rate determined by the temperature within limits which appear to be characteristic for a species.

TABLE 1
Effect of temperature on gonad development of North Carolina population of *Aequipecten irradians* in the absence of food.[*]

Reproductive stage	Temperature, °C	Gonad response
Vegetative stage with oogonia and oocytes prior to growth	10, 20, 30	Decrease in gonad index; resorption of oogonia and early oocytes
Gonads with a minimum accumulated reserve and with oocytes in the cytoplasmic growth phase	10 and 15	Cytolysis and resorption of oocytes
	20, 25, 30	Oocyte growth to maturation and spawning
Gonads with accumulated reserves and with oocytes in vitellogenesis	10 and 15	Cytolysis of oocytes
	20, 25, 30	Maturation of oocytes and spawning
Resting reproductive stage with neutral gonads	10, 15, 20, 25, and 30	No gonad growth or gametogenesis

[*] Based on data from Sastry, 1966a.

Food

The abundance of food has been generally correlated with the breeding period of marine invertebrates and is thought to ensure adequate nutritional availability for the planktonotrophic larvae (Thorson, 1950). Barnes (1957) found that larval release in

Balanus balanoides occurs during the spring diatom bloom; in the laboratory they are released earlier when the diatoms are supplied. Although a synchronization of the breeding period with the abundance of food is adaptive, this relationship is different from the food requirements for gonad development. Prolonged starvation of vegetative and resting stage animals prevents the gonad growth (Giese, 1959; Sastry, 1966a). The relationship between food and gonad growth is also indicated by the reciprocal relationship between the nutrient storage organ (i.e., the hepatic caeca in starfishes and the digestive gland in bivalve molluscs) and the gonad during the course of the reproductive cycle (Giese, 1959; Sastry, 1970a). Apparently, nutrients stored during the nonreproductive period are depleted with gonad growth. This relationship, however, is obscured in some other species where gonad development may directly depend upon food availability in the environment. The relationship between food and gonad growth seems to vary among species as well as in populations of a species occurring in different parts of the distributional range (Marshall and Orr, 1952; Giese, 1959; Barnes, Barnes, and Finlayson, 1963; Sastry, 1966a, 1970a).

Temperature and Food

As mentioned earlier, the period of gonad growth of *A. irradian* at Beaufort, North Carolina, coincides with the period of high phytoplankton production in the year. Since the sequential events in the reproductive cycle of this population closely correlate with the seasonal changes in food availability and temperature, it is likely these factors may influence differently the animals in different stages of the reproductive cycle. Sastry (1966a) showed the effects of food deprivation on the gonad growth of scallops in different stages of the reproductive cycle when they were maintained at a series of different temperatures. These studies indicate that the food requirement for gonad development varies in different stages of the reproductive cycle of animals held under temperature conditions optimal for growth of gonads to maturation. Starvation of vegetative and resting stage animals prevents gonad growth and oocyte growth, and also results in a decrease of digestive gland and gonad indices (Table 1). Under the stress of starvation, the reserves from these tissues seem to be utilized for maintenance metabolism rather than for gonad growth and gamete production. However, food deprivation has no effect on oocyte development to maturation in those animals having minimum reserves in the gonads. In these animals, the maximum amplitude of gonad growth is not reached as it is in those receiving adequate amounts of food. Barnes and Barnes (1967) also found that an adequate food supply is essential for the ovarian development of barnacles, *Balanus balanoides*. Interruption of food supply to these animals resulted in the regression of ovarian tissue. In *Mytilus edulis* a complex relationship was observed between the development of gametes and the readiness with which the carbohydrate and protein reserves were utilized by animals held under nutritional and temperature stress (Bayne and Thompson, 1970).

Sastry (1968) determined the interrelationship between food, temperature, and gonad development of an *A. irradians* population from Beaufort, North Carolina. Temperature and food levels seem to interact in the activation of gonad growth and gametogenesis. Resting stage scallops exposed to 20°C without food supply failed to initiate gonad and oocyte growth, but if they were provided with food, the oocytes entered the growth phase (Fig. 3). When the resting stage scallops were held at 15°C with adequate food supply, the primary germ cells and oogonia developed, but the oocytes did not enter the growth phase (Fig. 3). A minimum temperature is required for the initiation of oocyte growth, for at lower temperatures this phase of the reproductive cycle is not activated even if the food supply is adequate. In addition, temperatures exceeding a maximum oocyte growth do not occur even if the animals are supplied with adequate amounts of food. The increased metabolic demands on the available reserves for maintenance at the higher temperature may prevent channeling of reserves to the gonads and inhibit the oocytes from entering the growth phase. It appears that immediately before and after the initiation of gonad and oocyte growth, the temperature and food levels in the environment, feeding and

Fig. 3. The influence of food and temperature on oocyte growth in the population of *A. irradians* from Beaufort, North Carolina (from Sastry, 1968).

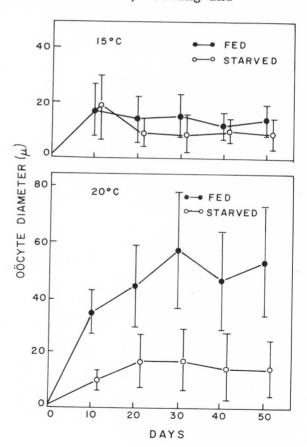

assimilation rates, and requirements for maintenance metabolism may influence in a complex manner the regulation of the utilization and distribution of reserves in the body tissues. Under optimal conditions for gonad growth, the nutrients are channeled to the gonads for gamete production.

The mechanisms regulating the transfer of reserves to the gonads and those activating and suppressing the oocyte growth are not well understood. The distribution of reserves to various body tissues may vary relative to the stage of the reproductive cycle and the seasonal changes in the environment. In some earlier studies, attempts to demonstrate the transfer of nutrients to gonad with the use of radiotracers were inconclusive. The sea urchins (Farmanfarian and Phillips, 1962; Boolootian and Lasker, 1964) and bivalve molluscs (Allen, 1962) fed with labeled food showed no rapid buildup of label in the gonadal tissue. In contrast, Marshal and Orr (1955, 1961) showed that 70% of the ^{32}P assimilated by the female copepod *Calanus finmarchicus* is transferred to the ova. Sastry and Blake (1971) showed that the transfer of reserves from the digestive gland to the gonads of *A. irradians* is regulated by the temperature and the stage of gametogenesis. The resting stage scallops injected with ^{14}C-leucine into the digestive gland and held at the threshold temperature with food incorporated a greater amount in the gonads than those held at a subthreshold temperature (Figs. 4 and 5). The uptake of ^{14}C-leucine into the gonad, digestive gland, and adductor muscle varied relative to the stage of

Fig. 4. Incorporation of ^{14}C-leucine into body tissues of resting stage *A. irradians* from Massachusetts held at 5°C with food (modified from Sastry and Blake, 1971, unpublished data).

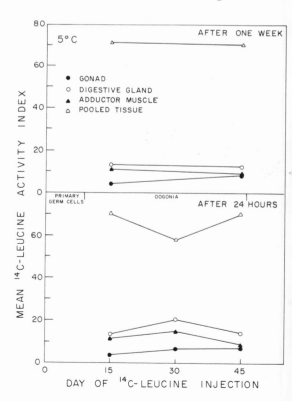

Fig. 5. Incorporation of
^{14}C-leucine into body
tissues of resting stage
A. irradians from
Massachusetts held at 15°C
with food (modified from
Sastry and Blake, 1971,
unpublished data).

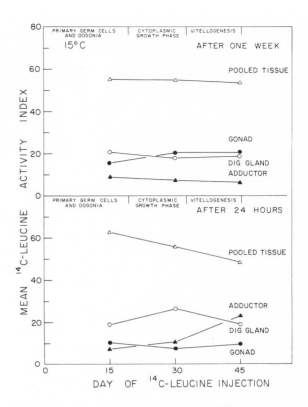

gonad development and the temperature. Incorporation of ^{14}C-leucine
into gonads of animals held at the threshold temperature increased
with time while a corresponding decrease occurred in the digestive
gland and the adductor muscle. The animals with quiescent gonads
at the subthreshold temperature incorporated ^{14}C-leucine more into
the adductor muscle than into the gonad, but less than into the
digestive gland. It would seem that the reserves from the digestive
gland are selectively transferred to the gonads or the adductor
muscle, depending on the activity of the gonads and the temperature.
In *Chlamys hercina*, gonadal lipid activity increases as the activity
in diverticular lipid decreases, which suggests transfer from the
digestive gland to the gonad (Vassallo, 1973). In *Mya arenaria* and
Venus striatula fed with labeled food, the incorporation of ^{32}P into
the gonads was greater in the animals with developing gonads than
in those with immature or mature gonads (Allen, 1970). These
results also suggest that nutrients are transferred to the gonads
when gonad growth and gametogenesis are activated.

Transfer of reserves to the gonads occurs slowly at a rate
regulated by the developing oocytes. This slow transfer of reserves
may produce temporal differences in the activation of the oocyte
growth phase, resulting in asynchrony in oocyte development within
an individual. Asynchrony in oocyte growth occurs in animals in
nature and also in those held under constant conditions in the
laboratory. The oocytes entering the growth phase may stimulate
gonad growth, with the rate of gonad development depending on the

rate at which nutrient reserves are utilized by the growing oocytes for synthesis of various biochemical constitutents. Both environmental factors and the stage of oocyte development seem to regulate the transfer of reserves to the gonads and control their growth activity.

INTERACTION BETWEEN EXOGENOUS AND ENDOGENOUS FACTORS

In addition to the environmental factors discussed above, a number of endogenous factors may also influence the course of the reproductive cycle of marine invertebrates. Detailed studies on the influence of endogenous factors or their interaction with the exogenous factors are relatively few. Thus, the knowledge of this aspect of reproductive physiology and ecology is still fragmentary. It is likely that an interaction between the events in the environment and endogenous factors occurs during the course of the reproductive cycle. From the available data, the environmental factors interacting with the age, metabolism, and neuroendocrines seem to regulate the course of events in the reproductive cycle.

Age

For most marine invertebrates, it is difficult to determine the influence of age on reproduction under experimental conditions in the laboratory. In many instances, however, the influence of age on reproduction can be inferred from observations made on a series of samples collected from the field. The influence of age on reproduction is most evident in some bivalve molluscs where the animals undergo sex change (Coe, 1943). The influence of age on reproduction of *A. irradians* can readily be determined for field-collected animals and also under experimental conditions, since a year class of this species lives only for 18 to 24 months and essentially undergoes a synchronized reproductive cycle in the first year. Gonad and oocyte growth begin in the population of *A. irradians* at the age of 9 to 10 months in Massachusetts, 8 to 9 months in North Carolina, and 4 to 5 months in Florida. All three populations mature and commence spawning at the age of 12 months.

Blake (1972) showed that the Massachusetts population of scallops at the age of 3 months does not initiate the gonad and oocyte growth even after holding them for 2 months at the threshold temperature with food. At the age of 6 to 7 months, the animals with primary germ cells and oogonia maintained at 10, 15, 20, and 25C with food initiated oocyte growth only at 15°C after 45 days (Fig. 6). The laboratory-held animals initiated the oocyte growth phase at about the same age as the population under natural conditions. After the initiation of gonad growth and oocyte growth, however, the gonad development can be advanced to maturation at appropriate temperatures much earlier than the 12 months required under natural conditions. Unlike the vegetative stage animals, the postspawning adults readily respond by initiating the oocyte growth within 7 to 10 days after maintaining them at the threshold temperature with food. Gonad growth in these animals can be also delayed by holding them at subthreshold temperatures. The response of *A. irradians* to the environmental factors for initiation of gonad growth and oocyte growth seems to vary in the vegetative and resting stage animals.

Fig. 6.   The influence of age on the initiation of oocyte growth phase in *A. irradians* from Massachusetts (Sastry, unpublished data).

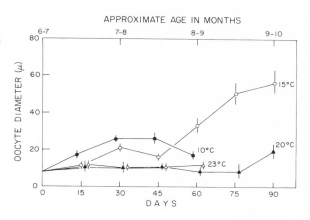

Metabolic Changes

Gonad growth during the course of the reproductive cycle may be regulated through a complex interaction between the food availability, the temperature, and the energy requirements for maintenance metabolism and growth. The energy requirements for maintenance and growth may have to be satisfied before the reserves can be mobilized to the gonads. In the absence of detailed data, it is impossible to understand how the metabolism of the organism is organized for seasonal gonad production relative to the food availability and temperature changes in the environment.

Seasonal changes in gonad growth of *Strongylocentrotus purpuratus* as determined by gonadal indices are much lower in the warmer areas of its distribution than in colder areas (Boolootian, 1966).   In *Mya arenaria* gonad development is interrupted when temperatures exceed the optimum for gonad development in the summer (Ropes and Stickney, 1965). A transplanted northern population of *A. irradians* fails to initiate gonad growth and oocyte growth during winter in warmer waters in the South (Sastry, unpublished data). As mentioned earlier, the gonad growth in *A. irradians* does not occur when the temperatures are beyond a certain range. At temperatures below the required minimum, the reserves accumulating in the digestive gland are transferred more into the adductor muscle than into the gonad.

Giese (1959, 1966, 1967) discussed the biochemical changes associated with the reproductive cycle of marine invertebrates. Gonad production involves intense biochemical synthesis with the formation of large amounts of nucleic acids for spermatozoa and mobilization of proteins and lipids for eggs. Barnes, Barnes, and Finlayson (1963) reported that seasonal variation in the oxygen consumption rate of isolated body tissues of *Balanus balanoides* is related to the changes in composition and activity of tissues. In *Mytilus edulis* seasonal changes in the oxygen consumption are related to the variation in the reproductive activity and accompanying changes in carbohydrate, protein, and lipid reserves in the mantle tissue (Bayne and Thompson, 1970). Metabolic rate has also been reported to vary with the stage of the reproductive cycle in some other species (Sparck, 1936; Thorson, 1936; Mori, 1968). The oxygen consumption rate measured on a seasonal basis showed a low Q_{10} of 1.6 between 18 and 22.5°C for the population of *A. irradians* at Beaufort, North Carolina (Kirby-Smith, 1970). The same temperature range is also required for the transfer of reserves to gonads (Sastry, 1968). The Q_{10} values were higher for the temperature range which the population experiences during the vegetative and resting stages of the annual reproductive cycle. Although changes in metabolism seem to occur with the season and the reproductive stage, it is not yet clear how these changes are regulated within the organism for seasonal gonad production.

Neuroendocrine Coordination

Neuroendocrine influences on reproduction of polychaetes (Clark, 1964), molluscs (Gabe, 1965; Lubet, 1966), and crustaceans (Adiyodi and Adiyodi, 1970) have been reviewed. In many cases the environmental interaction with neuroendocrine activity and reproductive activity has not been examined. In some bivalve molluscs, the seasonal changes in cyclical activity of neurosecretion based on histological examination have been related to reproduction (Gabe, 1965; Lubet, 1966). Neurosecretory products accumulate in the perikaryons during the period of gametogenesis and reach a maximum at the time of maturation. The neurosecretory products are emptied before spawning. Blake (1972) examined the changes in neurosecretory activity during the course of the reproductive cycle of populations of *A. irradians* from Massachusetts and Rhode Island. The neurons showed changes in size, granulation, and vacuolization to suggest a cycle of synthesis and release of secretory products. Based on these characteristics, the neurosecretory cycle has been divided into five stages and related to the stages of oocyte development. The neurosecretory cycle stages corresponding with the stages of oocyte development are as follows: Stage I, neutral or primary germ cells; Stage II, oogonia and early oocytes; Stage III, cytoplasmic growth phase; Stage IV, vitellogenesis and maturation; and Stage V, spawning. For the first 12 months, the neurosecretory cycle is highly synchronous for a population. The neurosecretory cycle and the reproductive cycle showed significant correlation with the seasonal changes in temperature, but not with each other.

Blake (1972) also determined the effects of subthreshold (5°C) and threshold (15°C) temperature on the neurosecretory activity and gonadal response of an *A. irradians* population from Massachusetts which was collected at different stages in the reproductive cycle. The vegetative stage scallops with primary germ cells held at the two temperatures for a period of 56 days showed no change in the gametogenic stage or neurosecretory cycle stage. When the scallops with oogonia and early oocytes were held at the subthreshold temperature, the neurosecretory cycle and the gametogenic stage remained unchanged. If the temperature was increased to the threshold level, the oocytes entered the growth phase and the neurosecretory cycle advanced to the next stage. When the animals at the threshold temperature for a period of 26 days were transferred to the subthreshold temperature, the neurosecretory cycle reverted to the earlier stage, but the oocyte size remained unchanged. Continued exposure of scallops to a temperature at or above the threshold level alone ensures oocyte growth with corresponding changes in the neurosecretory cycle. Progress of the neurosecretory cycle in the animals exposed to the threshold temperature may trigger the oocytes to enter the growth phase by initiating the transfer of reserves to gonads (Sastry, 1968, 1970b; Sastry and Blake, 1971; Blake, 1972).

After a certain period at the threshold temperature, in the scallops with oocytes in the growth phase, the stage of the neurosecretory cycle cannot be reverted to the earlier stage by lowering the temperature (Blake, 1972). The persistence of the neurosecretory cycle stage in scallops at subthreshold temperature during the course of gonad development does not ensure the normal development of oocytes to maturation. Normal growth of oocytes to maturation occurs only when there is a simultaneous change of neurosecretory cycle stages with the stages of oocyte development. Lubet (1966) found that the integrity of the cerebral ganglion in *Mytilus edulis* is essential for exhibiting the normal course of the reproductive cycle. Blake (1972) found that the neuronal degeneration in *A. irradians* is correlated with the extent of oocyte disintegration and resorption. Exposure of scallops to a certain minimum temperature after oocyte growth is required for maturation and spawning (Sastry, 1966a, 1968, 1970b).

In some bivalve molluscs, the neurosecretion is emptied before spawning (Lubet, 1966), but this stage was not reached in the scallops held under experimental conditions in the laboratory (Blake, 1972). Scallops collected with partially spent gonads have different neurosecretory cycle stages in each individual coinciding with the stages of oocytes present. After complete spawning, the neurosecretory cycle returns to a stage to coincide with the neutral stage of gonads. If these animals are held at the threshold temperature, the neurosecretory cycle remains in a stage which coincides with the oogonia and early oocytes. If the temperature is raised to the threshold level, the neurosecretory cycle advances to the next stage and oocytes enter the growth phase. The seasonal changes in temperature seem to regulate the neurosecretory cycle and the reproductive activity.

MECHANISMS REGULATING THE REPRODUCTIVE CYCLE

The ecological observations and laboratory results on reproduction of *A. irradians* discussed separately in the previous sections of this paper are synthesized here to consider some of the possible mechanisms regulating the pattern of the annual reproductive cycle (Fig. 7). During the vegetative stage, the population is refractory to the environmental stimuli for the initiation of gonad growth and oocyte growth. The inhibition of gonad development in this stage may have been due to the age and related energy requirements for maintenance metabolism and growth. The duration of the vegetative stage and the minimum age at which the gonad growth begins may be genetically fixed for a population.

As oogonia and early oocytes develop, with the corresponding change in the neurosecretory cycle stage, the scallops become responsive to the environmental stimuli for either initiation or delay of gonad and oocyte growth. If the temperature is above a certain minimum, gonad growth begins with a simultaneous change of neurosecretory cycle stage. Subthreshold temperature inhibits the progress of both neurosecretory cycle and gonad growth. At this stage, the scallops seem to have the ability for either suppressing or activating the gonad growth as influenced by the environmental temperature. The neurosecretory cycle stages II and III seem to act as an "on and off" mechanism to control the oocyte growth for some time before and after the initiation of gonad growth (Blake, 1972). A certain intensity and duration of temperature stimulus seem to be required for the entrainment and the continued progress of gonad growth and oocyte growth. With the initiation of oocyte growth, the reserves are transferred to the gonads. The transfer of reserves seems to be regulated by a complex interaction between temperature and food availability in the environment and neurosecretory cycle stage and gametogenic activity within the organism. The neurosecretion may act directly upon the gonad and/or it may influence the metabolism of the animal to allow channeling of reserves to the gonads. The normal course of gonad growth to maturation occurs only when there are corresponding changes in the neurosecretory cycle with the growth of oocytes. After certain minimum reserves are accumulated in the gonads, the oocyte development to maturation is independent of food supply, provided the temperatures are above the required minimum. If the temperature is below the threshold level during the course of gonad development, the neurosecretory cycle stage persists but the oocytes fail to mature. After vitellogenesis, maturation and spawning occur only in those animals exposed to another required minimum temperature (Sastry, 1970b). It would seem that a feedback control mechanism linking

293

the neurosecretory cycle with the gametogenic stages might be regulating the transfer of reserves to gonads and the normal growth of
oocytes to maturation (Bayliss, 1966).

   After maturation of oocytes, the spawning may be triggered
through an interaction between endogenous factors and environmental
factors inducing gamete release. Neurosecretory control of spawning has been suggested in some bivalve molluscs (Lubet, 1966).

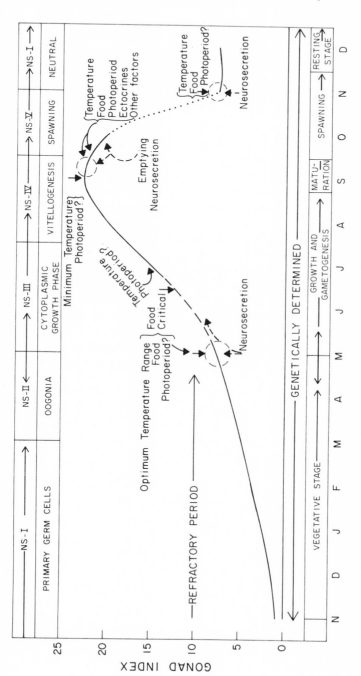

Fig. 7. Schematic drawing showing the interaction between exogenous and endogenous
factors in the regulation of the annual reproductive cycle of *A. irradians*.

The response to exogenous spawning inducers is maximal when the neurosecretory products decrease following the maximum coincident with maturation. Spawning may be induced by a number of environmental, biological, and hormonal factors (Giese, 1959; Sastry, in press).

At the end of the breeding period, the decrease of temperature to below threshold level, food availability, and the return of the neurosecretory cycle stage to coincide with the neutral stage of gonads may inhibit the initiation of gonad growth. The time course of the annual reproductive cycle seems to be regulated by a coordination of reproductive events with the seasonal changes in the environment through the mediation of neurosecretory activity. The characteristic pattern of annual reproductive cycle for the species may be a genetically controlled response to the environment.

GENETIC AND NONGENETIC VARIATION
IN THE REPRODUCTIVE CYCLES OF POPULATIONS

In species with wide geographical distributional range, the reproductive cycles of populations experiencing different environments may vary as a phenotypic response of a single genotype or it may be truly genetic, or both. The populations of sea urchins, *Arbacia punctulata*, from Massachusetts show two gonad index peaks during the year, while those from North Carolina have a single peak (Fig. 8). The breeding period of both populations is identical when Massachusetts sea urchins in the resting stage are transplanted to North Carolina and held under conditions similar to those of the local population (Sastry, 1966<u>b</u>). These population differences

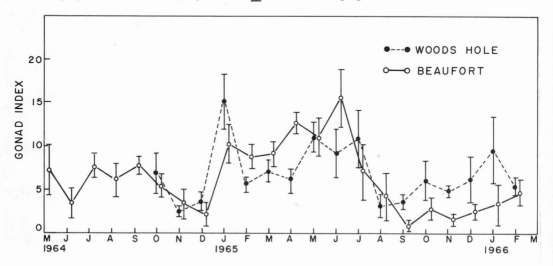

Fig. 8. Variation in the annual reproductive cycles of *Arbacia punctulata* from Massachusetts and North Carolina, which suggests phenotypic variation (Sastry, unpublished data).

reflect phenotypic variation in the reproductive cycles of the two populations. Latitudinally separated populations of the mud snail, *Nassarius obsoletus*, also show phenotypic variation in their reproductive cycles (Sastry, 1971). Recent electrophoretic studies of protein systems of *N. obsoletus* along the Atlantic Coast from Cape Cod, Massachusetts, to Beaufort, North Carolina, have shown a remarkable homogeneity of allele frequency, indicating genetic stability of the populations (Gooch, Smith, and Knupp, 1972). In some other species, the northern populations may exhibit an annual cycle while those in the southern areas have a semiannual cycle. The soft-shell clam, *Mya arenaria*, develops gametes in both autumn and spring in the areas south of Cape Cod, while those in the North have only one period of gametogenic activity in the summer (Ropes and Stickney, 1965; Pftizenmeyer, 1965). Summer temperatures in the South seem to exceed the optimum required for gametogenesis, while winter temperatures fall below this range, thus producing spring and autumn optima for gametogenesis in the year.

The variation in the reproductive cycles of geographically separated populations may result through genetic divergence. In this case, each population will have different environmental requirements for one or more events in the reproductive cycle. The populations of *A. irradians* from Massachusetts and North Carolina vary in the time of year when successive events in the reproductive cycle take place and also in the environment for these events (Fig. 9).

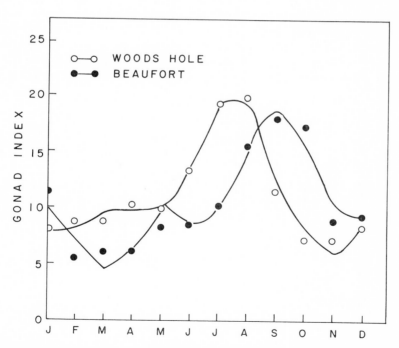

Fig. 9. Variation in the annual reproductive cycles of *A. irradians* from Massachusetts and North Carolina, which suggests genetic divergence (modified from Sastry, 1970a).

The populations have different temperature requirements for the initiation of gonad growth and oocyte growth which cannot be altered by thermal acclimation (Sastry, 1966a, b, 1968, 1970a; Sastry and Blake, 1971; Sastry, unpublished data).

GENERAL CONCLUSIONS

A considerable diversity exists in the patterns of the reproductive cycles of marine and estuarine invertebrates. These patterns have been generally associated with the environmental conditions to which the populations are exposed, but the causal relationships in determining these patterns have not been examined in any detail for most species. The reproductive cycle is influenced by a number of environmental factors, but also by endogenous factors. A coordination of the physiological processes within the organism with factors in the environment may regulate the pattern characteristic for a species. The reproductive cycle is a genetically controlled response to the environment. The time course of sequential events in the reproductive cycle occurs within the genetically determined environmental limits. The environmental limits and the duration for sequential events in the reproductive cycle may determine the synchrony and asynchrony of reproductive periodicity of a population in the year. In widely distributed species, the populations experiencing differences in the environment may adapt for reproduction by phenotypic variation or genetic divergence or both. The environment-organism interactions in reproductive physiological adaptation are significant for both basic studies in biogenography, evolutionary biology, and also to detect man-induced perturbations in the environment.

ACKNOWLEDGMENTS

The author's work reported in this paper was supported in part by grants from the National Science Foundation (GB-1356) and the office of Sea Grant Program NOAA.

LITERATURE CITED

Adiyodi, K. G. and R. G. Adiyodi. 1970. Endocrine control of reproduction in decapod crustacea. Biol. Rev. 45:121-65.
Allen, J. A. 1962. Preliminary experiments on the feeding and excretion of bivalves using *Phaeodactylum* labelled with ^{32}P. J. Mar. Biol. Ass. U. K. 42:609-23.

_____. 1970. Experiments on the uptake of radioactive phosphorous by bivalves and its subsequent distribution within the body. Comp. Biochem. Physiol. 36:131-41.

Barnes, H. 1957. Processes of restoration and synchronization in marine biology. The spring diatom increase and the spawning of the common barnacles, *Balanus balanoides*. Annéé Biol. 33: 67-85.

_____ and M. Barnes. 1967. The effect of starvation and feeding on the time of production of egg masses in the boreo-Arctic cirripede *Balanus balanoides* (L). J. Exp. Mar. Biol. Ecol. 1: 1-6.

_____, _____, and D. M. Finlayson. 1963. The seasonal changes in body weight, biochemical composition and oxygen uptake of two boreo-Arctic cirripedes, *Balanus balanoides* and *Balanus balanus*. J. Mar. Biol. Ass. U. K. 43:185-211.

Bayliss, L. E. 1966. Living Control Systems. London: The English University Press, Ltd.

Bayne, B. L. 1965. Growth and delay of metamorphosis of larvae of *Mytilus edulis* L. Ophelia 2:1-47.

_____ and R. J. Thompson. 1970. Some physiological consequences of keeping *Mytilus edulis* in the laboratory. Helgolander wiss. Meeresunters. 20:526-52.

Blake, N. J. 1972. Environmental regulation of neurosecretion and reproductive activity in the bay scallop, *Aequipecten irradians* Lamarck. Ph.D. dissertation, University of Rhode Island.

Boolootian, R. A. 1966. Reproductive physiology. In: Physiology of Echinodermata (R. A. Boolootian, ed.), pp. 561-613. New York: Interscience Publishers.

_____ and R. Lasker. 1964. Digestion of brown algae and the distribution of nutrients in the purple sea urchin *Strongylocentrotus purpuratus*. Comp. Biochem. Physiol. 11: 273-89.

Clark, R. A. 1964. Endocrinology and reproductive biology of polychaetes. In: Oceanography and Marine Biology, Annual Review (H. Barnes, ed.), pp. 211-55. London: George Allen and Unwin, Ltd.

Coe, W. R. 1943. Sexual differentiation in molluscs. I. Pelecypods. Quart. Rev. Biol. 18:154-64.

Crisp, D. J. 1957. Effect of low temperature on the breeding of marine animals. Nature 179:1138-39.

Farmanfarian, A. and J. A. Phillips. 1962. Digestion, storage and translocation of nutrients in the purple sea urchin, *Strongylocentrotus purpuratus*. Biol. Bull. 123:105-20.

Gabe, M. 1965. La Neurosecretion chez les mollusques et ses rapports avec la reproduction. Arch. Anat. Microsc. Morpho. Exp. 54:371-85.

Giese, A. C. 1959. Comparative physiology: Annual reproductive cycles of marine invertebrates. Ann. Rev. Physiol. 21:547-76.

_____. 1966. Lipids in the economy of marine invertebrates. Physiol. Rev. 46:244-98.

_____. 1967. Some methods for study of biochemical constitution of marine invertebrates. In: Oceanography and Marine Biology, Annual Review (H. Barnes, ed.), pp. 253–88. London: George Allen and Unwin, Ltd.

Gooch, J. L., B. S. Smith, and D. Knupp. 1972. Regional survey of gene frequencies in the mud snail, *Nassarius obsoletus*. Biol. Bull. 142:36–48.

Gunter, G. 1957. Temperature. In: Treatise on Marine Ecology and Paleoecology. I. Ecology (J. W. Hedgepeth, ed.), pp. 159–84. New York: Geological Society of America.

Kinne, O. 1970. Temperature, animals, invertebrates. In: Marine Ecology, A Comprehensive Treatise on Life in Oceans and Coastal Waters (O. Kinne, ed.), pp. 407–504. London: Wiley Interscience.

Kirby-Smith, W. W. 1970. Growth of the scallop, *Argopecten irradians concentricus* (Say) and *Argopecten gibbus* (Linne) as influenced by temperature and food. Ph.D. dissertation, Duke University.

Loosanoff, V. L. and H. C. Davis. 1950. Conditioning of *V. mercenaria* for spawning in winter and breeding its larvae in the laboratory. Biol. Bull. 98:60–65.

Lubet, P. 1966. Essai d'analyse experimentale des perturbations produites par les ablations des ganglions nerveux chez *Mytilus edulis* L et *Mytilus galloprovicialis* LMK (Mollusques Lamellibranches). Ann. Endoc. 27:353–65.

Marshall, S. M. and A. P. Orr. 1952. On the biology of *Calanus finmarchicus*. VII. Factors affecting egg production. J. Mar. Biol. Ass. U. K. 30:527–47.

_____ and _____. 1955. On the biology of *Calanus finmarchicus*. VIII. Food uptake and assimilation and excretion in the adult stage *V. Calanus*. J. Mar. Biol. Ass. U. K. 34:495–529.

_____ and _____. 1961. Studies on the biology of *Calanus finmarchicus*. XII. The phosphorous cycle: excretion, egg production and autolysis. J. Mar. Biol. Ass. U. K. 41:463–68.

Mori, K. 1968. Oxygen consumption and respiratory quotient in the tissues of oysters during the stages of sexual maturation and spawning. Tohoku J. Agric. Res. 19:136–43.

Orton, J. H. 1920. Sea temperature, breeding and distribution of marine animals. J. Mar. Biol. Ass. U. K. 12:339–66.

Pftizenmeyer, H. T. 1965. Annual cycle of gametogenesis in the soft-shelled clam, *Mya arenaria* at Solomons, Maryland. Chesapeake Sci. 6:52–59.

Ropes, J. W. and A. P. Stickney. 1965. Reproductive cycle of *Mya arenaria* in New England. Biol. Bull. 128:315–27.

Runnström, S. 1927. Über die Thermopathie der Fortpflanzung und Entwicklung mariner Tiere in Bezeitung zu ihrer geographischen Verbreitung. Bergens Mus. Arb. (Naturv. rekke) 2:1–67.

Sastry, A. N. 1963. Reproduction of the bay scallop, *Aequipecten irradians* Lamarck. Influence of temperature on maturation and spawning. Biol. Bull. 125:146–53.

_____. 1966a. Temperature effects in reproduction of the bay scallop, *Aequipecten irradians* Lamarck. Biol. Bull. 130:118-34.

_____. 1966b. Variation in reproduction of latitudinally separated populations of two marine invertebrates. Amer. Zool. 5:374-75.

_____. 1968. The relationships among food, temperature and gonad development of the bay scallop, *Aequipecten irradians* Lamarck. Physiol. Zool. 41:44-53.

_____. 1970a. Reproductive physiological variation in latitudinally separated populations of the bay scallop, *Aequipecten irradians* Lamarck. Biol. Bull. 138:56-65.

_____. 1970b. Environmental regulation of oocyte growth in the bay scallop, *Aequipecten irradians* Lamarck. Experentia 26:1371-72.

_____. 1971. Effect of temperature on egg capsule deposition in the mud snail, *Nassarius obsoletus* (Say). Veliger 13:339-41.

_____. 1974. Pelecypoda. In: Treatise on Physiology and Ecology of Reproduction in Marine Invertebrates (A. C. Giese and J. S. Pearse, eds.). New York: Academic Press. (In press.)

_____ and N. J. Blake. 1971. Regulation of gonad development in the bay scallop, *Aequipecten irradians* Lamarck. Biol. Bull. 140:274-83.

Sparck, R. 1936. On the relation between metabolism and temperature in some marine lamellibranchs and its zoogeographical significance. Biol. Medd. 13:1-27.

Thorson, G. 1936. The larval development, growth and metabolism of Arctic marine bottom invertebrates compared with those of other seas. Medd. Grønland 100:1-155.

_____. 1950. Reproduction and larval development of marine bottom invertebrates. Biol. Rev. 25:1-45.

Vassallo, M. T. 1973. Lipid storage and transfer in the scallop, *Chlamys hericia* Gould. Comp. Biochem. Physiol. 44A:1169-75.

Williams, R. B. and M. B. Murdoch. 1966. Phytoplankton production in the Beaufort Channel, North Carolina. Limnol. Oceanogr. 11:73-82.

The response of developmental stages of *Fundulus* to acute thermal shock

S. R. Hopkins and J. M. Dean

In his 1960 report J. R. Brett summarized the phases of research on temperature effects on fish. In the earlier studies, temperature was considered a lethal factor; subsequently, it was studied as a controlling factor; more recently, it was analyzed as a factor affecting cellular action, interaction, and stress. Brett concluded his article (Brett, 1960) by saying, "Temperature tolerance studies for every stage in the life cycle are sorely needed, particularly at the level of the developing embryo."

Developing embryos of estuarine animals are exposed to a wide range of environmental extremes. Generally, the embryonic stages of organisms are more sensitive to stress than are adults (Alderdice and Forrester, 1971), although the present study finds some evidence that at certain stages of development the embryos of estuarine fish are more tolerant of environmental stress than adults are. During other stages of development, however, environmental perturbations may be critical, resulting in abnormal growth or even death of the organism.

In an estuary the temperature, salinity, and level of dissolved oxygen are in a constant state of flux due to the daily tidal mixing of fresh and oceanic waters. Animals living in that changing environment are well adapted to the fluctuations encountered there, but man has added other stresses. One of these stresses is the potential threat of thermal shock from the heated effluents of industries and electrical generating stations. Despite marked differences in environmental tolerances of different stages in the

life cycle of estuarine animals, the data necessary for water quali‐
standards and for policy-making decisions on reactor sitings are
usually available only for the adult of a species. What happens to
a fish embryo subjected to higher than ambient temperatures for
short periods of time is unknown.

Most of the early research on fish embryos concerned tempera‐
ture as a lethal factor and dealt with defining the range of temper‐
ture for the normal development of various species (Gray, 1928;
Battle, 1929; Bonnet, 1939). Working with *Salmo fario*, Gray found
that embryos were larger when incubated at lower temperatures and
that precocious hatching occurred at higher temperatures. He rea‐
soned that higher temperatures increased the growth rate, but that
the embryos were smaller because more yolk was required for main‐
tenance, leaving less available for formation of new tissues.
Battle (1929), working on *Enchelyopus*, concluded that the thermal
limits of embryos are generally narrower than those of adults and
may vary from stage to stage. She suggested that the main factor
involved in lethal limits was not the effects on protein or lipids,
but that these effects may appear secondarily. Bonnet (1939), in
determining the highest temperature possible for development of the
cod, used four temperatures in the supra-optimal range. He found
an initial period of high mortality at all temperatures which
decreased with the closing of the blastopore. This period was
followed by another of low mortality until the embryo was three-
fourths the circumference of the egg membrane. At this point
mortality increased steadily again up to hatching. Other studies
of the range of tolerance for normal development include those of
Worley (1933), Price (1940), Piavis (1961), Kinne and Kinne (1962),
Hubbs (1965, 1966), and Alderdice and Forrester (1971). All agree
that temperature affects size and rate of growth as well as mortali‐
and that tolerance is dependent on both the temperature and the
stage of development.

Temperature as a controlling factor on the rate of development
was the second category of research. Worley (1933) noted that the
developmental time of the mackerel was increased from 49.5 to 207 h‐
when the temperature was lowered from 21 to 10°C. Hatching time was
also calculated at different temperatures for the whitefish *Coregonu*
by Price (1940). His results were similar to those of Worley.
Merriman (1935) did the same type of study on the cut-throat trout,
Salmo clarkii clarkii. He also found that embryos raised at differ‐
ent constant temperatures varied in size. His trout were smaller a‐
higher temperatures and larger at lower ones. He said that using
time-to-hatching for comparing eggs raised at different temperatures
may result in error because there are a number of other factors tha‐
affect hatching.

Blaxter (1956) found that herring raised in 3°C water required
36 days to hatch, while those in 18C water hatched in only 6.5 days
Kowalska (1957) found that with brook trout (*Salmo trutta* var. *faric*
the number of day-degrees at hatching was not a constant value. She
found that development took longer (a maximum of day-degrees) at

those temperatures near the optimal range for the species. Above and below that span the number of day-degrees decreased. Up to the "eyed-stage" the number of day-degrees needed for each consecutive developmental stage was approximately constant; therefore, the cause of early or late hatching must be looked for in the later part of development and especially at time of hatching (Kowalska, 1957).

Hubbs and his co-workers have done studies of temperature tolerance and rates of development on various species of freshwater fish (Hubbs, 1965; Hubbs, Peden, and Stevenson, 1969; Hubbs, Sharp, and Schneider, 1971; Wilson and Hubbs, 1972). For example, in work with *Etheostoma lepidum*, Hubbs, Peden, and Stevenson (1969) found that development is strongly affected by incubation temperature. The effects are greatest at the earlier stages and become less with increasing maturity. Kinne and Kinne (1962) found that there appears to be a period of low thermal stability during early development (i.e., fertilization to gastrulation) followed by a phase of increased stability and later, toward the end of development, by a second period of low stability. Other papers on developmental rate include those by Lasker (1964), Forrester (1964), Forrester and Alderdice (1966), and Garside (1959, 1966).

In many of the studies already mentioned, slight to severely abnormal embryos were noted at the extremes of the temperature range. For example, workers such as Hubbs (1922) found that variations in the numbers of vertebrae, scale-rows, and fin rays of *Notropis atherinoides* (shiner) and *Lepomis incisor* (sunfish) could be correlated with the temperature of the water during development. The numbers of these structures increased in colder water in both species. Gabriel (1942, 1944) showed that *Fundulus heteroclitus* embryos raised at colder temperatures had a higher average number of vertebrae than those raised at optimum temperatures. Tåning (1950) found that there was a "sensitive" period in the development of the embryos of teleostean fishes, during which the number of vertebrae was "determined." This was discovered by exposing eggs to a period of brief but considerable change in temperatures (10 to 13°C). The supersensitive period (145 to 165 day-degrees) corresponded to the period from the closure of the blastopore to eye pigmentation. If treated at this time there was not less than a 3.2 difference in the average number of vertebrae. Tåning also noted that embryos tolerated changes better in this supersensitive period in which the death rate was lower. This finding is in agreement with the results of Bonnet (1939) and Kinne and Kinne (1962).

In studies on vertebrae and fin rays Lindsey (1954) and Seymour (1959) also found that the lowest average number of vertebrae and the highest number of anal fin rays and dorsal fin rays occurred in the middle portion of the developmental temperature range for the species. At either extreme the opposite effect was observed. Itazawa (1959) agreed with Tåning and Lindsey when he raised *Channa argus* (snake-head fish) at different constant temperatures. In all of these papers it appeared that high numbers of vertebrae were produced at intermediate temperatures and low numbers of vertebrae were produced by high or low temperatures.

Garside and Fry (1959) assumed that there was a certain critic yolk size below which it became limiting, preventing the formation of the normal myomere complement. However, Garside (1959) also sai that "innocuous meristic changes appear to result from small change in developmental rate. Failure of whole organs and large portions of the body were due to temporary arrest of some or all of the differentiation processes." If, because of metabolic block, a structure was not formed in its normal position in the embryonic sequence, it would not develop at all. If fish embryos were subjected to temporary severe stress in the early gastrula stage, twinning or head distortions or both often resulted (Stockard, 1921 Stockard said that this type of deformity was not caused by hereditary factors, but by temporary inhibition of the rate of developmen a "developmental arrest," which deranged the relative gradient of certain processes. The deformity varied with the time of inhibitio because of the differences in relative metabolic gradients at the different stages of development. Other studies that also made note of deformities occurring at the upper thermal limits were those of Piavis (1961), Forrester (1964), and Wilson and Hubbs (1972).

Brett's third category, the effects of temperature on cellular action, interaction, and stress, is of primary concern to this stud The effect of short-term thermal shock on the developmental stages of animals is not as well documented as other temperature studies. Several investigators have used this type of stress on insects. Milkman (1962, 1967) found that a sudden and short thermal shock of a few minutes on the pupae of *Drosophila* caused deletions in the venation of the wings. The extent of the deletion was related to the time of treatment and the degree of temperature used. Horsfall Anderson, and Brust (1964), working with *Aedes sierrensis*, a mosqui found that larvae reared at 30°C during instars 1, 2, 3, and the first half of 4 had modifications in the palpi, antennae, oral stylets, and testes. Anderson and Horsfall (1965), using another species of *Aedes*, found that the primordia of organs of the male reproductive system developed abnormally when exposed to near-letha high temperatures during the cellular differentiation processes of embryogeny.

Frankhauser and Humphrey (1942), using cold shock, induced triploidy and haploidy in axolotl eggs. Heat shock caused polyploi in Black Sea sturgeon (Vasetskii, 1967). Bergan (1960) performed studies on heat-shocked embryos of the blue gourami, *Trichogaster trichopterus*. Depending on which phase of mitosis the egg was in when treated, Bergan found that the effects were: migration of centrioles stopped, definition of astral structure disappeared, the spindle shortened and then disappeared, and chromosomes changed int chromosomal vesicles. In all categories of mitotic blocking, cytokinetic anomalies were encountered in later development.

In earlier studies with Salmonids, we found that the effect of incubation at elevated temperatures, following the annual temperatu cycle of the river, did not alter the upper thermal limits of the juvenile fish. What was critical was the immediately previous thermal history.

During the course of this experiment, we observed that there were differences in the mortality of the fish at different developmental stages. We would also like to speculate on some thermal effects data we have that fits very well with the ecology of the beast.

In order to examine more critically some of the questions raised in considerations of these findings, we designed experiments using the embryo of *Fundulus heteroclitus*, the mummichug. The objectives of this study were twofold: (1) to determine whether stages in the development of *F. heteroclitus* are affected by acute temperature shock, and (2) to determine any effects of temperature shock on growth and development of the embryo and the larva.

Fundulus heteroclitus is one of the most common species of fish inhabiting the estuaries and *Spartina* marshes of the eastern United States (Cain, 1973). These fish typically inhabit inshore bays and inlets as well as shallow tidal creeks and pools. They are euryhaline and eurythermal and are found from the Gulf of St. Lawrence to Mexico (Bigelow and Schroeder, 1953).

F. heteroclitus has several features which make it useful for embryological research. The adults can be readily collected in minnow traps, and the eggs stripped and fertilized in the laboratory. The chorionic membrane is transparent, permitting a clear view of the developing embryo, the embryo is hardy, and the rapid rate of development is convenient for experiments. Also, the normal stages of its development have been thoroughly described and illustrated (Armstrong and Child, 1965).

MATERIALS AND METHODS

The parent stock of *F. heteroclitus* used in this study was collected from the North Inlet estuary near Georgetown, South Carolina, and from Johns Island, South Carolina. During winter months the fish were placed in 27°C water, a 16:8 LD photoperiod, and given 50 I.U. of human chorionic gonadotropin per day for 5 days. Spawning condition was achieved in about 3 wks. The gametes were removed from the fish following the technique of Strawn and Hubbs (1956). In this method eggs are squeezed from the females by applying slight pressure to the ventral region of the fish. Milt was removed from several males in the same way. Eggs and sperm thus removed were placed directly into fingerbowls containing filtered seawater. The salinity throughout the experiment was held at 20 o/oo. Several females and males were used to assure a random sample for viability. Rocking the bowls gently after the addition of the milt usually assured the fertilization of the eggs. The exact time of the mixing of gametes of each group was noted, and the treatment was begun at the same intervals so that fertilized eggs were the same age when treated.

The separation of the chorionic membrane from the egg to form the perivitelline space occurs within 15 min if the egg has been fertilized (Armstrong and Child, 1965). The eggs were examined at this time, and any eggs which were either damaged or not fertilized were removed. The remaining viable eggs were then placed in plastic dishes with bottoms made of nylon mesh and submerged in a constant temperature bath at 20°C until the desired stage of development was reached. Each bowl contained no more than 50 eggs to avoid over-crowding (New, 1966). The mesh covering the bottom of each dish was 1.5 mm while the diameter of the eggs was 2.0 mm. This assured rapid emersion for treatment without loss of eggs.

The test procedure consisted of producing a thermal shock to the embryos by exposure in a constant temperature bath at 40°C for 5 min, followed by an immediate return to the 20°C incubation temperature. This experimental design is similar to that of Milkman (1967) in his study of temperature adaptation in *Drosophila* pupae. The following stages after mixing of gametes were tested: ½, 1, 1½, 2, and 2½ hrs (1-cell stage; Stage 2 in Armstrong and Child, 1965). These five treatments were all prior to first cleavage. Other stages tested were: 10 hrs (64-cell stage, Stage 10), 15 and 20 hrs (early and late blastula, Stages 12 and 13), 33 and 37 hrs (early and late gastrula, Stages 17 and 18), 40 hrs (closure of the blastopore, Stage 19), 6 days (Stage 29), and 1 day prehatching (Stage 34). After the shock treatment the embryos were allowed to continue development at room temperature (about 22°C) in fingerbowls containing 20 o/oo filtered seawater to a depth of 3.5 cm. The water in each bowl was changed daily, and any nonviable (opaque) eggs were removed and recorded. Two kinds of controls were used in all tests: unhandled eggs for viability control (controls) and eggs handled in the same manner as the treated group to determine effects resulting from the manipulation of the eggs (handling controls).

To measure the effects of the treatment the following criteria were used: (1) percentage mortality after closure of blastopore (Stage 21 of Armstrong and Child, 1965), (2) percentage mortality at 8 days development (Stage 25), and (3) percentage hatch. Other aspects of development noted were formation and expansion of melano-phores, body movements, rate of development, and especially, abnormal morphological characteristics.

RESULTS

In a preliminary test to determine what the duration of the shock treatment should be, groups of eggs were exposed to 40°C in a timed series 1 hr after fertilization. The lengths of exposures were 5, 15, 30, 45, and 60 min. The percentage hatch in the treated groups ranged from 25 to 43%, while the hatch in the control groups was 67%. The group exposed for 5 min had a hatch of 31% with 5 out of 13 embryos deformed. Although the number of eggs treated was small, this test showed that even as short a time as 5 min was harmful.

The use of 40°C as a test temperature was chosen because it was environmentally realistic. Such a temperature could easily be encountered in habitats such as tide pools. As the preliminary test showed, 40°C was high enough to show an effect, but not high enough to cause excessive rapid mortality.

Early Cleavage

The test treatment was given to groups of eggs at ½-hr intervals after fertilization. Table 1 shows that all stages were sensitive except the 1-hr stage. The highest initial death (59.7%) and the lowest overall percentage hatch (14.9%) occurred in those eggs treated 2 hrs after fertilization. There was little mortality 8 days after treatment, but the hatch from eggs that were viable at 8 days was only 38.5%. The hatch for handling controls from this group was 68.2%.

TABLE 1
Results of heat shock treatment to embryos of *F. heteroclitus* during early cleavage

| | Time of Treatment Postfertilization | | | | |
	½ hr	1 hr	1½ hrs	2 hrs	2½ hrs
Number of eggs	369	347	280	67	520
% initial death	26.3	4.6	36.8	59.7	46.4
% viable at 8 days	73.4	95.4	45.4	38.8	52.5
% overall hatch	40.1	91.4	17.8	14.9	45.8
% hatch of 8-day eggs	54.6	95.8	39.4	38.5	87.2
Number deformed	82	1	46	6	9
Peak hatching days	15-18	16	16	16	16

The embryos treated at ½ hr and 1½ hrs, when compared to the 2-hr group, showed less initial death, 26.3% and 36.8%, respectively, but showed a higher incidence of malformations. The total hatch of the ½-hr group was 40.1%, compared to 77.8% for the handling controls. The 1½-hr group had a hatch of 17.8%, compared to 62.7% and 66.7% for handling controls and viability controls, respectively.

Those embryos treated at 2½ hrs showed, as did the 2-hr group, a high initial death rate but a low number of deformities. This group had a higher percentage hatch from eggs viable at 8 days posttreatment and overall. The group treated at 1 hr did not appear to be affected by the treatment, although some effect was seen in the preliminary tests. The overall hatch was 91.4%, compared to 94.6% for the handling controls. There was only one deformity out of the 347 eggs treated.

The treatment also appeared to shorten developmental time slightly. The average developmental time in the treated groups was

16 days, while eggs in the control groups took an average of 19 to 20 days to hatch. The total percentage hatch for all five groups was 67.4% for controls and 72.8% for handling controls. The percentage deformities for all groups combined is significant. The handling controls had 1.7% deformed embryos, the controls had 0.0%, while the treated group had 9.1%.

The abnormalities observed ranged from slight to severe. Those that were slight involved the malformation of the fins. In some cases the pectoral fins were narrow and pointed, and the caudal fin rays had a curled or crumpled appearance. At the other extreme, some embryos were not recognizable as fish. They appeared as a mass of tissue without symmetry, but with a feeble heartbeat and circulatory system.

Almost all abnormal embryos were undersized and slower in their rates of development. Generally the abnormal embryos were too weak or incompletely formed to break the chorionic membrane. The abnormalities shown in Figure 1 are typical of those found in all of the stages treated.

The 64-Cell to Late Gastrula Stage

Stages 10, 12, 13, and 15 did not appear to be sensitive to the shock treatment (Table 2). Stages 17, 18, and 19 were increasingly more sensitive as closure of the blastopore approached. Initial death was low in Stage 17 (4.9%), but the percentage deformities was approximately 10%, compared to between 1% and 2% in the other stages. At Stages 18 and 19, which are just prior to closure of the blastopore, there was higher initial death, but no abnormal embryos. The percentage hatch in all groups was not markedly different from that of the control groups.

When embryos in Stages 18 and 19 were examined immediately after being treated, it was observed that the yolk material in some eggs had expanded and was protruding through the opening of the blastopore. Within 30 min the symmetry of the embryonic shield was lost, and it appeared as an amorphous mass of tissue. The eggs were dead 24 hrs later.

These groups of eggs, obtained from fish injected with human chorionic gonadotropin hormone to induce spawning, did not differ in viability from the naturally spawned eggs. The percentage hatch of controls for all groups combined was 75.7% and 69.8% for the handling controls. The percentage deformities occurring in the controls and handling controls was 1.5% and 1.2%, respectively, with 2.6% in the treated groups.

Later Developmental Stages

Embryos shocked at Stage 29, or 6 days, had no initial death and an overall hatch of 68.9%. There were no deformities, and the peak hatch was at 21 days. The handling control group for these embryos had a hatch of 84.4% and also a peak hatch at 21 days.

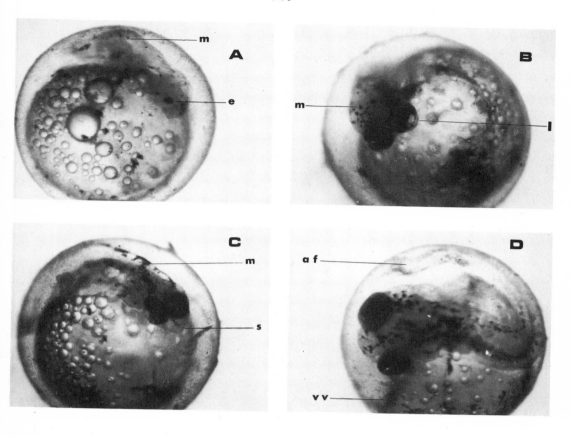

Fig. 1. Abnormal embryos of *F. heteroclitus* that were given a heat
shock at 1½ hrs postfertilization. A. The embryo in A was
one of the most severely deformed. There were only rudimentary eyes (e), almost no vitelline circulation, and no
melanophore formation (m). This embryo remained alive for
2 wks but never hatched. There was also no symmetry to the
body. The bubblelike structures are lipid droplets in the
yolk. B. The embryo in B showed further development in the
head region, but the posterior part of the body was a form-
less stump. The anterior part of the head was absent, and
the melanophores (m) were incompletely expanded. The eyes
were positioned at what is the front of the head, and the
lenses (l) protruded beyond the natural curvature of the
eyes. C. The embryo shown in C had a developed snout (s),
but it was too pointed. The posterior portion of the body
was incompletely formed as in B, but the melanophores (m)
showed some expansion. D. Embryo D showed more development
in the tail region, but there was curvature of the spine,
and the anal fin (af) was malformed. The eyes were in more
normal position but were unequal in size. The dark shadow
in the lower portion of the picture is the vitalline vein
(vv). In a normal embryo it would appear directly in front
of the head.

TABLE 2
Results of heat shock treatment to embryos of *F. heteroclitus* during development from 64-cell stage to closure of the blastophore

| | Stage at Time of Treatment | | | | | |
	10	12	13	15	17	18-19
Number of eggs	194	183	199	196	205	125
% initial death	7.7	8.2	7.7	8.7	4.9	16.0
% viable at 8 days	92.3	91.8	92.3	91.3	95.1	84.0
% overall hatch	78.4	65.6	67.8	84.7	63.4	70.4
% hatch of 8-day eggs	84.9	71.4	72.9	92.7	66.7	83.8
Number deformed	0	4	2	3	20	0
Peak hatching days	16-18	16	14	16	16-17	16-19

Another group of embryos just prior to hatching was given the usual 5-min shock treatment. In addition, two groups were held in the 40°C bath for 20 min and 3 hrs, respectively. The percentage hatch for the 5-min group was 50.7%; for the 20-min group it was 53.6%; and for the 3-hr group it was 36.0%. The number of embryos treated under each condition was 71, 28, and 25, respectively. Many of the remaining unhatched embryos appeared alive and normal, but failed to hatch and eventually died. Developmental time was slightly longer, taking 18 to 21 days (Table 3).

In a preliminary test with larvae, 41 fish were placed in 40°C for 3 hrs 15 min after hatching, then returned to 20°C water. There was initial loss of equilibrium by all fish from the heat shock, but some recovered. Twenty-four hrs after the test, 12 of the 41 were still alive. Adults exposed to 40°C were dead within 5 min.

DISCUSSION

The effects of thermal shock on the developing embryo can best be understood by examining the processes occurring in the egg at the time of treatment. The formation of the first and second polar body occurs at 3 and 4.5 to 5 min, respectively, postfertilization (Huver 1960); therefore, the maturation divisions are not involved in this study.

At $1\frac{1}{2}$ hrs postfertilization the protoplasm, which is distributed over the surface of the egg, "streams" and condenses in the submicropylar area (Armstrong and Child, 1965). This migration continues to some degree for $1\frac{1}{2}$ hrs. The eggs treated at $\frac{1}{2}$ hr showed a somewhat low initial death, but there were numerous deformities.

TABLE 3

Results of starvation death posthatch on heat shocked embryos of *F. heteroclitus*

	Hrs Postfertilization					Stage*							Prehatch Exposure†		
	½	1	1½	2	2½	10	12	13	15	17	18-19	29	5 min	20 min	3 hrs
Number of fish	148	147	106	137	139	35	36	36	36	48	26	27	21	8	9
Day of initial die-off	3	9	6	12	7	13	13	9	10	10	15	10	9	9	12
Day of 50% die-off	12	15	14	16	17	17	18	20	20	16	19	13	11	11	15
Day of 100% die-off	20	19	20	22	23	23	24	26	25	21	24	21	19	17	18

*Armstrong and Child, 1965.

†Period of exposure at 40 C.

Barnes (1953) found that eggs of *F. heteroclitus* had a sensitive period soon after fertilization. When eggs were centrifuged beginning from 10 to 90 min after fertilization, duplications occurred i centrifuged within the first 30 min. The deformities he found were true duplication, reduced and defective head structures, mesodermal deficiencies, and rudimentary "parasitic twin" structures. Althoug heat-shocked embryos showed none of the twinning phenomena, some of the deformities were similar to those reported by Barnes. Barnes suggested that "any agent capable of separating a presumptive organ zation center produces effects commensurate with the degree of redistribution of öoplasmic materials" (Barnes, 1953). The deformi ties noted in the present study could have been caused by a change in the rate of migration of the protoplasm.

At 1 hr postfertilization the 1-cell stage appears as a biconv lens-shaped structure. Since the eggs showed little initial death, high hatch, and low abnormalities, this stage is probably a relativ inactive period just prior to cleavage. The cytoplasmic materials had probably completed their migration, but the nucleus had not yet begun to form mitotic structures.

At 1½ hrs the protoplasm bulges above the curvature of the egg. Since each cleavage division took approximately 1 hr at 20°C the eggs at this time were beginning to undergo the mitotic processes, so that the first cleavage division was completed at 2 hrs 50 min.

Price (1940), Bergan (1960), and Vasetskii (1967) have shown that eggs are very sensitive during cleavage. Vasetskii said that mature eggs had a threshold exposure of 11 min, while eggs in cleavage had a threshold of 5 min when both were given a heat treat ment. He found that metaphase was most sensitive, especially at th second cleavage division.

Bergan's work with the blue gourami (1960) is very pertinent t this study. He found that the temperature interval in which mitoti blocking occurred was rather narrow. In his study, 39 to 40°C temporarily blocked mitosis, and morphological changes were slight. The mitotic process resumed immediately after the termination of th shock. At 43°C mitosis was blocked and no resumption of cell divisi occurred when the temperature shock was removed. At 40.5 to 42.5°C mitosis was blocked, and the mitotic structures were changed in various ways, but a resumption of mitosis generally took place; mortality of the eggs was not great. Bergan also found that at temperatures above 40°C, a 3 to 4 min shock blocked most mitotic stages, while a 4 to 7 min treatment caused blockage of all mitoti stages.

The results we have obtained on the effects of heat shock at or near the first cleavage could very well have resulted from mitot blocking. The optimum temperature for both species of fish was similar, and the temperature and duration used in this study were within the range used by Bergan. The differences shown between the 1½, 2, and 2½ hr stages are understandable for two reasons. First, since fertilization would not have occurred at exactly the same time for all eggs, they would not have undergone cleavage at the same time. Second, each stage in mitosis varies in its sensitivity to thermal shock, as has already been said. The percentage mortality and the extent of any deformities depend on the exact stage of mitosis of each egg when the heat shock is applied.

In Stages 10 through 19 many processes are occurring. There is flattening and expansion of the blastodisc in Stages 12 through 14. Thermal shock does not appear to be unduly harmful during these stages or at Stage 15, which is the beginning of gastrulation. Garside (1959) reported that if fish embryos were subjected to temporary severe stress in the early gastrula stage, twinning or head distortion or both occurred. The data from the present study show that middle and late gastrula are more sensitive than early gastrula.

The cause of death and deformity from thermal shock at Stage 17 was probably not the same as at early cleavage. Disruption of the induction processes probably accounts for the increased number of deformities over the previous four stages. Stages 18 and 19 occur just prior to closure of the blastopore. In Stage 18 the extra-embryonic ectoderm covers three-fourths of the surface of the yolk. At Stage 19 the blastopore is just a small opening, and the embryonic keel is prominent (Armstrong and Child, 1965). These two stages showed a high initial death over the previous five stages; in fact, it was twice as high, but there were no deformities recorded at this treatment time. The high initial death was caused by expansion of the yolk through the opening of the blastopore, causing rupture of the vitelline membrane. In addition, the combination of heat and mechanical injury destroyed the integrity of the embryonic shield. Those eggs observed immediately after treatment showed complete loss of symmetry of the embryonic shield and axis.

Rollefsen (1932, in Bonnet, 1939), Hayes and Armstrong (1942), Johnson and Brice (1953), and Hayes, Pelluet, and Gorham (1953) all reported high mortality rates in cod and salmon eggs when they were handled at the time of blastopore closure. *F. heteroclitus* eggs are not affected by handling to the degree that salmon eggs are. However, when high temperature was added, the vitelline membrane ruptured. Observations showed that those eggs in Stage 19 incurred more damage than those in Stage 18. Once the covering of the yolk is replaced by a layer of cells, stresses that would cause rupture are not effective. Tåning said that embryos tolerated change better after closure of the blastopore where the death rate was lower than in nearby stages.

When the embryo is six days old (Stage 29), many organ systems are functioning. Heat shock at this stage can be tolerated as shown by a hatch of 68.9% versus 84.4% for the handling controls. This compares favorably with those eggs treated at Stages 10 through 19. The main differences was that there was no initial death, and all of the eggs were still viable at 8 days. Thus, there is some development of thermal capacity by the fish at this time. Some of the viable embryos did not hatch, which would indicate there still was some negative response to the thermal insult.

Prehatched embryos gave similar results to those treated at Stage 29. There was no immediate death even in the group exposed for 3 hrs. The figures for percentage hatch are only for those fish that hatched successfully. Some appeared normal and remained alive for up to 10 days but never hatched. Price (1940) suggested that hatching enzymes are extremely sensitive to temperature. McCauley (1963) hypothesized that fish develop those mechanisms which enable them to resist high temperatures near or after hatching. If both of these suggestions are true, then the increased mortality reported by Kinne and Kinne (1962) and Bonnet (1939) could have resulted not from the temperature, but rather from failure of the fish to free itself from the egg.

McCauley's hypothesis is certainly valid concerning newly hatched larvae. Although our data are scanty, the capacity of newly hatched *Fundulus* for withstanding high temperature far exceeds that of the adults. This is an obvious ecological advantage, as a fish hatching out in tide pool where temperature may approach 40°C in the summer would be able to survive until the pool was flooded again on the incoming tide.

Ushakov (1968) said that the evidence on the heat resistance of cells and proteins indicates that at early stages of embryogenesis of multicellular animals, the cellular level of organization becomes "naked." The cells are subjected to the direct influence of stress and heat resistance of the majority of somatic cells increases as the animal develops. Changes in the cells' thermostabilities can no longer play a limiting role in regard to the resistant temperature adaptation of the whole organism. The specific level of cell thermostability is correlated with the reproductive temperature of the species and has an adaptive advantage. This interpretation would support the temperature tolerances of the embryos and larvae of *Fundulus* but would not explain the loss of tolerance in the adult.

Fundulus is well adapted to the environment where it is found. The eggs can stick to the *Spartina* stalks and remain as the tide ebb where the temperature and salinity in a tide pool or the marsh may reach 40°C and 40 o/oo; hours later, the temperature-salinity regime in the same pool may be 26°C and 2 o/oo if a summer thunderstorm occurs.

Fundulus embryos must be more heat resistant than adults as they cannot avoid a thermal stress. The fish does occupy a highly stressed environment and shows outstanding physiological lability while inhabiting this stressful but highly productive zone. The adult *Fundulus* can migrate into and out of the intertidal zone for feeding but can select an optimum or at least nonstressful thermal environment.

It is important to note that many species of fishes are not as eurythermal as *Fundulus*. In fact, Kinne and Kinne (1962) say that extreme conditions of temperature, salinity, and oxygen can induce developmental arrest. Such arrest may remain reversible if conditions are normalized within hours or a few days. Longer periods cause irreversible damage; they are "lethal." It now appears from this study and others that the time periods must be shortened from hours to minutes, and the developmental phase must be carefully defined, especially in light of a survey of 61 power reactor designs filed with the Atomic Energy Commission. This survey revealed that the average rise of condenser cooling water was 10.8°C. There are some electric generating plants anticipating increases ranging from 5.6 to 18°C (Coutant, 1970). The time for passage from intake head-work to the end of the discharge canal in tidal power stations in California ranges from 2 to 10 min (Adams, 1969). In the longer discharge canals the passage time may exceed 30 min (Hoss, personal communication). The impact of the effluents of these plants on the life stages of aquatic populations, especially embryos, could be highly significant.

SUMMARY

When a thermal shock of 40°C was applied for 5 min to the embryos of *F. heteroclitus*, death and malformations occurred. The effects of this thermal shock showed that several mechanisms were involved. At ½ hr postfertilization, disruption of the migrating protoplasm caused high mortality and numerous deformities. Blockage of mitosis at or near the first cleavage division also caused similar results. Mechanical injury and/or disruption of induction brought about deformities at Stage 17 and high initial death at Stages 18 and 19. The cause of death or failure to hatch of prehatched embryos is not fully understood at this time. These results are discussed as they relate to the use of water in the cooling systems of electrical generating plants.

ACKNOWLEDGMENTS

This work was supported by the Belle W. Baruch Foundation, Contribution #57, of the Belle W. Baruch Institute for Marine Biology and Coastal Research.

LITERATURE CITED

Adams, J. R. 1969. Ecological investigations around some thermal power stations in California tidal waters. Chesapeake Sci. 10: 145-54.

Alderdice, D. F. and C. R. Forrester. 1971. Effects of salinity, temperature, and dissolved oxygen on early development of the Pacific Cod. J. Fish. Res. Bd. Canada 28:883-902.

Anderson, J. F. and W. R. Horsfall. 1965. Thermal stress and anomalous development of mosquitoes (Diptera: Culcidae). V. Effects of temperature on embryogeny of *Aedes stimulans*. J. Exper. Zool. 158:211-21.

Armstrong, P. B. and J. S. Child. 1965. Stages of the normal development of *Fundulus heteroclitus*. Biol. Bull. 128:143-68.

Barnes, L. J. 1953. A further study of the effects of centrifugation and low temperature on the development of *Fundulus heteroclitus* (Abstr.). Biol. Bull. 105:370.

Battle, H. I. 1929. Effects of extreme temperatures and salinitie on the development of *Enchelyopus cimbrius* (L.). Contribs. Canadian Biol. and Fish. (n.s.) 5:109-92.

Bergan, P. 1960. On the blocking of mitosis by heat shock applied at different stages in the cleavage divisions of *Trichogaster trichopterus* var. *sumatranus* (Teleostei: Anabantidae). Nytt Mag. Zool. 9:37-121.

Bigelow, H. B. and W. C. Schroeder. 1953. Fishes of the Gulf of Maine. Fish. Bull. 53:1-577.

Blaxter, J. H. S. 1956. Herring rearing. II. The effects of temperature and other factors on development. Mar. Res. Dept. of Agri. and Fish. for Scot.

Bonnet, D. D. 1939. Mortality of the cod egg in relation to temperature. Biol. Bull. 76:428-41.

Brett, J. R. 1960. Thermal requirements of fish—three decades of study, 1940-1970. In: Biological Problems in Water Pollution Second Seminar, Robert A. Taft Sanit. Engng. Cent., Tech. Rep. W60-3:110-17.

Cain, R. 1973. The annual occurrence, abundance, and diversity of fishes in an intertidal creek. Master's thesis, University of South Carolina.

Coutant, C. C. 1970. Biological aspects of thermal pollution. I. Entrainment and discharge canal effects. CRC Critical Review in Environ. Control 1:341-81.

Forrester, C. R. 1964. Laboratory observations on embryonic development and larvae of the Pacific Cod (*Gadus macrocephalus* Tilesius). J. Fish. Res. Bd. Canada 21:9-16.

_____ and D. F. Alderdice. 1966. Effects of salinity and temperature on embryonic development of the Pacific Cod (*Gadus macrocephalus*). J. Fish. Res. Bd. Canada 23:319-40.

Frankhauser, G. and R. R. Humphrey. 1942. Induction to triploidy and haploidy in axolotl eggs by cold treatment. Biol. Bull. 83:367-74.

Gabriel, M. L. 1942. The effect of temperature on vertebral numbers in *Fundulus*. The Collecting Net 17:85-86.

_____. 1944. Factors affecting the number and form of vertebrae in *Fundulus heteroclitus*. J. Exp. Zool. 95:105-47.

Garside, E. F. 1959. Some effects of oxygen in relation to tempera ture on the development of Lake Trout embryos. Canadian J. Zool. 39:689-98.

_____. 1966. Effects of oxygen in relation to temperature on the development of embryos of Brook Trout and Rainbow Trout. J. Fish. Res. Bd. Canada 28:1121-34.

_____ and F. E. J. Fry. 1959. A possible relationship between yolk size and differentiation in trout embryos. Canadian J. Zool. 37:383-86.

Gray, J. 1928. The growth of fish. III. The effect of temperature on the development of the eggs of *Salmo fario*. J. Exp. Biol. 6:125-30.

Hayes, F. R. and F. H. Armstrong. 1942. Physical changes in the constituent parts of developing salmon eggs. Canadian J. Res. D 20:99-114.

_____, D. Pelluet, and E. Gorham. 1953. Some effects of temperature on the embryonic development of the salmon *(Salmo salar)*. Canadian J. Zool. 31:42-51.

Horsfall, W. R., J. F. Anderson, and R. A. Brust. 1964. Thermal stress and anomalous development of mosquitoes (Diptera: Culicidae). III. *Aedes sierrensis*. The Canadian Entomol. 96: 1369-72.

Hubbs, C. 1965. Developmental temperature tolerance and rates of four southern California fishes, *Fundulus parvipinnies*, *Atherinops affinis*, *Leuresthes tenuis*, and *Hypsoblennius* sp. Calif. Fish. and Game 51:113-22.

_____. 1966. Fertilization, initiation of cleavage, and developmental temperature tolerance of the cottid fish, *Clinocottus analis*. Copeia 1966:29-42.

_____, A. E. Peden, and M. M. Stevenson. 1969. The developmental rate of the Greenthroat Darter, *Etheostoma lepidum*. Am. Midl. Nat. 81:182-88.

_____, B. Sharp, and J. F. Schneider. 1971. Developmental rates of *Menidia audens* with notes on salt tolerance. Trans. of the Am. Fish. Soc. 100:603-10.

Hubbs, C. L. 1922. Variations in the number of vertebrae and other meristic characters of fishes correlated with temperature of water during development. Amer. Nat. 56:360-72.

Huver, C. W. 1960. The stage at fertilization of the egg of *Fundulus heteroclitus*. Biol. Bull. 119:320.

Itazawa, Y. 1959. Influence of temperature on the number of vertebrae in fish. Nature 183:1408-9.

Johnson, H. E. and R. F. Brice. 1953. Effects of transportation of green eggs, and of water temperature during incubation on the mortality of Chinook salmon. Prog. Fish-Cult. 15:104-8.

Kinne, O. and E. M. Kinne. 1962. Rates of development in embryos of a cyprinodont fish exposed to different temperature-salinity-oxygen combinations. Canadian J. Zool. 40:231-53.

Kowalska, A. 1957. On the influence of temperature on the embryonic development of the Brook Trout *(Salmo trutta* var. *fario* L.). Przegląd Zoologiczny 3:253-59.

Lasker, R. 1964. An experimental study of the effect of temperature on the incubation time, development, and growth of Pacific Sardine embryos and larvae. Copeia 1964:399-405.

Lindsey, C. C. 1954. Temperature-controlled meristic variation in the Paradise Fish, *Macropodus opercularis* L. <u>Canadian</u> <u>J</u>. <u>Zool</u> 32:87-98.

McCauley, R. W. 1963. Lethal temperatures of the developmental stages of the sea lamprey, *Petromyzoa marinus* L. <u>J</u>. <u>Fish</u>. <u>Res</u> <u>Bd</u>. <u>Canada</u> 20:483-90.

Merriman, D. 1935. The effect of temperature on the development of the eggs and larvae of the Cut-throat Trout (*Salmo clarkii clarkii* Richardson). <u>J</u>. <u>Exp</u>. <u>Biol</u>. 12:297-305.

Milkman, R. 1962. Temperature effects on day old *Drosophila* pupae <u>J</u>. <u>Gen</u>. <u>Physiol</u>. 45:777-99.

_____. 1967. Kinetic analysis of temperature adaptation in *Drosophila* pupae. In: <u>Molecular</u> <u>Mechanisms</u> <u>of</u> <u>Temperature</u> <u>Adaptations</u> (C. Ladd Prosser, ed.), pp. 147-62. Washington, D. C.: American Association for the Advancement of Science.

New, D. A. T. 1966. Fish. In: <u>New</u>, <u>The</u> <u>Culture</u> <u>of</u> <u>Vertebrate</u> <u>Embryos</u> (D. A. T. New, ed.), pp. 187-93. New York: Academic Press.

Piavis, G. W. 1961. Embryological stages in the sea lamprey and effects of temperature on development. <u>Fish</u>. <u>Bull</u>. <u>of</u> <u>the</u> <u>Fish</u> <u>and</u> <u>Wildlife</u> <u>Ser</u>. 61:111-43.

Price, J. W. 1940. Time-temperature relations in the incubation of the whitefish, *Coregonus clupeaformis* Mitchill. <u>J</u>. <u>Gen</u>. <u>Physiol</u>. 23:449-68.

Rollefsen, G. 1932. The susceptibility of cod eggs to external influences. <u>J</u>. <u>de</u> <u>Conseil</u> <u>perm</u>. <u>Internat</u>. <u>pour</u> <u>l'explor</u>. <u>de</u> <u>la</u> <u>Mer</u> 7:367-73.

Seymour, A. 1959. Effects of temperature upon the formation of vertebrae and fin rays in young Chinook salmon. <u>Trans</u>. <u>Am</u>. <u>Fish</u>. <u>Soc</u>. 88:59-69.

Stockard, C. R. 1921. Developmental rate and structural expressio An experimental study of twins, 'double monsters' and single deformities, and the interaction among embryonic organs during their origin and development. <u>Am</u>. <u>J</u>. <u>Anat</u>. 28:115-266.

Strawn, K. and C. Hubbs. 1956. Observations on stripping small fishes for experimental purposes. <u>Copeia</u> 2:114-16.

Tåning, A. V. 1950. Influence of environment on number of vertebrae in Teleostean fishes. <u>Nature</u> 165:28.

Ushakov, B. P. 1968. Cellular resistance adaptation to temperature and thermostability of somatic cells with special reference to marine animals. <u>Mar</u>. <u>Biol</u>. 1:153-60.

Vasetskii, S. G. 1967. Changes in the ploidy of sturgeon larvae induced by heat treatment of eggs at different stages of development. <u>Doklady</u> <u>Akademic</u> <u>Nauk</u> <u>SSSR</u> 172:1234-37.

Wilson, S. and C. Hubbs. 1972. Developmental rates and tolerances of the plains killifish, *Fundulus kansae*, and comparison with related fishes. <u>Texas</u> <u>J</u>. <u>Sci</u>. 23:371-79.

Worley, L. G. 1933. Development of the eggs of the mackerel at different constant temperatures. <u>J</u>. <u>Gen</u>. <u>Physiol</u>. 16:841-57.

Some physical and nutritional factors which affect the growth and setting of the larvae of the oyster, *Crassostrea virginica*, in the laboratory

J. L. Dupuy

Due to diseases, pollution, and natural catastrophes in the southeastern and Gulf states, the availability of seed oysters for planting by commercial operations has greatly decreased. In turn, this has increased the interest in the building of commercial oyster hatcheries designed to "efficiently" culture American oyster larvae. In addition, other organizations are also culturing oyster larvae for their research programs in order to obtain disease-resistant, fast growing, and well-shaped oysters through breeding selection. One of the prime parameters in successfully rearing large numbers of oyster larvae to metamorphosis is the types and the quantities of food used during the larval growth period.

Since Cole (1937) first demonstrated that pure cultures of naked flagellates could be used to produce growth of *Ostrea edulis* larvae under laboratory conditions, many other investigators (Bruce, Knight, and Parke, 1940; Davis, 1950, 1953; Walne, 1956, 1963, 1965, 1966; Davis and Guillard, 1958; and Ukeles and Sweeney, 1969) have shown that the best single foods to rear the larvae of *Ostrea edulis* and *Crassostrea virginica* to metamorphosis were the chlorophyte, *Pyramimonas grossi*, and the chrysophytes, *Isochrysis galbana* and *Monochrysis lutheri*. In contrast to the usage of single algal species as foods for growing oyster larvae, the Glancy Method and the Hidu Method (Hidu et al., 1969) have been used extensively and successfully by commercial hatcheries and some research operations. These methods were based on the utilization of selectively filtered or centrifuged bay water, allowing the smaller forms of phytoplankton to be grazed on by the oyster larvae in the static cultures. Mackie (1969) demonstrated that the larvae of *Crassostrea virginica*, from Delaware Bay littoral waters, selected phytoplankters 1 to 30μ in size. She further stated that as the oyster larvae grew in size from straight hinge to eyed larvae, the size of the phytoplankton they selected also increased. She cautioned, however, that there appeared to be a qualitative selection with respect to phytoplankter of comparable size and that in each size category, several species of algae were utilized by the oyster larvae. In addition, Davis and Guillard (1958) concluded that combinations of *Isochrysis galbana* and *Monochrysis lutheri* with *Plátymonas* sp. and *Dunaliella tertiolec* provided better growth of oyster larvae than did any of these foods singly.

The utilization of unialgal cultures of *Isochrysis galbana* and *Monochrysis lutheri*, singly or in combination, as a larval food to obtain consistent and rapid setting of the Chesapeake Bay oyster *Crassostrea virginica* on a year-round basis for experimental laboratory breeding has left much to be desired. During the three-year period (1968 through 1971), attempts to obtain set from genetic crosses for the breeding program resulted in long larval periods (18 to 25 days, but usually 25 days) and also inconsistent setting success. Experiments with inbreeding and sibling crosses from different genetic parent stock lines indicated that different geneti parent stock lines and individuals within parent stock lines varied in their ability to produce vigorous larvae, relative to growth rates, and to setting intensity. Furthermore, the inability to obtain consistent reproducible results from the bioassay of new algal species isolated from the York River estuary as a food for oyster larvae prompted the investigation of other parameters which could affect the growth and setting of *Crassostrea virginica* larvae.

Preliminary experimentation utilizing different sizes of larval containers indicated that depth and total volume affected the growth rate and setting success of oyster larvae. Walne (1958) and other investigators (Castagna, Hidu, and Powell, personal communications) also reported faster growth rates and better success in setting oyster larvae when utilizing larger larval culture containers with any given set of their experimental conditions. The purpose of this paper is, then, to report on: (1) the effects of different types of algal foods on the growth of larvae, and (2) the effects of the size of larval containers upon the growth of oyster larvae.

METHODS

Conditioning and Spawning

The methods used for conditioning and spawning oyster brood-stocks were those of Dupuy and Rivkin (1972). Briefly, selected broodstocks were conditioned in newly designed flumes. Each oyster received 10 1 of 18 o/oo raw river water per hour at 24°C, containing 0.2 ppm cornstarch and 5×10^6 cells of *Phaeodactylum tricornutum*. After 5 wks, these oysters were spawned in a small flume with running filtered (1μ Cuno Cotton Filter) river water at 30°C. Once spawning began, the males and females were individually placed in dishes to allow for controlled fertilization. The fertilized eggs were then put into 250-1 containers at 50 larvae per ml of filtered river water (18 to 19 o/oo at 27°C) until straight hinge larvae were observed (36 hrs). The larvae were then distributed to the 250-1 culture containers and the 4-1 containers at a concentration of 15 larvae per ml of 18 to 19 o/oo river water.

Larval Culture

A standard procedure of feeding algae every day and renewal with 1μ filtered river water (Cuno Cotton Filter) every third day was followed. All cultures were aerated and kept at 27°C with aquarium heaters. The fiber glass culture tanks used were 130 cm deep by 55 cm in diameter, with a conical-shaped bottom. The fiber glass culture tanks were coated with a gel coat. Samples of larvae were collected every 2 days and preserved in neutralized formalin.

Measurements of the longest dimension of the larvae for two sets of bioassay and spatial experiments reported on were accomplished by the use of a πMC Particle Analyzer (Millipore Filter Corporation). For each sample, 100 larvae chosen at random were measured individually. Recalibration of the instrument for the longest dimension was accomplished by the use of a special Millipore slide containing squares of known sizes ranging from 5μ to 500μ. Maximum error of the πMC Particle Analyzer, as checked by the use of a filar ocular micrometer for measurement of larvae, ranged no greater than 3%.

In the final series of experiments, the end point for bioassay was chosen arbitrarily when 30% of the population of an experimental set (one larval container) reached the eyed larval stage (290μ). Observations of larval size each day were made to determine approximately the general size distribution of that particular population and that point where eyed larvae appeared (ca. 30%) in each culture container.

Algal Culture

Four new algal species were isolated and cultured from the York River estuary. They were two species of Chlorophyceae, *Nannochloris oculata* (Fig. 1) and *Pyramimonas virginica* (Fig. 2), and two species of Chrysophyceae, *Pseudoisochrysis paradoxa* (Fig. 3) and *Chrysophaeropsis planktonicus* (Fig. 4). The Chrysophytes *Monochrysis lutheri* and *Isochrysis galbana* were cultured as the standard algal food for all experimental controls.

All algal species were grown in heat-sterilized, enriched seawater medium in 18- and 40-1 Pyrex carboys in semicontinuous and batch unialgal cultures (not bacteria-free).

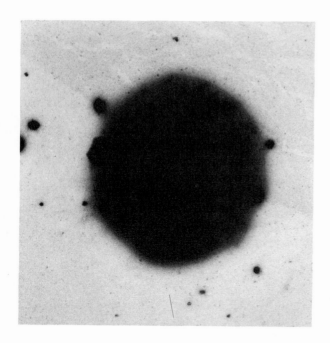

Fig. 1. *Nannochloris oculata* (1 to 2μ), 33,750x.

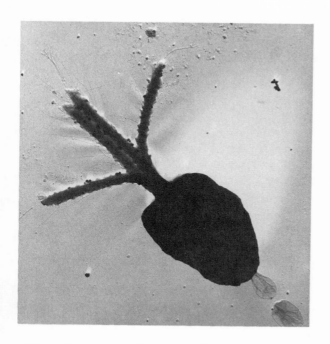

Fig. 2. *Pyramimonas virginica* nom. prov. (3 to 4μ), 23,750x.

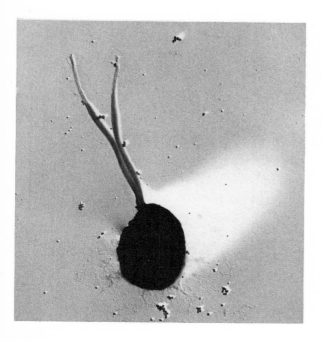

Fig. 3. *Pseudoisochrysis paradoxa* nom. prov. (4 to 6μ), 18,750x.

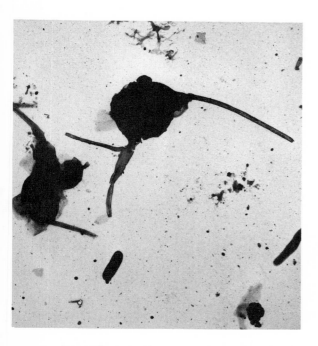

Fig. 4. *Chrysophaeropsis planktonicus* nom. prov. (8 to 10μ), 6,250x.

The preparation of the single stock solution for the N_2M enrichment of the seawater requires seven basic stock solutions. Their preparation is as follows:

(1) Sodium Silicate Solution

$Na_2SiO_3.9H_2O$	4.66 g
Dist. H_2O	to 100 ml

(2) Ketchum and Redfield's Solution "A"

KNO_3	20.2 g
Dist. H_2O	to 100 ml

(3) Ketchum and Redfield's Solution "B"

$Na_2HPO_4.7H_2O$	3.0 g
$CaCl_2.2H_2O$	2.7 g
$MgSO_4$ (anhyd.)	2.9 g
$FePO_4$	0.5 g
HCl (conc.)	2.0 ml
Dist. H_2O	to 100 ml

(4) Sodium Molybdate Solution

$Na_2MoO_4.2H_2O$	0.0119 g
Dist. H_2O	to 100 ml

(5) Arnon's Micronutrient Solution (mod.)

H_3BO_3	0.286 g
$MnCl_2.4H_2O$	0.181 g
$ZnSO_4.7H_2O$	0.0222 g
$CuSO_4.5H_2O$	0.0079 g
$CoCl_4.6H_2O$	0.004 g
Dist. H_2O	to 100 ml

(6) Soil Extract Solution

This solution must be made up immediately before mixing into the N_2M stock solution. Mix 1 kg of top soil with 2 l of distilled water and sterilize in a large flask 1 hr at 15 lbs pressure. Allow to settle for several days, decant the liquid, pass it through a layer of cotton, then filter through a Whatman No. 1 filter paper. The soil extract is now ready to use.

(7) Vitamin Mix

Thiamine HCl	200 mg
Biotin	1 mg
B_{12}	1 mg
Dist. H_2O	to 100 ml

This enrichment is added to the filtered seawater via a single stock solution, the composition of which is as follows:

Sodium Silicate Solution	100 ml
Ketchum and Redfield's Solution "A" . . .	200 ml
Ketchum and Redfield's Solution "B" . . .	100 ml
Sodium Molybdate Solution	50 ml
Arnon's Micronutrient Solution (mod.) . .	50 ml
Soil Extract Solution	200 ml
Vitamin Mix	175 ml

This enrichment is dispensed into bottles, plugged, sterilized, and stored at 0-5°C. It is used at the rate of 2 ml/l of filtered seawater.

Withdrawal for each species was accomplished daily with the increments from each culture carboy for the one species being combined and filtered through a 50μ stainless steel filter screen. The cell counts were obtained by using a hemocytometer. Total cell volume for each species, and for any combination fed, was then calculated.

RESULTS AND DISCUSSION

Of the thirty-one experiments performed, only two will be described in any detail. Until 1970, many experiments involving the bioassay of new species of algae as food for American oyster larvae appeared to be unsuccessful. Preliminary trials with different sizes of culture containers indicated that the size of the container did affect the growth of oyster larvae.

In addition, since this laboratory had not been having consistent results by using *Monochrysis lutheri* and *Isochrysis galbana* as the standard food for all breeding experiments to obtain disease-resistant oysters, preliminary observations on the selective grazing by the oyster larvae in relation to the size of the cells within the populations of *I. galbana* and *M. lutheri*, which they were fed, were initiated. Observations during these preliminary trials indicated that the oyster larvae (straight hinge to early umbo) fed primarily on the smaller sized cells of the two species of algae used as food. Furthermore, the utilization of this algal culture (*M. lutheri* and *I. galbana*) in different growth phases as a food indicated that mass algal cultures which were 3 to 4 wks old (semicontinuous culture) · yielded a smaller proportion of smaller algal cells (3 to 4μ) than the newly inoculated mass cultures (1 to 2 wks old). Slower growth rates or a longer larval period to the eyed stage were observed when utilizing older mass cultures.

In one of the bioassay trials, two types of larval culture containers were used to compare the growth rates of oyster larvae utilizing a combination of *M. lutheri* and *I. galbana*. The 250-1 culture containers received a total combination cell volume of $1,800μ^3 \times 10^9$ per 3.5×10^6 oyster larvae per day (Table 1). The 4-1 culture containers received a proportional total combination cell volume per 0.06×10^6 larvae per day. The experiment was terminated in this case when the first eyed larvae were observed in the 250-1 culture containers. Each point in Figure 5 represents the mean for 200 measurements made from the replicate sets, at that particular time, from the two types of culture containers. The results indicate that the first eyed larvae were observed at 18 days in the 250-1 culture containers though the mean size of the population was only 225μ. In contrast, those larvae in the 4-1 culture

TABLE 1
Sequential feeding protocol (per 3.5 x 10.6^3 oyster larvae per day)

Larval Size Range	Species	Total Cell Volume Fed	Total Combination Cell Volume Fed
70μ to 105μ	*Nannochloris oculata*	588μ^3 x 10^9	
	Pyramimonas virginica	332μ^3 x 10^9	= 920μ^3 x 10^9
105μ to 150μ	*Pyramimonas virginica*	581μ^3 x 10^9	
	Nannochloris oculata	368μ^3 x 10^9	= 949μ^3 x 10^9
140μ to 250μ	*Pyramimonas virginica*	596μ^3 x 10^9	
	Nannochloris oculata	318μ^3 x 10^9	
	Pseudoisochrysis paradoxa	1,284μ^3 x 10^9	= 2,198μ x 10^9
250μ to pediveliger larvae	*Pyramimonas virginica*	298μ^3 x 10^9	
	Pseudoisochrysis paradoxa	963μ^3 x 10^9	
	Chrysophaeropsis planktonicus	1,512μ^3 x 10^9	= 2,933μ^3 x 10^9

Standard and control feeding protocol (per 3.5 x 10^6 oyster larvae per day)

70μ to pediveliger	*Monochrysis lutheri*	900μ^3 x 10^9	
	Isochrysis galbana	900μ^3 x 10^9	= 1,800μ^3 x 10^9

Note: (1) Total single species cell volume fed, is calculated from the volume of a single cell of: *Nannochloris oculata*, 4.2μ^3; *Pyramimonas virginica*, 36μ^3; *Pseudoisochrysis paradoxa*, 45μ^3; *Chrysophaeropsis planktonicus*, 140μ^3; *Monochrysis lutheri*, 115μ^3; *Isochrysis galbana*, 115μ^3.

(2) Total single species cell volume fed = volume of one cell (μ^3) x number of cells/ml of culture x number of ml fed.

beakers only reached a mean size of 152μ, with the largest oyster larvae measuring 176μ. The utilization of 420-1 culture containers for rearing five different larval broods yielded eyed larvae in 15 days when fed the same combination of *M. lutheri* and *I. galbana* (3,100μ^3 × 10^9 total cell combination volume) per 6.3 × 10^6 larvae per day.

Fig. 5. Differential growth of oyster larvae (*C. virginica*) in two types of culture containers.

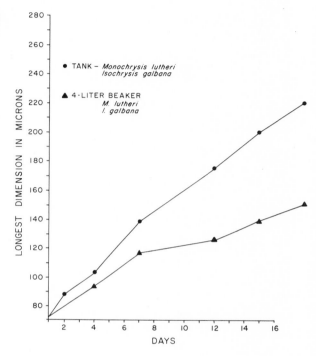

The second series of experiments (the results of one are shown in Figure 6) were designed primarily to establish the advantage, if any, of utilizing three new species, *Pyramimonas virginica*, *Nannochloris oculata*, and *Pseudoisochrysis paradoxa*, in various combinations with the standard foods, *Monochrysis lutheri* and/or *Isochrysis galbana*. With the exception of *Pseudoisochrysis paradoxa*, the new species utilized in addition to *M. lutheri* and *I. galbana* as food for the larvae in the 250-1 culture containers ranged from 1 to 4μ in size. The replicate combinations and the control or standard (*M. lutheri* and *I. galbana*) were fed a total combination cell volume of 1,800μ^3 × 10^9 for each experimental and control set (culture container of 3.5 × 10^6 larvae per day). The results indicate that, regardless of the combination algal food for American oyster larvae (addition of one or two new species to *M. lutheri* and/or *I. galbana*), growth rates are increased to yield eyed larvae in 15 days, whereas *M. lutheri* and *I. galbana* alone or in combination did not. It is interesting to note that major differential growth occurred during the first two days (from straight hinge to early umbo) when contrasting the experimental combinations with the standard foods. Although the combination of *M. lutheri*, *P.*

virginica, and *N. oculata* yielded 41% eyed larvae in 15 days compared to other experimental combinations, no conclusion can be rendered as to the optimal combination needed to obtain optimal growth and eyed larvae within the experimental combination presented.

Fig. 6. Differential growth of oyster larvae *(C. virginica)* fed five different combinations of algae.

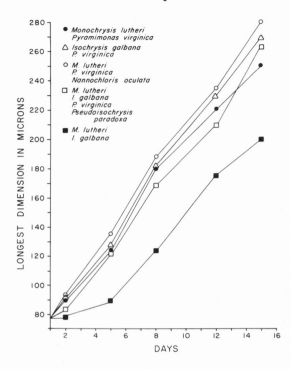

The last series of experiments (Tables 1 and 2) reflect the sequential feeding protocol that has been established to yield eyed larvae (ca. 30% of the larval population) within a 11- to 14-day period. It is important to note that there is a gradation or increase in the size of algal cells fed as well as the total combination cell volume fed. Table 1 is presented as a guide for laboratory use where an additional algal species, *Chrysophaeropsis planktonicus*, is utilized from the 250µ to the pediveliger larvae stage. *Chrysophaeropsis planktonicus* appeared to yield better growth from 250µ to the pediveliger larvae. However, the erratic growth and culture of this algal species precluded its consistent utilization as larval food. Table 2 represents a sequential protocol with algal species that can be consistently grown in standard enrichment media and still yield eyed larvae (30% of larval population) in 12 to 14 days. It is quite apparent that further bioassay is necessary to establish optimal algal food concentrations during the 70µ to 105µ larval stages since the work of Ukeles and Sweeney (1969) indicates much lower cell concentrations and cell volumes to obtain optimal growth and yields for the setting of these American oyster larvae.

Only two factors, space and nutrition, have been briefly considered in the culture of the larvae of the American oyster, *Crassostrea virginica.*

TABLE 2
Sequential feeding protocol (per 3.5×10^6 oyster larvae per day)

Larval Size Range	Species	Total Cell Volume Fed	Total Combination Cell Volume Fed
70µ to 105µ	*Nannochloris oculata*	$588µ^3 \times 10^9$	
	Pyramimonas virginica	$332µ^3 \times 10^9$	$= 920µ^3 \times 10^9$
105µ to 150µ	*Pyramimonas virginica*	$581µ^3 \times 10^9$	
	Nannochloris oculata	$368µ^3 \times 10^9$	$= 949µ^3 \times 10^9$
140µ to 250µ	*Pyramimonas virginica*	$596µ^3 \times 10^9$	
	Nannochloris oculata	$318µ^3 \times 10^9$	
	Pseudoisochrysis paradoxa	$1,284µ^3 \times 10^9$	$= 2,198µ^3 \times 10^9$
250µ to pediveliger larvae	*Pyramimonas virginica*	$900µ^3 \times 10^9$	
	Pseudoisochrysis paradoxa	$2,000µ^3 \times 10^9$	$= 2,900µ^3 \times 10^9$

Note: (1) Total single species cell volume fed is calculated from the volume of a single cell of: *Nannochloris oculata*, $4.2µ^3$; *Pyramimonas virginica*, $36µ^3$; *Pseudoisochrysis paradoxa*, $45µ^3$.

(2) Total single species cell volume fed = volume of one cell $(µ^3)$ x number of cells/ml of culture x number of ml fed.

Major factors such as the genetic and environmental parameters, as they affect the vigor of larvae and, in retrospect, gamete quality, have not been considered. Results from breeding experiments with wild populations and inbred lines indicate that the interrelated genetic and environmental parameters of parent broodstocks as they affect gamete quality, larvae vigor, and, finally, setting intensity may camouflage the nutritional bioassay of oyster larvae.

SUMMARY

1. Four new isolated algal species from the York River Estuary, *Nannochloris oculata*, *Pyramimonas virginica*, *Pseudoisochrysis paradoxa*, and *Chrysophaeropsis planktonicus* when fed in combination as a food, yielded eyed oyster larvae (290μ) in 11 to 14 days.
2. *Nannochloris oculata*, *P. virginica*, and *P. paradoxa*, when fed in combination singly with *Monochrysis lutheri* and/or *Isochrysis galbana* as a larval food, yielded eyed larvae in 15 to 16 days.
3. The growth of oyster larvae was affected by the size of the larval culture container with an increase in growth of oyster larvae when the size of the culture container was increased (from 4 1 to 420 1).

ACKNOWLEDGMENTS

I wish to thank Dr. Franklyn Ott, who provided the phytoplankton; Dr. Frank Perkins and Dr. Fred Kazama, who provided the electron photomicrographs of the new algal isolates; and Mr. Samuel Rivkin and Miss Nancy Troneck, who helped with the culture and measurements of the oyster larvae.

We gratefully acknowledge support provided by the Department of Commerce under Sea Grant P. L. 89-688, Grant GH-67.

This publication is Contribution No. 532, Virginia Institute of Marine Science, Gloucester Point, Virginia.

LITERATURE CITED

Bruce, J. R., M. Knight, and M. W. Parke. 1940. The rearing of oyster larvae on an algal diet. J. Mar. Biol. Assoc. U. K. 24:337-74.

Cole, H. A. 1937. Experiments in the breeding of oysters (*Ostrea edulis*) in tanks, with special reference to the food of the larva and spat. Fish. Invest., London (Series 2) 15:1-28.

Davis, H. C. 1950. On the food requirements of larvae of *Crassostrea virginica*. Anat. Rec. 108:132-33.

_____. 1953. On food and feeding of larvae of the American oyster, *C. virginica*. Biol. Bull. 104:334-50.

_____ and R. R. Guillard. 1958. Relative value of ten genera of micro-organisms as foods for oyster and clam larvae. U. S. Fish and Wildlife Service, Fish. Bull. 58:293-304.

Dupuy, J. L. and S. Rivkin. 1972. The development of laboratory techniques for the production of cultch-free spat of the oyster, *Crassostrea virginica*. Chesapeake Science 13:45-52.

Hidu, H., K. G. Drobeck, E. A. Dunnington, Jr., W. H. Roosenburg, and R. L. Beckett. 1969. Oyster hatcheries for the Chesapeake Bay region. NRI Special Report No. 2 Natural Resources Institute, University of Maryland Contr. No. 382.

Mackie, G. 1969. Quantitative studies of feeding in the oyster, *Crassostrea virginica*. Nat. Shellfish. Assoc. Proc. 59:6-7.

Ukeles, R. and B. M. Sweeney. 1969. Influence of dinoflagellate trichocysts and other factors on the feeding of *Crassostrea virginica* larvae on *Monochrysis lutheri*. Limnol. Oceanogr. 14:403-10.

Walne, P. R. 1956. Experimental rearing of the larvae of *Ostrea edulis* L. in the laboratory. Fish. Invest., London (Series 2) 20:1-23.

_____. 1958. The importance of bacteria in laboratory experiments on the rearing of the larvae of *Ostrea edulis*. J. Mar. Assoc. U. K. 37:415-26.

_____. 1963. Observations on the food value of seven species of algae to the larvae of *Ostrea edulis* L. I. Feeding experiments. J. Mar. Biol. Assoc. U. K. 43:767-84.

_____. 1965. Observations on the influence of food supply and temperature on the feeding and growth of the larvae of *Ostrea edulis* L. Fish. Invest., London (Series 2) 24:1-45.

_____. 1966. Experiments in the large scale culture of the larvae of *Ostrea edulis* L. Fish. Invest., London (Series 2) 25:1-53.

Physiological and behavioral evidence for color discrimination by fiddler crabs (Brachyura, Ocypodidae, genus *Uca*)

G. W. Hyatt

Decapod Crustacea show varied behavioral responses toward visual stimuli under natural and laboratory conditions (see Wulff, 1956, and Waterman, 1961, for reviews). The ability to discriminate colors seems to be common among the Crustacea. Generally, investigators have attributed the behavioral and ecological relevance of color discrimination by Crustacea to cryptic coloration (Koller, 1927), food acquisition (e.g., "color dances" by *Daphnia*, Smith and Baylor, 1953), or perhaps to shelter identification (Koller, 1928). In addition to wavelength, the other physical parameters of light, viz., intensity (Schöne, 1961) and polarization plane (Horridge, 1967; Waterman and Horch, 1966) are discriminated by representative decapod species. Shape and form can be identified by *Pagurus* (leg and claw models, Hazlett, 1969, 1972) and by *Ocypode saratan* (sand "pyramid" landmarks, Linsenmair, 1967). *Goniopsis cruentata* demonstrates a time-compensated celestial compass which enables it to orient itself perpendicular to its home beach (Linsenmair, 1967). Thus, color, intensity, polarization plane, shape, and celestial cues all provide decapods with usable visual information. The basic characteristic of crustacean compound eyes is truly one of "cellular parsimony with functional versatility" (Waterman, 1961).

The visual sense of fiddler crabs (genus *Uca*) may be quite germane to their overt behavior and survival, especially during their reproductive activities. The striking coloration of many species, especially tropical ones, suggests that color vision might play an important role in conspecific identification and the further

stimulation of courtship or other behaviors. Crane (1944) lists
five pigments (black, red, yellow, white, and a nonchromatophoric
blue) that contribute to coloration in *Uca*. She identifies four
conditions of color changes demonstrated by fiddler crabs in the
field: (1) diurnal darkening, (2) submergence darkening,
(3) darkening upon capture, and (4) display brightening during
which some species bleached to a "dazzling" white. Zucker (personal
communication) has observed that during display, the major cheliped
and ambulatory legs of *U. terpsichores* become a vivid purple. She
also suggests that the elevated shelters built by this species may
act as dark backgrounds against which a bleached, displaying male
might further emphasize his presence.

Uca males possess a single enlarged (major) cheliped which is
used during the daytime in a courtship display ("waving") as well
as in ritualized combat (Crane, 1941, 1966, 1967). Females do not
wave and both their claws are small. Waving appears to attract
conspecific females to the male's burrow where mating occurs. All
male fiddlers show a species-specific temporal pattern in their
waving displays. Crane (1941) was the first to describe in detail
the waving behavior of a series of tropical, and later (1943),
temperate fiddlers.

Despite the obvious relevance of the visual sense to fiddler
crabs (Crane, 1957, 1967; Salmon, 1965; Salmon and Atsaides, 1968;
Langdon, 1971; Herrnkind, 1968b, 1972), little has been done to
test their responses to light. Holmes (1908) described a positive
phototaxis to strong light by *U. pugnax*. The tactic response broke
down when the eyestalks were fashioned into an "X" by crossing and
tying them with thread, or when the animals were frightened by
sudden movement. Schwartz and Safir (1915) tested *U. pugilator* and
U. pugnax in a box "maze" and found a persistent preference for a
particular corner. They concluded that sight and touch were the
senses used in learning, with the former considered more important.

Dembowski (1926) studied the burrowing behavior of *U. pugilator*
and tried to relate phototaxis and thigmotaxis to the position and
time of burrowing. He concluded that neither was sufficient
explanation and that digging was initiated by many factors.
Matthews (1930), after making field observations of *U. leptodactyla*,
stated, "The crabs are very alert; a shadow falling across the
colony causes them all to bolt into their burrows. . . ." Clark
(1935) used optomotor methods to test visual acuity in *U. pugnax*.
He derived an acuity value of 0.0042 in the laboratory, with values
derived in the field being of the same order of magnitude.

Salmon and Stout (1962), using dead crabs, and von Hagen (1961)
using various models, demonstrated that the presence of a major
cheliped is the main stimulus for sexual discrimination by male
U. pugilator and *U. tangeri*, respectively. When the major cheliped
is lacking, the model elicits courtship display from the male;
otherwise, aggressive behavior results. Males of each species
demonstrate a very characteristic pattern of courtship display with
regard to color, type, and rate of claw waving. However, except

for *U. pugnax* (Aspey, 1971), they apparently do not discriminate between different species of females. Presumably, then, it is the females which respond to the species-typical patterns of their conspecific males (Crane, 1941; Hartnoll, 1969).

Altevogt (1963) tested the visual discrimination ability of *U. tangeri*. The choices were: light of different intensity, of different polarization plane, and of different colors (648 nm red, 547 nm yellow, 546 nm green, 430 nm blue). Of the 23 crabs trained, 5 learned to discriminate relative brightness in 50 to 75 trials over 3 to 5 days (77 to 99% criterion). Training to polarized light was unsuccessful for 3 males (6 days, 124 trials per animal). A series of 719 choices by 6 crabs over 29 days did not show discrimination between red/green and blue/yellow pairs of equal intensity. He cautions, however, that this result should not lead one to conclude categorically that color vision is absent in *Uca*. He feels that brightness differences may supercede color differences in *Uca* visual discrimination, but that color could play an integral role.

Altevogt and von Hagen (1964) and Korte (1965) demonstrated polarized light perception by *U. tangeri*. Korte (1966) used an optomotor apparatus to test responses under various light intensities and eye positions. By using colored stripes in place of black ones on the inner surface of the optomotor cylinder, Korte was able to demonstrate that *U. tangeri* could detect red, blue, green, and possibly ultraviolet wavelengths.

Palmer (1964) used an automatic activity recorder to monitor the photic responses of *U. pugnax*. He found a persistent light preference rhythm with maximal attraction in the early morning hours. The response turned to dark preference slightly after 6:00 P.M. Herrnkind (1968a) demonstrated a positive phototaxis for stages II, III, and IV zoeae of *U. pugilator*. However, they became negatively phototactic when the light intensity was reduced.

More recently, Langdon (1971 and personal communication) has investigated visual shape discrimination and learning by *U. pugilator*. In both laboratory and field experiments he has demonstrated behaviors toward particular shapes that could be involved in the identification of three classes of significant stimuli: (1) conspecifics, (2) harborage elements, and (3) predators. Field experiments with real crabs and models reaffirmed that the major cheliped is the releaser for aggressive behavior, and that movement of a female model is a necessary factor for eliciting male courtship behavior. In general, objects with curved outlines, or with low perimeter-squared to area ratios (P^2:A) were avoided; those with high P^2:A ratios were approached. Vertical rectangles were preferred over all other shapes tested, with the suggestion that the generally vertical grasses (*Spartina* spp.) and trees which fiddlers approach for shelter in the wild may lend ecological relevance to this behavior. Field and laboratory experiments with bird-shaped, ellipsoid, and spherical models showed aversion to all three. Avoidance of the sphere and ellipsoid was thought to be due to the resemblance of these shapes, respectively, to a head-on or

quartering predatory bird. Training experiments suggested a learned
preference for vertical, vegetationlike stimulus objects, modifiable
to adapt to changing environmental characteristics. Young crabs
learned a new preference faster than adults, indicating the possibility that preferences are established most easily when the animals
are first introduced to the habitat.

The specific goals of the present study were: (1) to define
the spectral sensitivities of the eyes of Uca in physiological terms
(2) to use the technique of selective adaptation to test for a
multiple visual pigment system that would, if found, suggest color
vision; and (3) to carry out tests for behavioral evidence of color
discrimination.

MATERIALS AND METHODS

General

Female animals were used in this research, since it was
assumed that females respond to the species-specific displays of
their conspecific males as outlined by Crane (1941) and Hartnoll
(1969). Mature female Uca pugilator, U. minax, and U. pugnax
were collected at the Duke University Marine Laboratory (DUML),
Beaufort, North Carolina, and U. pugilator were collected near
Flagler Beach and Steinhatchee, Florida. The animals were kept at
DUML in Fiberglas tanks (66 × 119 × 22 cm) with running natural
seawater at a depth of 1 cm over a shallow layer of sand. At
Illinois, these animals were maintained in aquaria (59 × 72 × 22 cm)
with a sand-shell-brick substrate under 1 cm of artificial seawater
(Rila Marine Mix, 25 ppt) filtered through sand, charcoal, and
glass wool. Laboratory temperatures were 22-24°C at DUML and
Illinois. The lighting at DUML was natural (June-September), through
windows in the animal room, while the photoperiod at Illinois was
maintained at 14 L: 10 D using 40-watt "Naturescent" fluorescent
tubes 32 cm above the water surface. The animals were fed daily
with a variety of foods. Oatmeal, chopped fresh shrimp, chopped
raw liver, and fresh fish or a prepared fish food (Tetramin flakes)
were accepted readily. Algae, growing naturally in the tanks, were
also consumed. The artificial seawater in the Illinois aquaria was
changed at irregular intervals, and animals thrived in the tanks
for over a year. Experiments were performed from 1970 through 1972.

Electrophysiological Experiments

Recording and stimulation. Crabs were dark-adapted in a bowl of
seawater 2 to 24 hrs. Before testing, they were wrapped loosely in
cheesecloth under dim light so that the left eye remained upright
through the weave. The wrapped animal was placed in a Plexiglas
holder containing a pool of seawater. A sheet of rubber dam with

an eyehole was placed over the animals and held down with rubber bands. A small slit was then made laterally in the cornea with a fine dissecting needle.

All recording was done with glass pipette electrodes filled with seawater and broken off to 40 to 60μ tip diameters. The pipette was inserted into the slit in the cornea and connected via an Ag-AgCl wire to the input of a Tektronix (Type 122) preamplifier (0.2 and 50.0 Hz bandpass settings). A ground wire was placed in the seawater pool. The electroretinogram (ERG) signals were displayed on an oscilloscope (d - c = coupled) and their deflections in millivolts recorded directly from the screen by eye or photographed with a Polaroid camera.

Light stimuli were generated by a Bausch and Lomb (Model 33-86-02) grating monochromator (350 to 700 nm) using a 150-watt xenon arc lamp. The output beam was focused on the eye with a series of quartz lenses. Intensity was modulated with Kodak #96 Wratten neutral density (N.D.) filters individually calibrated with a Yellow Springs Instruments radiometer (Model 65, with 6551 probe, reported to have linear output from 280 to 2600 nm) at wavelengths below 450 nm. Visible wavelengths were blocked with a Corning 5970 filter when wavelengths from 350 to 490 nm were presented. Stimuli were given every 10 nm from 350 to 490 nm and every 20 nm from 500 to 640 nm. Flashes (0.1 sec) were controlled with a Compur shutter operated by a solenoid. One minute was allowed for recovery between flashes.

Dark-adaptation time. *U. pugilator* were prepared for ERG recording and dark-adapted for 2 hrs. The dark-adapted ERG amplitude was determined using 0.1 sec flashes of 470 nm light (2.2 mw/cm^2). The eye was then light-adapted for 0.5, 5.0, 10.0, or 15.0 min (four separate animals) using a microscope illuminator focused so that all the ommatidia were flooded with white light (180 mw/cm^2). The ERG's produced by 0.1 sec flashes of 470 nm light were measured at the end of the adaptation period with the adapting lamp still on, and then at the following times after the light was turned off: once per minute for the first 6 min, once every 2 min for 7 to 20 min, and once every 5 min for 25 to 60 min. The percentage of dark-adapted response achieved was then plotted against time in the dark.

Dark-adapted spectral sensitivity. Crabs were prepared for ERG recording as above and dark-adapted. Five flash intensities were presented at 0.3 log unit increments and a stimulus-response curve drawn for each test wavelength. A 1.0 millivolt (mv) criterion response (below saturation within the range of test flashes) was interpolated from each curve and the logarithm of the relative energy necessary to produce this response was plotted against wavelength to yield a spectral sensitivity curve. Twelve *U. pugilator*, six *U. pugnax*, and six *U. minax* were tested in this manner. The mean and standard error were calculated for the relative energies at each wavelength.

Light-adapted spectral sensitivity. Sensitivity curves were obtained
for *U. pugilator* both in the dark- and light-adapted states. Light
adaptation was achieved using a microscope illuminator and a Kodak
#21 Wratten filter (transmission > 540 nm; yellow-orange), or a
GE G5S11 ultraviolet lamp with an output peak near 365 nm in conjunction with a Kodak #18A filter (near-UV-adapted). After allowing
30 to 60 min for light adaptation, ERG data were recorded as above.
Four test animals were yellow-orange adapted, and six were UV-
adapted. The means and standard errors were calculated as above.

Behavioral Experiments

Equipment, control measures, and procedures. All behavioral results
described below are based upon experiments performed during the
summer of 1971 and the spring of 1972 at DUML, using freshly caught
female *U. pugilator*. The mean carapace width of a random sample of
57 animals was 13.9 ± 1.8 mm SD; median = 13.5 mm; range = 10.0 to
20.0 mm.
    A circular choice chamber (Fig. 1), made of ¼" thick black and
white Plexiglas with a central start chamber and tightly fitting
cover, was used to present paired light stimuli to single animals.
The chamber was 19.4 cm in diameter by 7.8 cm high. The start chamber was 3.2 cm in diameter and 3.8 cm deep, with a plunger inserted
through the bottom to elevate the animal to the level of the chamber

Fig. 1. View of the
    choice chamber
    apparatus without the
    cover. Individual
    crabs were placed in
    the start chamber. A
    trial began when the
    crab was elevated to
    the floor level and
    simultaneously exposed
    to two light sources
    at opposite sides.

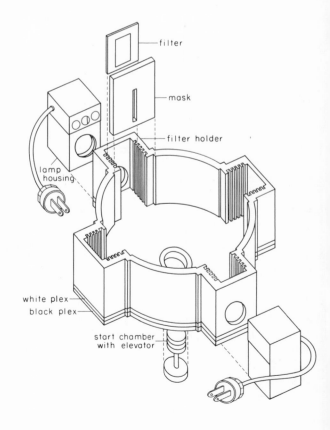

floor. The inner surfaces of the apparatus were uniformly roughened with fine-grained sandpaper to limit internal reflections. Paired lights, presented from opposite directions, shone through a mask with a vertical slit (33 × 5 mm). Slots milled into the light holders accepted Kodak Wratten (#18A near UV, #48 blue, #72B red-orange) and neutral density filters and waxed paper diffuser-depolarizers mounted in plastic 35 mm slide holders. The paired light sources were GE 15-watt tungsten filament bulbs operated independently from a pair of Variacs. Since only two lights were used, the alternate lamp housings shown in Figure 1 were fitted with solid black Plexiglas masks. Light intensities were measured at the entrance slit with a Yellow Springs Instruments (Model 65) radiometer. The masks, lamp housings, and bottom were removable for cleaning.

　　During the experiments the animals were held in plastic basins with 1 cm of seawater and light-adapted with a 60-watt incandescent lamp placed 40 cm above. Except for the adapting lamp, which was shielded from the choice chamber, the room was darkened. Before trials were begun the chamber was leveled using a small "line" level. A trial consisted of placing an animal in the center well, fitting the cover to the chamber, waiting 5 secs, elevating the animal, and immediately removing the cover to see which light the animal had approached. A choice was defined when the animal came to rest on either side of the diameter of the chamber, perpendicular to the light paths. No response (NR) was scored if the animal failed to leave the elevator surface or if it lay on the diameter. The numbers of animals approaching each light, or failing to respond, were recorded.

　　To control for potential orientational cues provided by the presence of the experimenter or the test apparatus itself, animals were run through the circular chamber with masks in place, but with no stimulus lights present. To observe the paths of the animals, the fitted top was removed and a 60-watt incandescent lamp with an aluminum painted reflector was placed 20 cm above the center of the chamber. With the rest of the room darkened, the experimenter was invisible to the crab looking out of the chamber, as evidenced by the lack of startle responses to a hand passed over the chamber on the observer's side of the 60-watt lamp.

　　Two types of choice experiments were run at ± 2 hrs of times corresponding to both high and low diurnal tides at the subject's collection site. The first series (Hyatt, in press) matched lights of the same wavelength composition and was designed to define the sign of phototaxis by varying intensity only. In the second series, paired lights of different wavelength were used to investigate true color discrimination (color vision). In both series, presentations consisted of holding one light at a given intensity while changing the intensity of the other with neutral density filters (1.0, 2.0, and 3.0 N.D. values, which reduced light intensities by a factor of 10, 100, and 1000). Thus, the first experiments served as additional controls to the second series, in that the distribution of choices could be compared to those derived from heterochromatic (two-color) experiments as a model of what a simple intensity response should be. In all experiments, each half of an individual test pairing served as a control on the other half in that equivalent stimulus pairs were presented from opposite light sources.

In all, seven separate color discrimination experiments were performed, involving 2,834 trials (not including NR's) with approximately 2,000 individuals. The five percent significance criterion was considered in testing H_o:P (C_1) = P (C_2) = 0.5, where C_1 and C_2 represent numbers of choices to the two lights. Each animal was used only once within any C_1:C_2 comparison. The probabilities for these comparisons indicated in Table 3 (e.g., p = .010 for C_1:C_2 = 62:38 at attenuation 1.0 in experiment A) were obtained by binomial exact tests. Of necessity some of the 2,000 animals used in this study appeared in more than one of the 2,834 trials (not including NR) so that all observations in the column of totals in Table 3 are not entirely independent. The probabilities (in parentheses) from the corresponding binomial exact tests of these totals are thus suspect. No conclusion need be based on these totals, however, that is not inherent in the individual comparisons, and they are included only for their values as summaries of the equally attenuated comparisons in the two directions.

Definition of photostimuli. The behavioral experiments were carried out after the electrophysiological data were gathered and analyzed, so that light stimuli in the choice chamber were adjusted and presented with a prior understanding of the animal's physiological spectral sensitivity characteristics. The series of experiments in which each animal was presented with lights of different wavelength was based upon the following rationale. Given knowledge of the relative spectral sensitivity of the eye, the stimulus lights can be adjusted to produce equivalent physiological responses. With radiometric equivalence easily measured, the occurrence of statistically significant choice behaviors at both physiological and radiometric equality would thus be interpreted as an ability to discriminate two lights based upon wavelength differences alone, since both subjective and objective intensity cues were known and controlled.

To define the light stimuli, calculations were made by using information from both the electrophysiological data and the physical parameters of the light source and filters. Table 1 shows the differences in relative spectral sensitivity derived for *U. pugilator* at three wavelengths selected as representative of the transmission wavelengths of the Kodak Wratten filters used in the two-color tests (see the arrows on Fig. 2). In order for a 600 nm light to produce an ERG response equal in amplitude to one from a 460 nm light, the former must be adjusted approximately 0.8 log units brighter than the latter. Likewise, the other two intensity relationships have been calculated and shown. These values were used for experiments A through D in Table 3.

Figure 2 shows the relationship of the tungsten output spectrum (LeGrand, 1957) to the spectral sensitivity curve of *U. pugilator*.

TABLE 1
Differences in *Uca pugilator* relative spectral sensitivity*

Experiment	Kodak Wratten filter	Wavelength Max.(nm)	Log relative energy	Sensitivity diff. in log units
A	18A	370	.335	A minus B = .186
B	48	460	.149	C minus A = .548
C	72B	600	.983	C minus B = .834

*Taken from Fig. 6.

Fig. 2. Scheme for deriving relative intensity values (see Table 2) for white/colored light experiments using information from the *U. pugilator* spectral sensitivity curve and the energy distribution curve for a tungsten source.

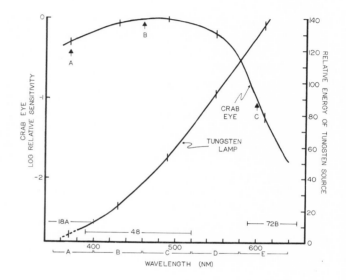

Table 2, calculated from the figure, shows the results of equating monochromatic with panchromatic stimuli. These values were used for experiments E through G in Table 3. Computation was as follows. The X-axis was divided arbitrarily into five equal segments, A through E, and the midpoints of each curve segment within a section

TABLE 2
Relationship of tungsten spectrum and *Uca pugilator* spectral sensitivity

Midpoint of section	Relative energy (tungsten)	Log Relative Sensitivity (eye)	Relative Sensitivity (eye)	RE x RS (tungsten) x (eye)
A	7.0	.74	.55	3.8
B	24.0	.92	.83	20.3
C	53.5	.98	.95	50.7
D	93.0	.76	.57	53.0
E	136.0	.78	.06	8.2
	313.5			136.0

shown. Two values, one a sensitivity value and one a light energy value, were thus determined from the curves. The difference in relative stimulus intensities can thus be expressed by the equation

$$DI = \log \left(\frac{RET_{\lambda T} \times RSE_{\lambda I}}{RSE_{\lambda T}} \right)$$

where DI = difference in light intensities for the two lights,
    RET = relative energy of tungsten source,
    RSE = relative spectral sensitivity of the eye,
    λT = total spectrum; λI = individual wavelength.

The cumulative relative energy for the tungsten spectrum (RET) over A through E is 314.0. The total subjective brightness to the crab, considering its spectral sensitivity, is 136.0 (= $RSE_{\lambda T}$). In order for a red-orange light to be radiometrically equivalent to the white in a stimulus pair, the red-orange light must be of relative brightness 313.5. To a crab, however, this brightness is a function of its red sensitivity which, from Table 2, is 0.06 (= $RSE_{\lambda I}$). Therefore, the red-orange light would appear 313.5 × 0.06 = 18.8 in relative brightness to white light. When compared by the crab, the ratio of red-orange to white would be 18.8:136.0 or approximately 1:7 = .0.14. The logarithm of 0.14 is approximately 0.8. Therefore in order for a red-orange light to be subjectively equivalent to a panchromatic light of equal initial relative intensity (313.5), the red-orange light must be 0.8 log units brighter radiometrically. In a similar manner the relationship for blue:white can be calculated, only in this case, since the transmission characteristics of the blue filter span over sections B and C, the average of the blue sensitivity over B and C (0.89) is used to find the relative brightness of blue: 313.5 × 0.89 = 279.0. The ratio of blue:white is then 279.0:136.0 or approximately 2:1 = 2.0. The logarithm of 2.0 is about 0.3; therefore, the white light in this case must be 0.3 log units brighter than the blue to achieve physiological equality. By similar calculations, white must be approximately 0.1 log unit brighter than UV.

While all of the above values are within one of the intensity increments used throughout the experiments, knowing where they fall dismisses the possibility that the demonstrated responses could be due to gross subjective intensity inequalities. With these methods the assumption that the stimulus lights are producing equal physiological responses is more firmly based than if the lamps were simply adjusted equal to a radiometer. At the same time it is possible to identify where the lights are equal to the radiometer by interpolation on the attenuation axes of Figures 9 and 10.

RESULTS

Part I: Electrophysiological Experiments

The electroretinogram. Electroretinograms of *U. pugilator*, *U. pugnax*, and *U. minax* at a series of wavelengths, intensities, and states of adaptation are shown in Figure 3, A through C. The ERG characteristics of all three species showed a negative on-effect, followed by a return to baseline during illumination and ending with a positive off-effect when the stimulus was removed. Both *U. pugilator* and *U. minax* showed a small negative deflection immediately after lights out. This effect was absent in *U. pugnax*, but can probably be explained as a difference in electrode geometry or placement. The latencies for all three species remained at about 20 msec throughout, except at neutral density 2.4 (.0087 mw/cm^2) where the latency was approximately 40 msec. The 20 msec latency was maintained for 1.0 sec flashes of 370, 600, and 640 nm light.

Fig. 3. Electroretinogram traces for the three species of fiddler crabs. CW = carapace width. The left column shows ERG traces resulting from 1.0 sec flashes of 470 nm light at the N.D. values shown (at N.D. 0.0, intensity = 2.2 mw/cm^2). The right column shows ERG traces at different wavelengths. The light intensities used at these wavelengths are shown.

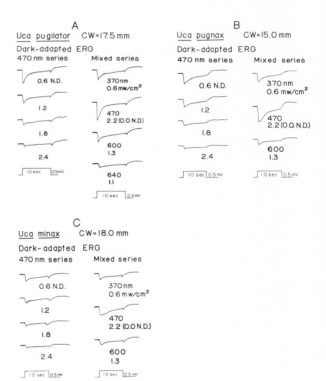

The stimulus-response function. Figure 4 A through C shows typical response (ERG "a" wave) amplitudes vs stimulus intensity for the three species of fiddler crabs presented with 0.1 sec flashes of 380, 480, and 640 nm light. The response varied from 0.3 to 3.0 mv over approximately 3.0 log units of intensity. Saturation was not

achieved within this range. The 1.0 mv criterion amplitude used
for determining the spectral sensitivity curves falls near the
bottom of the linear portion of the stimulus-response curves, while
being as close as possible to threshold. Therefore, linear interpolation to this criterion was justified for determining spectral
sensitivity curves.

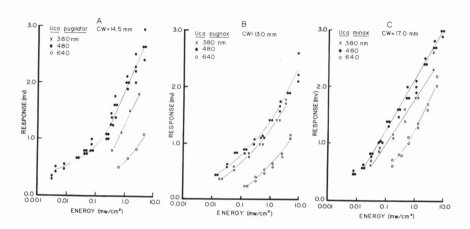

Fig. 4.   Stimulus-response curves (fitted by eye) for three indi-
           vidual fiddler crabs. X-axis = light energy in mw/cm^2
           (log scale); Y-axis = ERG "a" wave amplitude in mv.   S-R
           functions for three wavelengths from the middle and
           extremes of the tested spectrum are shown. CW = carapace
           width.

Dark-adaptation experiments.   Time courses for dark-adaptation are
shown in Figure 5. A sharp rise in sensitivity was apparent within
2.0 min of extinguishing the adapting light. After 4 min the
sensitivity changed more slowly. The zero time points on the graph
were determined while the adapting light was still on and they

Fig. 5.   Dark-adaptation
    (D.A.) curves for four
    Uca pugilator after
    different periods of
    light adaptation (white
    light, 180 mw/cm^2).
    CW = carapace width.

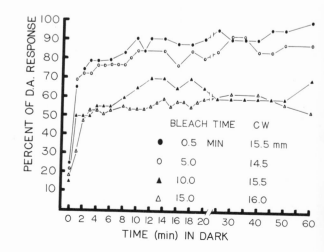

indicate the relative depression of the ERG. Longer periods of
light adaptation resulted in slightly more depressed sensitivities
and less gain in sensitivity over the test interval. Although there
was some fluctuation, the slow trend toward full dark-adaptation
was evident over the 60 min that animals were tested. The dark-
adaptation curves justified the dark-adaptation periods (2 to 24 hrs)
allotted before spectral sensitivity experiments were performed.

Dark-adapted spectral sensitivity experiments. Broad sensitivity
curves with maxima between 450 and 500 nm characterized each species
(Fig. 6). *U. pugilator* and *U. minax* maintained a slightly higher
sensitivity into the short wavelengths than did *U. pugnax*. However,
for all three species the sensitivity at 350 nm was about a log unit
higher than at 640 nm. Sensitivity decreased sharply beginning at
520 nm for the three species, so that the sensitivity at 640 nm was
1.2 to 1.7 log units down from 500 nm.

In selective adaptation experiments with *Uca pugilator*, spectral
sensitivity curves for female *U. pugilator* adapted with near-UV and
"long" (> 540 nm) wavelengths are compared with the dark-adapted
curve in Figure 7. For both adaptation experiments there was a
1.0 to 1.4 log unit depression in overall sensitivity compared to
the dark-adapted curve. If the two light-adapted curves are the
same, then a measure of the sensitivity depression (UV depressed
minus long wavelength depressed sensitivity values) should yield
a line whose slope does not differ from zero. The computation was
performed and the slope of the resultant curve differed significantly
from zero $\left(p(r = 0) < .0005\right)$; therefore, there was a difference
between the two light-adapted curves.

Part II: Behavioral Experiments

Behavior in the absence of paired stimuli. When trials were run
with no stimulus lights present, 70 animals followed almost direct
paths from the elevator platform to the walls of the chamber
(Fig. 8). Little hesitation was observed. Once the animals reached
the level of the chamber floor, they proceeded immediately to the
wall. In addition to the linear pathways, very little lateral
movement occurred after the animals reached the wall. Rather, they
attempted to climb the sides or to depress their bodies for as
close a fit as possible into the intersection of the floor and wall.
The absence of curved pathways suggests very few mid-course corrections were made once a direction was chosen. Figure 8 also shows
the tallies of choice scores when the chamber is divided in the
manner shown. None of these arbitrary divisions shows statistically
significant departures from a random distribution. Therefore, the
conclusion was made that orientational cues, if any, provided by
the experimenter or the choice chamber topography had no effect on
the animals.

Fig. 6. Spectral sensitivity curves (fitted by eye) for the three species of fiddler crabs. The bars represent standard errors of the mean and are shown for only a few points. (Standard errors for the other points approximate the ones shown.)

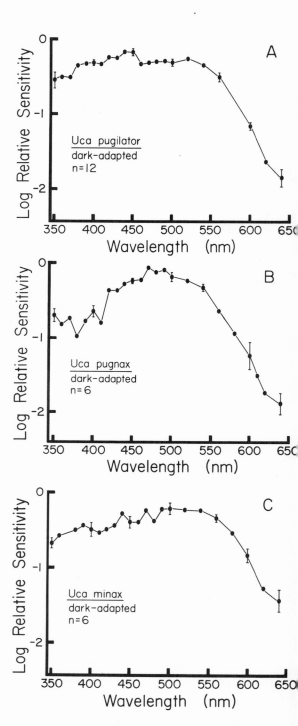

Fig. 7. Selective adaptation of *Uca pugilator* eyes with UV and long wavelength lights (Kodak #18A and #21 Wratten filters; shaded areas show transmission characteristics). Bars represent standard errors of the mean.

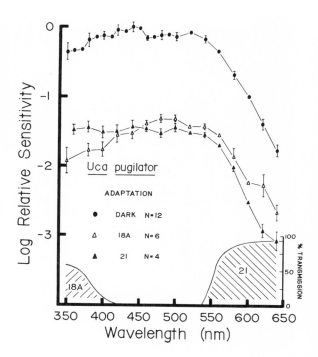

Fig. 8. Paths shown by 70 individual *Uca pugilator* (each crab tested once) in the choice chamber when paired light stimuli were absent. By arbitrarily dividing the chamber into the sections shown, the hypothesis of random choice distribution cannot be rejected at the 0.05 level by chi-square.
L = light housings.

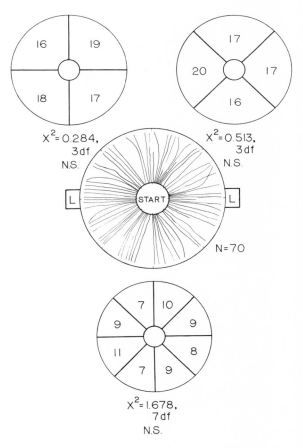

Phototaxis experiments with paired lights at high and low tide.
The results of phototaxis experiments are described elsewhere
(Hyatt, in press). In summary, when presented with paired white
lights at low tide, *U. pugilator* showed a negative phototaxis over
a three log unit span of initial light intensities. The sign of
phototaxis changed to positive when *U. pugilator* were presented wit
paired lights composed entirely of blue or red-orange wavelengths.
The taxis disappeared when identical experiments were run at high
tide.

Discrimination experiments with paired colors at low tide. Figure
(A and B) and Table 3 (A through D) show the results of pairing blu
and red-orange wavelengths. On the X-axis, where $C_1 = C_2$ is indicated, the intensity of the 600 nm light is actually 0.8 log units

TABLE 3
Summary and comparison of choice tallies for the seven behavioral experiments

Experiments	DISCRIMINANDA[a] (C_1/C_2)		C_1 + N.D. 3.0 C_1 C_2	2.0 C_1 C_2	ATTENUATION 1.0 C_1 C_2	= C_1 C_2	1.0 C_1 C_2	C_2 + N.D. 2.0 C_1 C_2	3.0 C_1 C_2	TOTAL ΣC_1
A	B/RO 1.0/6.6	N	– –	42 38	62 38	67 33	64 36	43 37	– –	278 1
		P[b]	–	.3687	.0104	.0004	.0033	.2882	–	(<.00
		NR[c]	–	–	–	–	–	–	–	
B	B/RO 2.2/15.0	N	– –	44 36	49 31	49 31	49 31	43 37	– –	234 1
		P	–	.2170	.0283	.0283	.0283	.2882	–	(.000
		NR	–	10	12	11	16	10	–	59
C	B/UV 1.5/2.2	N	– –	65 35	62 38	60 40	64 36	57 43	– –	308 1
		P	–	.0017	.0104	.0284	.0033	.0966	–	(<.00
		NR	–	33	34	44	31	30	–	172
D	RO/UV 3.5/1.0	N	– –	45 35	60 40	70 30	61 39	63 37	66 34	365 2
		P	–	.1571	.0284	<.0005	.0176	.0060	.0008	(<.00
		NR	–	16	24	20	19	27	28	134
E	W/B 77.0/77.0	N	11 29	3 21	30 40	10 30	13 27	22 38	27 33	116 2
		P	.0032	.0001	.1409	.0011	.0192	.0259	.2594	(<.00
		NR	6	8	16	8	9	16	8	71
F	W/RO 75.0/75.0	N	22 18	25 35	27 43	18 32	12 28	18 32	22 28	144 2
		P	.3179	.1225	.0361	.0324	.0082	.0324	.2399	(<.00
		NR	7	13	15	10	17	23	8	93
G	W/UV 0.2/0.2	N	– –	20 20	21 19	27 13	27 13	24 16	– –	119
		P	–	.5626	.4373	.0192	.0192	.1340	–	(.004
		NR	–	14	18	15	9	15	–	71

[a] W=panchromatic (white); B=blue; RO=red-orange; UV=ultraviolet light. Numbers
refer to light intensities (in mw/cm^2) at $C_1 = C_2$.
[b] Binomial probability that the departure from $p(C_1) = p(C_2) = 0.5$ is random.
H_o: $p(C_1) = p(C_2) = 0.5$.
[c] NR=nonrespondents.

greater than the 460 nm light. But, the two lights are presented at equivalent physiological intensity as measured from the ERG response. Attenuation then proceeded as in the previous experiments. At physiological equality, and at 1.0 log unit attenuation of both lights on each side of equality, a nonrandom orientation toward blue wavelengths was shown (p ≤ .0104). This distribution maintained itself even when the initial intensities were approximately doubled (Fig. 9 B, p = .0283). The small arrows on the X-axis of the figures show the region where the red-orange and blue lights reversed their absolute intensity relationship according to the radiometer. The statistical significance of the response toward

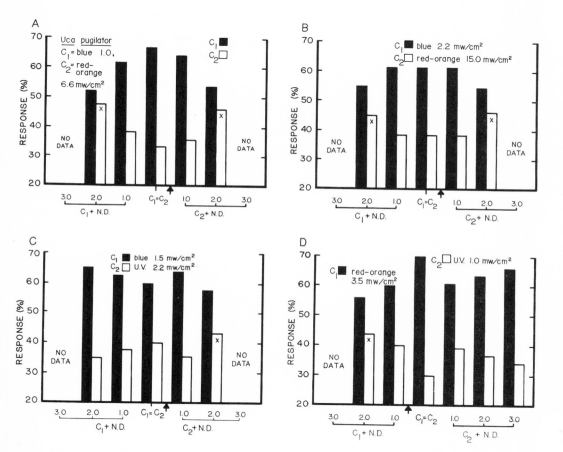

Fig. 9.   Results of heterochromatic light choice experiments with
          Uca pugilator. In A and B, blue vs red-orange; in C, blue
          vs UV; in D, red-orange vs UV. Where $C_1 = C_2$, the light
          stimuli are paired at amplitudes which give similar magni-
          tudes of ERG response. The response is toward blue in A
          through C and toward red-orange in D. The arrows on the
          X-axes indicate where the intensity relationships of the
          two lights reversed themselves according to the radiometer.
          See also Table 3 (A through D) for further details.

the blue was maintained to the left of the arrow. The total choice toward blue in both tests (Table 3) were highly significant (p < .0001; p = .0003), indicating that blue was consistently approached regardless of its intensity relationship to red-orange.

A similar format was followed in pairing near-UV with blue (Fig. 9 C) and near-UV with red-orange lights (Fig. 9 D). Figure 9 shows a nonrandom response toward blue at physiological equality as well as when the blue was attenuated by 2.0 log units (p ≤ .0284). A significant response was apparent to the right of the arrow (p = .0033), and overall (p < .0005). The discrimination broke down when the UV was attenuated by 2.0 log units. Figure 9 D shows a differential response toward red-orange light at both physiological and beyond radiometric equality (≤ .0284), and overall (p < .0005). The arrow again shows the transition zone where the stimulus intensities reversed their relationship according to the radiometer.

Discrimination experiments with colored vs white lights at low tide
Figure 10 A and Table 3 E show the results of pairing blue with white light, both set at 77.0 mw/cm^2. Blue was approached at all attenuation levels of white light except one, and the total choices toward blue showed high significance (p < .0005). The response persisted through 2.0 log units of attenuation of the blue light and degenerated only after the level of the blue light reached approximately 0.7-0.07 mw/cm^2.

Figure 10 B and Table 3 F show the result of pairing red-orange with white light. The response was not as clear-cut as in the white vs blue experiment. It is apparent, however, that red-orange was approached when it was radiometrically equal to white, and when it was attenuated by 1.0 log unit (p ≤ .0324). The overall number of choices toward the red-orange remained highly significant (p < .0005).

Figure 10 and Table 3 G display data from a UV-white pairing. Only two of the pairings showed a nonrandom distribution, but the overall trend was to approach the white light more than the UV (p = .0043). The distribution turned random at the extremes of attenuation (as in the red-orange vs white experiment above), and suggested an intensity effect. The arrows on the X-axes were placed to indicate the level of attenuation where the energy of the panchromatic (tungsten) spectrum was approximately at subjective equality (for the animals) to the energy of the transmission spectrum of the colored filters. The values were calculated as outlined in the METHODS section (Table 2 and Fig. 2).

Peripheral vs central processing of the behavioral response.
Comparison of the physiological and behavioral data offered an opportunity to establish an index of the relative importance of peripheral and central nervous system (CNS) processing of the visual stimulus. Table 4 equates the relative physiological sensitivity differences (column P) to the behavioral sensitivity differences (column B) in intensity units. Column B was derived by

calculating the light intensity difference between the $C_1 = C_2$
pairing condition and that condition of pairing where random
$(p(C_1) = p(C_2) = 0.5)$ behavior occurred (refer to Fig. 9 and Table 3,
A through D). The difference (P minus B) between these two was
thus an indication of how much processing of the stimulus was done
centrally (i.e., by the CNS) in contribution to the overt behavioral
response. This difference as a percentage of P is shown in the

Fig. 10. Results of white
vs monochromatic light
choice experiments with
U. pugilator. A, white
vs blue; B, white vs
red-orange; C, white vs
UV. In A and B the
response is toward blue
or red-orange. In C
the response is toward
the white light. The
arrows on the X-axes
indicate where the
intensity relationships
of the two lights reversed
themselves according to
the radiometer. See
also Table 3 (F through G)
for further details.

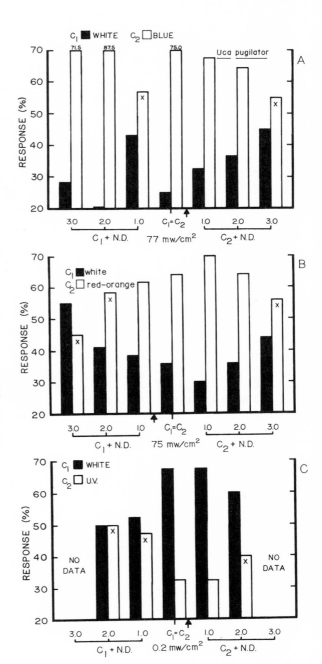

last column of Table 4. Thus, it appears that less than 20% of the neural integration required to initiate the locomotory response toward or away from the stimulus lights occurs in the CNS.

TABLE 4
Summary of peripheral vs central processing of the behavioral response

Light Pairing	Log sensitivity dif. (ERG)	P Arithmetic sensitivity dif. (ERG)	B Arithmetic intensity dif. at behavioral equality[*]	P-B	(P-B) as % of P
UV/Blue	.186	1.5 I units	1.4 I units	0.1	6.6
UV/Red	.648	4.4	3.5	0.9	20.0
Red/Blue	.834	6.8	6.5	0.3	4.4

[*]Calculated as the intensity difference between $C_1 = C_2$ and the first indication of random response in the choice experiments (Choice complement scaled to 1.0).

The relationship of the magnitude of choice response to the number of nonrespondents. In the course of the choice experiments, three pieces of information were made available for comparative analysis: the choice tallies to C_1 and to C_2, and the number of animals that were nonrespondent (NR). The NR tallies are included in Table 3 along with the choice tallies. Upon analysis of the data, it was noticed that the number of nonrespondents rose in the heterochromati pairings when the choice became more clear-cut, i.e., when one light was chosen more frequently.

Least-squares regression analyses were performed on the absolute values of the differences between C_1 and C_2 ($|C_1 - C_2|$) and the NR tallies. Two separate analyses were performed, one with the experiments wherein intensity was the only variable (Hyatt, in press and one with experiments in which wavelength differences existed between C_1 and C_2 (A through G in Table 3). Figure 11 A shows the analysis of the taxis experiments. Where intensity difference is the only factor eliciting the response, the number of NR is poorly correlated with the magnitude of choice difference $(r = +.1206,$ 33 df; $p(r = 0) = .4901)$.

Figure 11 B shows the analysis of experiments A through G, wherein at least one of the lights is monochromatic. Where wavelength differences are present, the correlation between the magnitude of choice difference and the number of nonrespondents is substantially higher $(r = +.5057, 32$ df; $p(r = 0) = .0023)$.

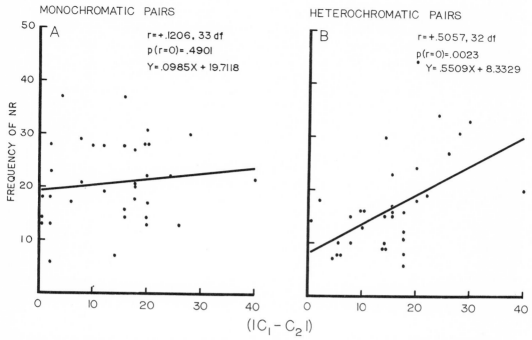

Fig. 11. Linear regression analyses of the frequency of non-respondents (NR) vs absolute difference in choice tallies ($|C_1 - C_2|$). A. Analysis of NR vs taxis experiments. B. Analysis of NR vs A-G.

DISCUSSION

Study of the fiddler crab as an "experimental system" has yielded a gradual synthesis of structure, function, and ecological relevance with its demonstrated behavioral capabilities. The original quantification of waving and acoustic signals and social interactions in the field by Crane and Salmon raised questions directed toward stimulus filtration, receptor processes, and response in the receiver, as well as toward an analysis of signal production.

The response of *Uca* toward visual stimuli was reviewed in the INTRODUCTION section of this paper. The only prior electrophysiological examination of vision in *Uca* was done by Altevogt (1957). The electrical responses of fiddler crab eyes appear to follow the pattern of some other Crustacea. Similar ERG responses were shown for *U. marionis* and *U. annulipes* by Altevogt (1957), for the isopod *Ligia occidentalis* by Ruck and Jahn (1954), and for *Panulirus*, *Callinectes*, and *Eupanopeus* by Waterman and Wiersma (1963). All are cornea negative responses with a positive off response, and latencies are within the same order of magnitude as for *U. pugilator*, *U. pugnax*, and *U. minax*.

Inspection of the dark-adaptation curves for *U. pugilator* (Fig. 5) indicates that the rate of dark-adaptation is independent of the period of light-adaptation. Under all conditions of previous light-adaptation, the animals showed 50 to 80% recovery within the first 4 min. Short dark-adaptation times are characteristic of "fast"eyes (Autrum, 1950), and the data agree favorably with Altevogt's (1957) results with other *Uca* species.

Broad spectral sensitivity curves sometimes suggest the presence of a heterogeneous population of receptors or multiple pigment with sensitivity maxima in different spectral regions (Post and Goldsmith, 1969). An alternative question that may be asked, especially where UV sensitivity is demonstrated, is whether some eye part may be fluorescing at a longer wavelength upon exposure to UV, thus shifting the sensitivity into the UV region. This is probably not the case for *Uca pugilator* (Fig. 6 A), since fresh frozen eye sections viewed by UV-dark field microscopy failed to show anything but a slight cast after a 20-min exposure on high-speed Ektachrome film (pilot observation, this research). Pigment data corroborate the 450 to 500 nm sensitivity maxima found for *U. pugilator* and *U. pugnax* (Goldsmith and Fernandez, 1966).

Selective adaptation (Fig. 7) is a method of determining whether an eye contains classes of pigments or receptors that differ in the wavelengths of light to which they are maximally sensitive. It is also a method of demonstrating a potentiality for color vision (Wald, 1968b; Post and Goldsmith, 1969). Using this procedure, Post and Goldsmith presented clear physiological evidence for two classes of color receptors in the eye of the butterfly *Colias eurytheme*. The selective adaptation curves for *U. pugilator* (Fig. more closely resemble those of the green crab, *Carcinus maenas*, and the spider crab, *Libinia emarginata*, derived under similar conditions by Wald (1968b). The very close agreement of the *Uca* curves with those of *Libinia* perhaps allows some of the same interpretations. Adaptation of the *U. pugilator* eye with yellow-orange light tilts the curve toward shorter wavelengths with little change between 400 and 500 nm. Wald attributed his similar results with *Libinia* to the presence of at least two visual systems apparently differentiated for use in intensity discriminations, and secondarily for color vision. The shift upon light-adaptation corresponds to a reverse Purkinjee phenomenon. Most probably the dark-adapted spectral sensitivity curve of *U. pugilator* is maintained by at least two separate pigment or receptor systems, one maximally sensitive in the long (yellow-red), and the other in the short (violet-blue) wavelength region of the spectrum. For *U. pugilator*, which behaviorally show color discrimination, the presence of a two-pigment system suggests the possible physiological mechanism.

The methods used to investigate color vision in arthropods (Wulff, 1956) include the use of a conditioned response, observation of spontaneous responses to stimulation, optomotor responses, selective adaptation, and measurement of electrical responses. Crustaceans have been conditioned with differing degrees of success and effort by the investigator (see the INTRODUCTION). Altevogt (1963) failed to demonstrate conditioned responses to colored lights by *U. tangeri*. The method used in this research precluded the necessity of conditioning and exploited a spontaneous response toward light.

In order to avoid confusion regarding the behavioral data, two major operating definitions are offered here which qualify the behavioral observations presented. Fraenkel and Gunn (1961) identify a taxis as a directed response during which, with a single source of stimulation, the long axis of the body is oriented in line with the source, and locomotion is toward (positive) or away (negative) from it. More specifically, they identify telotaxis as a response only toward light stimuli, wherein attainment of orientation is direct, without lateral deviations. Movement is toward a light source as if it were a goal, even if two or more stimuli are present. Tropotaxis is also a direct response, but orientation occurs between two stimulus sources, probably because of a more or less differential stimulation of paired receptors and the simultaneous comparison of stimuli by the two. They further outline a major difficulty in the identification of taxis behaviors that involve crabs. Because of the varied methods of locomotory progression shown by these decapods (e.g., forward, backward, "crabwise") there is no tidy relationship between the angle of orientation to a stimulus source and the main body axis. Crabs orient to the stimulus equally well regardless of locomotor style. Usually the behavior is characteristic of telotaxis, but occasionally tropotaxis-like patterns are observed which, because of the generous eye morphology of most crabs, cannot be easily tested via unilateral eye ablation (which results in characteristic "circus" movements if tropotaxis is the mechanism).

Figure 12 is a pictorial representation of pathways taken by animals during one of the white light taxis series. The paths to the dim light were straight as in the control (Fig. 8), while a group of ten curved trails was apparent in the series of choices toward the brighter light. Whereas the linear pathways toward the dim light support negative photo-telotaxis, the curved ones toward the bright light illustrate the type of observation described by Fraenkel and Gunn (1961) that suggests a tropotaxis component. For the purposes of this research, photo-telotaxis was assumed to be the major orientational method.

Mazokhin-Porshnyakov (1969) defines color vision as the ability to distinguish radiations by their wavelength characteristics at any value of intensity. If an animal responds differentially to lights of different spectral composition regardless of their relative intensities, it possesses color vision. Mazokhin-Porshnyakov cautions, however, that because an animal demonstrates a differential response to, for example, blue and red lights, this does not mean that the animal recognizes "red" or "blue" as colors, since "color" does not fully characterize the physical composition of the radiation (e.g., "yellow" can be a monochromatic 585 nm or a combination of red and green wavelengths). Fiddler crabs show the ability to discriminate between lights based upon the wavelength of the radiation, but it is unknown if the animal perceives these radiations as colors. In the descriptions of the behavior, therefore, references to "red" or "blue" or "color," rather than "640 nm," "480 nm," or "radiation" was simply an aid to the reader and not an implication or simplification of the subjective perceptual capabilities of the animals.

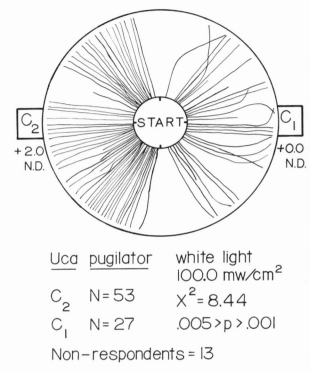

Fig. 12. Choice pathways observed during white light taxis experiment.

$\underline{Uca}$ $\underline{pugilator}$ white light
100.0 mw/cm^2

C_2 N = 53 x^2 = 8.44

C_1 N = 27 .005 > p > .001

Non-respondents = 13

The effort spent in equating light stimuli for the color discrimination tests represents a different approach to the investigation of color responses in Uca. Altevogt (1963) did not consider the visual spectral sensitivity function of the Uca eye. With the present methods it became possible to adjust the stimulus lights in the choice chamber so that they produced equivalent responses in the animal's eye, and not just equal responses to a radiometer with some entirely irrelevant spectral sensitivity. (No Crustacea are known to have linear spectral sensitivity functions.) Radiometric equality did, however, remain identifiable by interpolation. Given this information, the logic of the experiments was as follows. If an identical response was apparent at both physiological and radiometric equality, then the argument that intensity was the cue for making a choice could be dismissed on the grounds that there was, in fact, no difference in "perceptual" intensity under the former condition. Since wavelength was the only other independent

variable, the hypothesis that the crabs have the ability to discriminate two lights based only upon their wavelength differences
(color vision according to Mazokhin-Porshnyakov, 1969) is supported.
In the heterochromatic experiments A through D, the animals were
demonstrating a persistent orientation toward one light regardless
of the conditions of intensity, thus fitting the description for
the demonstration of color vision. Similar behavior patterns and
results occurred in experiments E through G when panchromatic lights
were paired with monochromatic lights. This suggests that *Uca*
pugilator not only discriminates two different monochromatic lights,
but also can discriminate a monochromatic light from a panchromatic
light that includes wavelengths of the former.

The role of peripheral and central nervous system processing
of visual stimuli outlined in the RESULTS section warrants further
investigation. The findings suggest that most of the analysis
(80% or more) of the visual input occurs peripherally, probably at
the level of the optic ganglia. Morphologically, *U. pugilator* optic
tracts are constructed in a manner which supports the hypothesis of
major peripheral processing (Nunnemacher, 1966). Nunnemacher's
comparative studies showed that *U. pugilator* possesses the lowest
fiber/facet ratio of the seven species of reptantian decapods
examined. The *Uca* ratio (2.1:1), compared to the maximum (9.1:1 in
Pagurus) and the mean (5.4:1), led Nunnemacher to suggest that there
was a possibility of peripheral integration, and that the lower
order processing may be associated with the active scanning habits
and terrestrial habitat of the crab. Some further experimentation
is possible, incorporating more extensive anatomical studies, and
electrophysiological examination of the optic ganglia (with
ablations), coupled with behavioral observations on operated animals.

An interesting finding, and one for which the full meaning has
probably not been understood, is the correlation of the increase in
NR with the increasing nonrandomness of the choice scores for the
heterochromatic tests. For lack of further data or comparable
results from other research, four possible reasons for the correlation are presented here. (1) "Ambivalence" arises when stimuli are
afforded that make a choice mandatory within the framework of the
Uca CNS. Such ambivalence is manifested by remaining stationary.
(2) The choice differential presumably reflects the degree of
"attractiveness" or "repulsiveness" of one stimulus over the other.
At high stimulus efficiencies, "startle" responses could preclude
or confound coordinated orientation movements. (3) In clear-cut
stimulus situations, the response latency increases to beyond the
time allotted for a choice to be made (5 sec), thereby increasing
the number of apparent NR. If this is the case, the demonstrated
correlation may be artifactual, and the critical test would be to
perform trials allowing a longer time to make a choice. (4) When
presented with stimuli which involve a wavelength discrimination,
some other cue (e.g., movement, sound) facilitates final orientation. Stimulus summation is well known as a basic receptor phenomenon, and it is possible that an animal, such as *U. pugilator*, which

responds to a group of relevant stimuli through several modalities may well be programmed with a specific "combination" of necessary stimulus prerequisites that unlock its behavior.

The fiddler crabs, along with a very few other crustaceans, occupy a unique position in the continuum of arthropod evolution in that they are semiterrestrial. Along with special physiological prerequisites for life on land, problems of communication arise that must be solved in ways different from obligate aquatic crabs. Courtship communication is especially critical. As far as is known, two methods of courtship communication are used by fiddler crabs: (1) acoustical signals transmitted through the substrate and (2) visual signals sent through the air. Both types show stereotyped, species-specific format and temporal patterning that probably serve as cues for species identification. Other visual stimuli, such as shape, size, and color, may be equally efficient identifiers Indeed, the work of Langdon (1971) clearly demonstrated a marked ability for *U. pugilator* to discriminate shapes. The obvious (to humans) differences in coloration among fiddler crabs suggest that color may also be a cue. The behavioral demonstration of color vision by this research makes color a favorable candidate for a communicative modality. Adult *U. pugilator* and *U. pugnax* show distinct blue areas on the carapace. *U. pugnax* always possesses an area on the rostrum (about 2×3 mm) of about the same shade of "aqua" blue from animal to animal. *U. pugilator* has an area around the cardiac depression that varies from violet to "robin's egg" blue. There is little doubt that the colored areas are visible to wandering females when the male is displaying during daytime courtship. The spectral sensitivity curves suggest the presence of the physiological mechanism(s) for perceiving blue light wavelengths more efficiently than, for example, yellow or red. The demonstrated behavorial response toward blue over near-UV and red-orange lights coincides with the physiological data.

The response toward blue over red-orange has further implications. An observer of a colony of *U. pugilator* can easily identify immature males (see Herrnkind, 1972, for developmental criteria) by the size and shape of the major chelae (smaller and rounder). These males perform courtship waving that appears identical in format and vigor to that of adult males (unpublished observations by the author) A major difference is that small males invariably lack the blue area on the carapace. At the same time their coloration is obviously more rusty-orange than adult males, including a rust-orange shading of the major chela. Although no comparisons have been made of the courtship success of immature vs mature males, the pigmentation differences and the behavorial data described above suggest that the immature males might indeed be less conspicuous to females or, if not less conspicuous, at least less preferred or approached less frequently. Ignoring immature males in preference for mature males confers an obvious selective advantage on the species, since it raises the probability of producing offspring via copulation with more "experienced" males. Color may therefore act as an isolating mechanism which effectively keeps immature males outside the breeding population.

Although color may be used in conspecific identification in the wild, it may not elicit behavior as shown in the laboratory choice chamber. In order to be effective, a sign stimulus must be presented while an animal is in the appropriate motivational state (Hinde, 1966). A safe guess is that a fiddler crab, once handled by a human and placed in a plastic chamber, is not in a motivational state conducive to courtship or the identification of any of its sign stimuli (but escape responses may not be affected). However, Langdon (1971) found that shape discrimination by *U. pugilator* was involved in identifying three important classes of environmental stimuli (conspecifics, shelter, and predators). He concluded that "crabs avoided circles and bird shapes in the apparatus just as in the field. These similarities in response, in addition to the similarity in response toward numerous vertical rectangles in the laboratory and grass-like vegetation in the field, suggest that much of the information obtained using these [choice chamber] techniques can be applied to situations in the natural environment." Likewise, Muntz (1962, 1963) demonstrated an ontogeny of phototaxis in the laboratory for the frog *Rana temporaria* that he related to the natural conditions of the animal. Whether the responses observed during the present research would change if experiments were carried out in the field (very difficult to execute and control using colors) cannot be foreseen. If the laboratory responses are not like the "real" ones, then the chance of error is one that all investigators take when they remove an animal from its natural environment.

Uca pugilator and *U. pugnax* coexist in some areas. In the tropics, upward of a dozen species can be found on the same mud flat or beach. In areas of sympatry, the problem of maintaining species distinctiveness arises. Differential responses to "releasers" or "sign" stimuli (movements, colors, and patterns) have been demonstrated for numerous animals, under various conditions, for different behavioral functions (Hinde, 1966). Of the thirty-two sign characters of major importance enumerated by Hinde, twenty-one involve the visual sense of the receiver. Of the eighteen types of released behavior listed, five involve reproductive behavior or its initiation, and of the sixteen species, six are invertebrates. Hinde states that "since relational properties are usually important, few sign stimuli can readily be specified quantitatively along a single physical scale. . . . Where a response is influenced by stimulus characters acting through more than one sensory modality, or by more than one stimulus character in any one modality, their effects supplement each other." In all probability, conspecific identification by fiddler crabs is a multistimulus, if not a multimodality process that involves higher level neural integration and evaluation. Strictly behavioral (as opposed to morphological) constraints in the form of differential responses to sign stimuli, perhaps built into the genotype, probably contribute heavily to prevent hybridization and gamete wastage.

To date, the characteristics of visual stimuli that are presented and/or detected by *Uca* are: (1) intensity (Hyatt, in press),
(2) wavelength (this research), (3) polarization plane (Altevogt and
von Hagen, 1964; Herrnkind, 1968b), (4) spatial pattern (Langdon,
1971), and (5) temporal pattern (Crane, 1941, 1943, 1944, 1957, 1958,
1966, 1967; Salmon, 1965, 1967). On the same scale, a fiddler crab
communicating over a distance of 0.5 m is approximately the same as
a human transmitting visual signals over 70 m. Conspicuousness is
an asset that is obviously enhanced by every means possible. Likewise, in a general sense, long distance signals that are liable to
interference from the environment (e.g., grass, other animals,
topography) must be different enough from one another so that they
are quickly and accurately identified (Marler, 1968). Marler
states that "at long range, a respondent will be unable to perceive
subtle variations in signal structure; there is likely to be a trend
toward highly stereotyped signals. In other situations, small
variations in signal structure can have great value for communicating
certain types of information and may be exploited when circumstances
permit." The spatial and temporal patterning of *Uca* waving displays
are highly stereotyped and could easily serve as long distance signals. Coloration is more subtle, and for close-up identification,
color cues may be ideal.

SUMMARY

The results of general physiological experiments show that the
fiddler crabs *Uca pugilator*, *U. pugnax*, and *U. minax* have a diphasic
cornea-negative electroretinogram characteristic of "fast" eyes.
They show response functions that vary as the logarithm of the light
stimulus intensity. In dark-adaptation experiments, *U. pugilator*
returns to 50 to 80% of maximum sensitivity within 4 min of extinguishing the adapting light. In general, the *Uca* eye appears to
have physiological characteristics similar to certain other
crustacean eyes.

All three species show broad spectral sensitivity curves
(criterion amplitude method) with sensitivity maxima between 450
and 500 nm. The curves drop sharply beginning at 510 to 520 nm, so
that the sensitivity at 640 nm is approximately 1.2 to 1.7 log units
down from 500 nm. Sensitivity is maintained (about 0.5 log units
down) out to 350 nm (near-UV). Selective adaptation experiments
with *U. pugilator* suggest at least a two-pigment/receptor system
with one pigment/receptor sensitive in the blue-violet and the other
in the red-orange region of the spectrum. Published pigment data
corroborate the 450 to 500 nm sensitivity maxima found for *U.
pugilator* and *U. pugnax*.

Two types of behavioral (choice) experiments were run at times
corresponding to ± 2 hrs of both high and low diurnal tides at the
subject's collection site. The first series matched lights of the

same wavelength composition and was designed to define the sign of phototaxis by varying intensity only. In the second series, paired lights of different wavelength were used to investigate true color discrimination.

When presented with paired white lights at low tide, *U. pugilator* shows a negative phototaxis over a three log unit span of initial light intensities. The sign of phototaxis changes to positive when *U. pugilator* is presented with paired lights composed entirely of blue or red-orange wavelengths. The taxis disappears when identical experiments are run at high tide.

When lights of different wavelength are paired, *U. pugilator* shows a hierarchy of choice behavior from most to least frequently chosen as follows: blue, red-orange, white, UV. The two-color experiments demonstrate that *U. pugilator* can discriminate two lights based upon wavelength differences alone (intensities closely controlled and related to the spectral sensitivity function) and thus possess color vision. The demonstration that *U. pugilator* shows a negative phototaxis in panchromatic light differs from observations made by previous workers. The reversal of sign in blue and red-orange lights is an enigma, but perhaps suggests that color cues might act as close-range identifiers and that the approach made by females toward males may be affected by color. The disappearance of the response during high tides is less puzzling, since the animals are not normally active at these times.

The choice of blue over red-orange may act as an isolating mechanism which keeps immature males out of the breeding population. (Immature *U. pugilator* males are redder than mature ones, yet they display with equal vigor.) The spatial and temporal patterning of *Uca* displays are stereotyped and could easily serve as long distance signals. Coloration is more subtle and varied and could ideally serve as a second-stage, "close-up" identifier to initiate or perpetuate courtship or other behaviors.

ACKNOWLEDGMENTS

This study was done in partial fulfillment of the requirements for the degree of Doctor of Philosophy at the University of Illinois at Champaign-Urbana. The work was supported by NSF GB-20766X to Michael Salmon. This paper is a contribution of the Duke University Marine Laboratory.

I am grateful to William Mead and C. Ladd Prosser for thought-provoking discussions. I wish to thank Michael Salmon and Howard Schein for comments on the manuscript, and John D. Costlow, Jr. and the staff of the Duke University Marine Laboratory for incomparable southern hospitality.

LITERATURE CITED

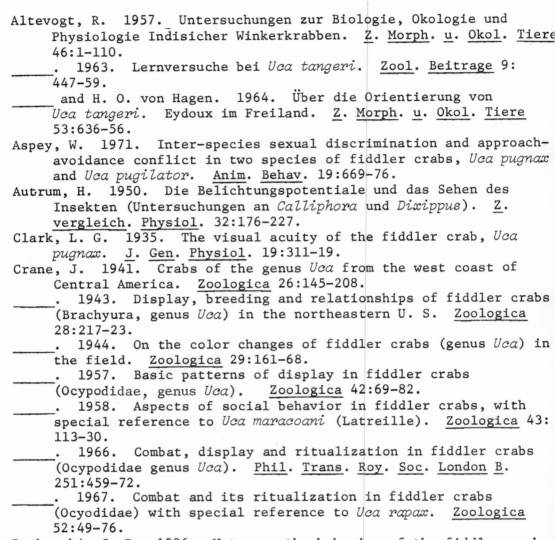

Altevogt, R. 1957. Untersuchungen zur Biologie, Okologie und Physiologie Indisicher Winkerkrabben. *Z. Morph. u. Okol. Tiere* 46:1-110.

_____. 1963. Lernversuche bei *Uca tangeri*. *Zool. Beitrage* 9: 447-59.

_____ and H. O. von Hagen. 1964. Über die Orientierung von *Uca tangeri*. Eydoux im Freiland. *Z. Morph. u. Okol. Tiere* 53:636-56.

Aspey, W. 1971. Inter-species sexual discrimination and approach-avoidance conflict in two species of fiddler crabs, *Uca pugnax* and *Uca pugilator*. *Anim. Behav.* 19:669-76.

Autrum, H. 1950. Die Belichtungspotentiale und das Sehen des Insekten (Untersuchungen an *Calliphora* und *Dixippus*). *Z. vergleich. Physiol.* 32:176-227.

Clark, L. G. 1935. The visual acuity of the fiddler crab, *Uca pugnax*. *J. Gen. Physiol.* 19:311-19.

Crane, J. 1941. Crabs of the genus *Uca* from the west coast of Central America. *Zoologica* 26:145-208.

_____. 1943. Display, breeding and relationships of fiddler crabs (Brachyura, genus *Uca*) in the northeastern U. S. *Zoologica* 28:217-23.

_____. 1944. On the color changes of fiddler crabs (genus *Uca*) in the field. *Zoologica* 29:161-68.

_____. 1957. Basic patterns of display in fiddler crabs (Ocypodidae, genus *Uca*). *Zoologica* 42:69-82.

_____. 1958. Aspects of social behavior in fiddler crabs, with special reference to *Uca maracoani* (Latreille). *Zoologica* 43: 113-30.

_____. 1966. Combat, display and ritualization in fiddler crabs (Ocypodidae genus *Uca*). *Phil. Trans. Roy. Soc. London B.* 251:459-72.

_____. 1967. Combat and its ritualization in fiddler crabs (Ocyodidae) with special reference to *Uca rapax*. *Zoologica* 52:49-76.

Dembowski, J. B. 1926. Notes on the behavior of the fiddler crab. *Biol. Bull.* 50:179-201.

Fraenkel, G. S. and D. L. Gunn. 1961. *The Orientation of Animals*. Dover, N. Y.: Dover Press.

Goldsmith, T. and H. R. Fernandez. 1966. Some photochemical and physiological aspects of visual excitation in compound eyes. In: *The Functional Organization of the Compound Eye*, Wenner-Gren Center International Symposium Series, Vol. 7 (C. G. Bernhard, ed.), pp. 125-43. New York: Pergamon Press.

Hartnoll, R. G. 1969. Mating in the Brachyura. *Crustaceana* 16: 161-81.

Hazlett, B. A. 1969. Further investigations of the cheliped presentation display in *Pagurus bernhardus* (Decapoda, Anomura). Crustaceana 17:31-4.

_____. 1972. Stimulus characteristics of an agonistic display of the hermit crab *(Calcinus tibicen)*. Anim. Behav. 20:101-7.

Herrnkind, W. F. 1968a. The breeding of *Uca pugilator* (Bosc) and mass rearing of the larvae with comments on the behavior of the larval and early crab stages (Brachyura, Ocypodidae). Crustaceana, Suppl. II. 214-24.

_____. 1968b. Adaptive visually-directed orientation in *Uca pugilator*. Am. Zool. 8:585-98.

_____. 1972. Orientation in shore-living arthropods, especially the sand fiddler crab. In: Behavior of Marine Animals, Vol. I (H. E. Winn and B. L. Olla, eds.). New York: Plenum Press.

Hinde, R. A. 1966. Animal Behavior. A Synthesis of Ethology and Comparative Psychology. New York: McGraw-Hill.

Holmes, S. J. 1908. Phototaxis in fiddler crabs and its relation to theories of orientation. J. Comp. Neurol. Psychol. 18:493-97.

Horch, K. W. and M. Salmon. 1969. Production, perception and reception of acoustic stimuli by semiterrestrial crabs (Genus *Ocypode* and *Uca*, family Ocypodidae). Forma et Functio 1:1-25.

Horridge, G. A. 1967. Perception of polarization plane, colour and movement in two dimensions by the crab, *Carcinus*. Z. vergl. Physiol. 55:207-24.

Hyatt, G. W. Behavioral evidence for light intensity discrimination in the fiddler crab *Uca pugilator* (Brachyura, Ocypodidae). Anim. Behav. In press.

Koller, G. 1927. Über Chromatophorensystem, Farbesinn und Farbwechsel bei *Crangon vulgaris*. Z. vergl. Physiol. 5:191-246.

_____. 1928. Versuche über den Fasbensinn der Eupaguriden. Z. vergl. Physiol. 8:337-53.

Korte, R. 1965. Durch polarisiertes Licht hervorgerufene Optomotorik bei *Uca tangeri*. Experientia 21:98.

_____. 1966. Untersuchungen zum Sehvermogen einer Dekapoden, inbesondere von *Uca tangeri*. Z. Morph. u. Okol. Tiere 58:1-37.

Langdon, J. W. 1971. Shape discrimination and learning in the fiddler crab *Uca pugilator*. Ph.D. dissertation, Florida State University.

LeGrand, Y. 1957. Light, Colour and Vision. New York: Wiley.

Linsenmair, K. W. 1967. Konstruktion und Signal Funktion der Sandpyramide der Reiterkrabbe *Ocypode saratan* Forsk (Decapoda, Brachyura, Ocypodidae). Z. Tierpsychol. 24:403-56.

Marler, P. 1968. Visual systems. In: Animal Communication; Techniques of Study and Result of Research (T. Sebeok, ed.), pp. 103-26. Bloomington, Ind.: Indiana University Press.

Matthews, L. H. 1930. Notes on the fiddler crab *Uca leptodactyla*, Rathbun. Ann. Mag. Nat. Hist., Ser. 10, 5:659-63.

Mazokhin-Porshnyakov, G. A. 1969. Insect Vision. New York: Plenum Press.

Muntz, W. R. A.   1962.   Effectiveness of different colors of light
in releasing the positive phototactic behavior of frogs, and
a possible function of the retinal projection to the diencepha-
lon.   J. Neurophysiol. 25:712-20.
_____.   1963.   The development of phototaxis in the frog (Rana
temporaria).   J. Exp. Biol. 40:371-79.
Nunnemacher, R. F.   1966.   The fine structure of optic tracts of
decapoda.   In: The Functional Organization of the Compound
Eye, Wenner-Gren Center International Symposium Series, Vol. 7
(C. G. Bernhard, ed.), pp. 363-75.   New York: Pergamon Press.
Palmer, J. W.   1964.   A persistent light-preference rhythm in the
fiddler crab, Uca pugnax and its possible adaptive significance.
Am. Nat. 98:431-34.
Post, C. T. and T. H. Goldsmith.   1969.   Physiological evidence for
color receptors in the eye of a butterfly.   Ann. Ent. Soc. Am.
62:1497-98.
Ruck, P. and T. L. Jahn.   1954.   Electrical studies on the compound
eye of Ligia occidentalis. (Dana) (Crustacea: Isopoda).
J. Gen. Physiol. 37:825-49.
Salmon, M.   1965.   Waving display and sound production in the
courtship behavior of U. minax and U. pugnax.   Zoologica 50:
123-50.
_____.   1967.   Coastal distribution, display and sound production
by Florida fiddler crabs (genus Uca).   Anim. Behav. 15:449-59.
_____ and S. P. Atsaides.   1968.   Visual and acoustical signalling
during courtship by fiddler crabs (genus Uca).   Am. Zool. 8:
623-39.
_____ and J. F. Stout.   1962.   Sexual discrimination and sound
production in Uca pugilator Bosc.   Zoologica 47:15-20.
Schöne, H.   1961.   Learning in the spiny lobster Panulrius argus.
Biol. Bull. 121:354-65.
Schwartz, B. and S. R. Safir.   1915.   Habit formation in the
fiddler crab.   J. Anim. Behav. 5:226-39.
Smith, F. E. and E. R. Baylor.   1953.   Color responses in the
Cladocera and their ecological significance.   Am. Nat. 57:
49-55.
Stevcic, Z.   1971.   The main features of Brachyuran evolution.
Systematic Zool. 20:331-40.
von Hagen, H. O.   1961.   Experimentele Studien zum Winken von Uca
tangeri in Sudspanien.   Ver. Deutsch. Zool. Ges. in
Saarbrücken.   Leipzig: Geest and Portig.
Wald, G.   1968a.   Oscillations of potential in the electroretinogram
of the lobster.   J. Gen. Physiol. 51:261-71.
_____.   1968b.   Single and multiple visual systems in arthropods.
J. Gen. Physiol. 51:125-56.
Waterman, T. H.   1961.   The Physiology of Crustacea, Vol. 2.
New York: Academic Press.
_____ and K. W. Horch.   1966.   Mechanism of polarized light
reception.   Science 154:467-75.

_____ and C. A. G. Wiersma. 1963. Electrical responses in decapod crustacean visual systems. J. Cell. Comp. Physiol. 61:1-16.

Wulff, V. J. 1956. Physiology of the compound eye. Physiol. Rev. 36:145-63.

Zucker, N. Shelter building by the fiddler crab *Uca terpsichores* (Crustacea: Ocypodidae). Submitted for publication.

Dinoflagellate phototaxis: Pigment system and circadian rhythm as related to diurnal migration

R. B. Forward, Jr.

Studies of the daily vertical distributions of marine planktonic dinoflagellates both in nature (Hasle, 1950, 1954) and in
experimental situations (Eppley, Holm-Hansen, and Strickland, 1968)
show that the cells rise to the surface waters during the day and
descend or disperse at night. Probably the most critical environmental cue for the upward-oriented movement as the sun rises is the
increase in light intensity (Hasle, 1950, 1954). Laboratory investigations of dinoflagellates indicate that these organisms have a pronounced phototactic response to light (Halldal, 1958; Hand, Forward,
and Davenport, 1967). At high intensities the behavioral response
consists of an initial cessation of movement (i.e., stop-response),
followed by swimming in the direction of the stimulus (i.e., phototaxis) (Hand, Forward, and Davenport, 1967). Other flagellated phyto-
planktonic organisms show a similar cessation of movement upon stimulation with light, e.g., *Chlamydomonas* (Feinlieb and Curry, 1971)
and *Volvox* (Huth, 1970). In all cases this initial stop-response is
considered an integral part of the total phototactic response, and
thus is an indicator of phototactic light reception.

Assuming that the diurnal early morning ascent by dinoflagellates represents a simple phototactic response to increasing over-
head light intensity, then ideally the cells should perhaps exhibit
diurnal changes in sensitivity to light, with the greatest sensitivity
occurring around dawn. The following discussion of physiological
aspects of phototaxis is directed at supporting this hypothesis.

INSTRUMENTATION AND METHODS

The estuarine dinoflagellates *Gyrodinium dorsum* Kofoid and *Gymnodinium splendens* Lebour were cultured in an enriched seawater medium (Hand, Forward, and Davenport, 1967) in constant temperature growth chambers at 18 to 19°C. Growth lights providing light:dark cycles of 12:12 hr were used for the circadian rhythm experiments, and 16:8 hr for all other experiments. Since photoresponsiveness varies with culture age, only 5 to 6 day-old cultures were used in experiments in which action spectra were determined; all tests were begun 3 to 4 hr after the onset of the light period.

The equipment for monitoring photoresponses of individual cells consisted of either a flying spot scanning microscope (Forward and Davenport, 1968), a simple dark-field inverted microscope (Forward, 1973), or a combination microscope-closed circuit television system (Fig. 1). Stimulus lights consisted of either a No. 33-86-02 gratin

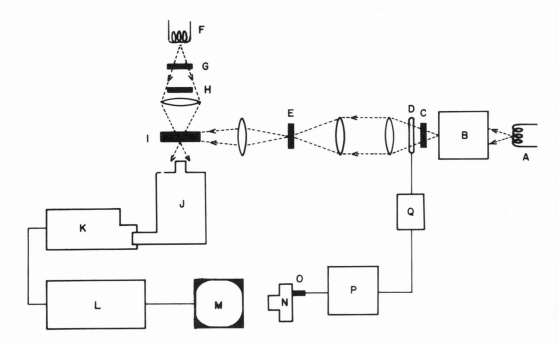

Fig. 1. Schematic diagram of light stimulus system and combination microscope-closed circuit television system for monitoring phototaxis. (A) Stimulus light; (B) monochromator; (C) wide band pass light filter; (D) shutter; (E) neutral density filters; (F) microscope illumination light; (G) interference filter; (H) dark-field condenser; (I) sample cuvette; (J) microscope objective; (K) television camera; (L) video tape recorder; (M) video monitor; (N) 35 mm still camera; (O) camera solenoid; (P) timing control for camera solenoid; and (Q) timing control for shutter.

monochromater (Bausch and Lomb) with a Tungsten Halide light source or a model f/3.5 monochromater (Farrand Optical Co.) coupled to a 150w Xenon arc lamp. Lights were focused onto the experimental chamber (lucite or quartz cuvette) by appropriate lenses, and the stimulus duration was controlled by a timed shutter assembly.

Both the initial stopping and the subsequent directional phototactic movement upon light stimulation can be quantitatively measured with the described monitoring systems. Phototaxis was measured either by recording photographically the individual cell movement relative to the stimulus light (± 10° was considered a directional response), or by cell accumulation at the window of the cuvette through which the stimulus light enters. The initial stop-response was also measured photographically. Since the stop-response latency was 0.4 to 0.6 sec, a photograph of the cells was taken by a solenoid-driven 35 mm camera about 0.6 sec after the beginning of light stimulation. In this way the largest number of responding cells was recorded. By using either a 0.25 or 0.5 sec camera exposure, stopped cells were recorded as round or oval dots and moving cells as long blurry lines. The response criterion, which was arbitrarily established, was based on the finding that the percentage of cells stopping upon stimulation (after being placed in darkness following illumination by the growth lights), decreased over time in a consistent manner (Forward and Davenport, 1968). A response in which more than 50% of the cells ceased moving upon stimulation is arbitrarily designated as a "positive stop-response;" a level of stopping from 35% to 50% is a "response drop-off." Subsequently, both the change in the time course of the response to a constant stimulus intensity (i.e., the time in darkness until a stop-response drop-off occurs) and the light energy threshold necessary for initiating a positive stop-response were used as behavioral assays for light sensitivity.

ACTION SPECTRA FOR PHOTOTAXIS

The visible action spectrum for phototaxis by *Gyrodinium dorsum*, as measured by the accumulation of cells at the stimulus window upon irradiation with wavelengths set at equal intensities, indicates a maximum responsiveness to 470 nm (Fig. 2). This result is further verified by determining the light intensity threshold necessary to initiate maximum phototaxis for wavelengths between 460 and 490 nm (insert, Fig. 2).

The action spectrum for the stop-response by *G. dorsum* was determined by measuring the light intensity threshold at selected wavelengths for a positive stop-response. Prior to stimulation, cells were either exposed to the culture chamber lights or to a red (620 nm) irradiation (1 min at an average intensity of 4.92×10^{15} hv cm^{-2} sec^{-1}). Following these irradiations, cells were left in darkness for 1 min, and then the light intensity threshold was determined. Figure 3 indicates that maxima occur at 470 and 280 nm.

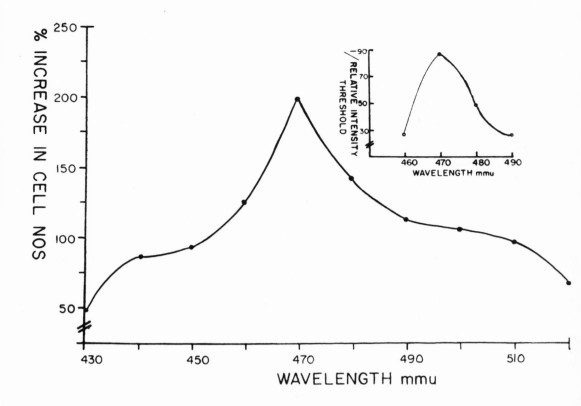

Fig. 2. The action spectrum for positive phototaxis by *Gyrodinium dorsum* as determined by measuring the percentage increase in number of cells (accumulation) at the stimulus window after 1 min stimulation with light at selected wavelengths of equal energies. Insert: the action spectrum for positive phototaxis in the spectral region of the maximum. Ordinate: the inverse of the lowest relative light intensity that would induce maximum orientation during a 10 sec stimulation (from Hand, Forward, and Davenport, 1967).

The visible action spectra for both the initial stopping and the subsequent phototactic movement are identical, having a single maxima at 470 nm. This would indicate that the same photoreceptive pigment is participating in both behavioral responses. Thus, the stop-response, which is technically easier to measure quantitatively is a valid indicator of phototactic light reception and was measured in subsequent experiments to indicate phototactic sensitivity.

For comparison, Figure 4 shows the action spectrum for the stop-response by *Gymnodinium splendens*, as determined immediately after removal from the culture chamber. The visible maximum at 450 nm is somewhat different from that for *Gyrodinium*, but the ultraviolet

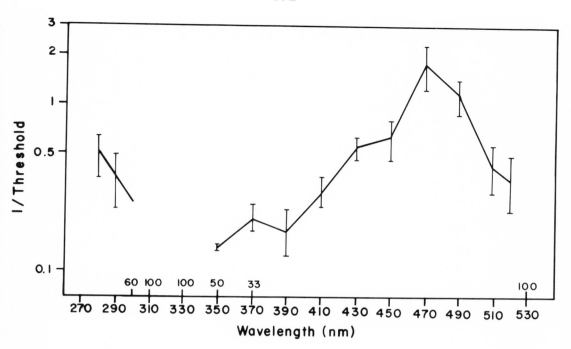

Fig. 3.   Ultraviolet and visible action spectra for the stop-response
by *Gyrodinium dorsum*. Ordinate: inverse of the intensity
threshold (1/threshold x 10^{15} hv cm^{-2} sec^{-1}) for a positive
stop-response. The vertical lines indicate the standard
deviation. The numbers directly above the various test wavelengths (abscissa) indicate the percentage of nonresponsive
cultures tested. A minimum of 5 cultures was tested at each
wavelength. Responsiveness to wavelengths below 280 nm was
not tested (from Forward, 1973).

Fig. 4. Ultraviolet and
visible action spectrum
for stop-response by
Gymnodinium splendens.
The actual plot is as
described for Figure 3.

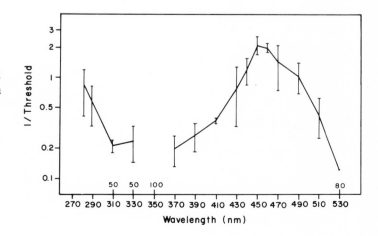

maximum is again at 280 nm. For both dinoflagellates, then, maximum visible sensitivity occurs to blue light, the region of the visible spectrum that penetrates the clear ocean to the greatest depths. In coastal and estuarine waters where these species are encountered, however, the wavelengths of maximum penetration are in the green and yellow regions, not the blue, due to the presence of suspended matter (Jerlov, 1968). A clear correlation between light penetration and sensitivity by the cells is not apparent. Dinoflagellate diurnal migrations, however, are confined to shallow depths of less than 15 m (Hasle, 1950, 1954; Eppley, Holm-Hansen, and Strickland, 1968). Furthermore, the intensity of light in the region 450 to 470 nm in coastal waters at shallow depths (Jerlov, 1968) is probably not sufficiently decreased to prevent phototaxis.

PIGMENTS INVOLVED IN PHOTOTAXIS

By comparing the above action spectra with those for other phytoflagellates, it is possible to determine the probable identity of the photoreceptor pigment (Forward, 1973). The loss of responsiveness at longer wavelengths above 530 nm and the large peaks at 470 nm (Fig. 3) and 450 nm (Fig. 4) resemble responses of other dinoflagellates, *Peridinium trochoidium* and *Gonyaulax catenella* (Halldal 1958). of *Euglena gracilis* (Diehn, 1969), and of volvocales (Halldal 1961). An additional pronounced maximum occurs at about 365 to 375 nm in action spectra for phototaxis in *Euglena* (Diehn, 1969) and for chloroplast movements in the chlorophycean alga *Vaucheria* (Haupt and Schönbohm,1970); this is considered supporting evidence for a flavin as the photoreceptor pigment. Since such an ultraviolet peak is absent from the action spectrum for *Platymonas*, Halldal (1961) suggests that the pigment here is a carotenoid.

The action spectra for *Gyrodinium* and *Gymnodinium* also lack near-ultraviolet maxima, so a carotenoid is likewise the probable photoreceptor in both organisms. The strong responses at 280 nm (Figs. 3, 4) resemble that by *Platymonas* and probably correspond to absorption by aromatic amino acids associated with a protein (Halldal 1961), suggesting that the photoreceptor pigment in both *Gyrodinium* and *Gymnodinium* is a carotenoprotein. Pigments of this type are known to be present in a related dinoflagellate, *Gyrodinium resplendens*, in which the major carotenoids are β-carotene, peridinin, dinoxanthin, and diadinoxanthin (Loeblich and Smith, 1968). Further experiments, however, are necessary to identify the responsible carotenoid in *Gyrodinium dorsum* and *Gymnodinium splendens*.

The first suggestion that a pigment other than the carotenoprotein was involved in phototaxis was the observation that *Gyrodinium dorsum* responded phototactically only when cultured under a combination of incandescent and fluorescent lights and not when the former were absent. Since these two types of lights differ in their output in the red, a pigment which absorbs in this region and is

capable of activation of a physiological system was suspected.
Through a series of experiments it was found that a 4-min exposure
to blue (470 nm) light abolished phototactic responsiveness to this
wavelength in succeeding darkness. However, subsequent exposure to
red light produced a return of sensitivity to blue. The sensitizing
effects of red light could be reversed by irradiation with far-red
light. This indicates that a second light-receptive pigment, probably
a phytochrome, is involved in phototaxis.

The action spectrum maxima for the red-absorbing form of the
phytochrome (P_R) are at 320 and 620 nm (Fig. 5) and for the far-red-
absorbing form (P_{FR}) they are at 360 to 390 nm and 700 nm (Fig. 6).
The fact that the effects of red and far-red light are mutually reversible is evidence that this pigment is indeed a phytochrome (Table 1).
Furthermore, since the action spectrum for phototaxis is distinctly
different from that for the phytochrome, two distinctly different
pigments are participating in phototaxis by *Gyrodinium dorsum*.

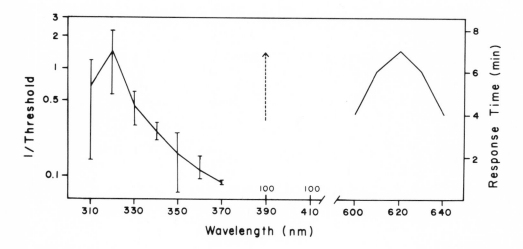

Fig. 5. Action spectrum for conversion of P_R to P_{FR} in the ultraviolet and visible by *Gyrodinium dorsum*. Left ordinate: the
inverse of the threshold for a positive stop-response to 470
nm for cells after being irradiated with blue 470 nm (12.4 x
10^{15} hv cm^{-2}s^{-1}) for 4 min to abolish responsiveness, then
exposed for 3 min to the test wavelengths (abscissa) set at
equal quanta (e.g., 350 nm = 8.7 x 10^{15} hv cm^{-2}s^{-1}) to reacti-
vate responsiveness, and then 1 min in darkness. A minimum
of 5 cultures was tested at each wavelength. The actual plot
is as described in Figure 3 (replotted from Forward, 1973).
Shown on the right is the visible action spectrum for P_R as
replotted from earlier work (Forward and Davenport, 1968).
The vertical arrow indicates the direction of the change in
1/threshold and response time upon greater conversion of
P_R to P_{FR}.

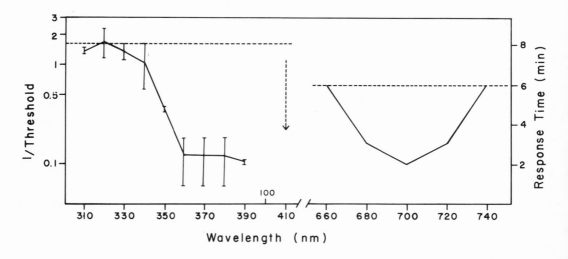

Fig. 6. Ultraviolet and visible action spectrum for conversion of
P_{FR} to P_R by *Gyrodinium dorsum*. Left ordinate: inverse of
the threshold for a positive stop-response to 470 nm after
the cells were irradiated with red (620 nm) light (6.5 x 10^{15}
hv $cm^{-2}s^{-1}$). The actual plot is as described in Figure 3.
The horizontal line indicates the control response (red light
plus 4 min in darkness). A minimum of 3 cultures was tested
at each wavelength (from Forward, 1973). On the right is
replotted the visible action spectrum from earlier work
(Forward and Davenport, 1968). The response of the control
is indicated by the horizontal line. The center vertical
arrow shows the direction of change in 1/threshold and
response time upon greater conversion from P_{FR} to P_R.

TABLE 1
Reversibility of stop-response activation and inactivation by red
(620 nm) and far-red (700 nm) irradiation.*

Combinations of red(R), far-red (FR) after 4 min exposure to blue	Time in darkness until response drop-off (min)
FR	0
R	7
R, FR	2
FR, R	7
FR, R, FR	2
R, FR, R	7
R, FR, R, FR	2
FR, R, FR, R	7

*From Forward and Davenport, 1968.

Curiously, experimental evidence indicates that a phytochrome is not associated with phototaxis by *Gymnodinium splendens*, so this pigment cannot be involved in phototaxis by all dinoflagellates. Thus, at least two different pigment systems (carotenoid alone; carotenoid-phytochrome) are involved in phototaxis by dinoflagellates.

Phytochrome from higher plants has red and far-red absorption maxima at about 660 to 665 and 730 nm, respectively (Butler, Hendricks, and Siegelman, 1964). West (1968) has argued that a red/far-red reversible photoperiodic process could not reasonably be expected in sublittoral algae, because neither type of light is present at sufficient intensities at depths where they grow. Dring (1970, 1971), however, presents convincing underwater energy measurements for red (660 nm) light, which indicate that within the photic zone of most types of oceanic and coastal waters the intensity of 660 nm is sufficient to activate a phytochrome system. A similar argument for 730 nm is perhaps untenable. Thus, 730 nm, which is important in establishing P_R/P_{FR} ratios, may be the real limiting wavelength. Although Dring's work (1967) with the red alga *Porphyra* indicates that phytochrome action spectrum maxima are at 660 and 730 nm, this situation is perhaps atypical for algae. Isolated phytochrome from the green alga *Mesotaenium* has maxima at 649 and 710 nm (Taylor and Bonner, 1967) and in a blue-green alga a pigment resembling phytochrome has maxima at 520 and 650 nm (Scheibe, 1972). Finally, in *Gyrodinium dorsum*, a representative of brown-colored algae, maxima are at 620 and 700 nm (Figs. 5, 6). Thus, in algae other than Rhodophyta, the red and far-red maxima for absorption by phytochrome-like pigments are at shorter wavelengths than those for higher plants (Butler, Hendricks, and Siegelman, 1964). An intriguing theory is that algal phytochrome absorption maxima are adapted to light transmission chracteristics of seawater which allows greater penetration by wavelengths below 730 nm.

CIRCADIAN RHYTHMS OF PHOTORESPONSES

Phototactic responsiveness by the phytoflagellates *Euglena* (Pohl, 1948) and *Chlamydomonas* (Bruce, 1970) is not uniform over a 24-hr light-dark (LD) period, but rather shows circadian rhythmicity with maximum responsiveness during the light phase. Similar rhythms in bioluminescence (Hastings and Sweeney, 1958, 1960), division (Sweeney and Hastings, 1958), and photosynthesis (Sweeney, 1960) have been found in dinoflagellates. Since these organisms are known to show diurnal migration presumably in response to light, the presence of a circadian rhythm in photoresponsiveness and the timing of maximum responsiveness may be very important as related to the daily movements.

If *Gymnodinium splendens* is grown on a 12:12 LD cycle which is representative of LD cycles in nature, and then placed under constant low level light (two cool white fluorescent lamps, intensity at

surface of culture vessels = 2.0 watts/m^2) and its responsiveness to
a constant intensity of 450 nm light is monitored over the next 24
hrs, a clear rhythm in sensitivity is evident (Fig. 7). Interestingly

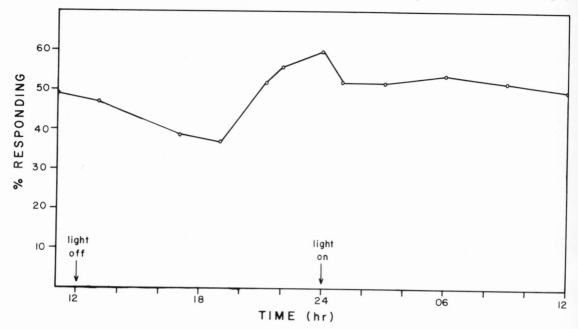

Fig. 7. Circadian rhythm in responsiveness by *Gymnodinium splendens*.
Ordinate: percentage of cells showing a stop-response to
stimulation with 450 nm light at an intensity of 5.3 x 10^{14}
hv cm^{-2}s^{-1}. Abscissa: time through day; time for the onset
and end of the light phase indicated by "light on" and "light
off." Average cell sample size at each test time was 148
cells.

the cells are maximally responsive at the time for the beginning of
the light phase. The same pattern of responsiveness is observed
under constant dark conditions.

In contrast, if *Gyrodinium dorsum* is grown on a 12:12 LD cycle
and placed in constant darkness, no rhythm in responsiveness to 470
nm light is evident (Fig. 8a). Instead, the cells show increasing
sensitivity to blue light with longer time in darkness. *Gyrodinium*,
however, does have a phytochrome pigment which can alter phototactic
responsiveness upon irradiation.

The test for a rhythmic interaction of phytochrome with phototaxis was to grow cells on a 12:12 LD cycle, place them in constant
darkness and over time, test for responsiveness to blue (470 nm)
light after the phytochrome was irradiated for 1 min with red (620
nm) light (intensity = 8 x 10^{15} hv cm^{-2}s^{-1}). In this experiment the
light intensity available at 470 nm for stimulation was much lower
than that necessary for initiating a response to 470 nm light alone
(Fig. 8a). Thus, any response to 470 nm would occur only if the

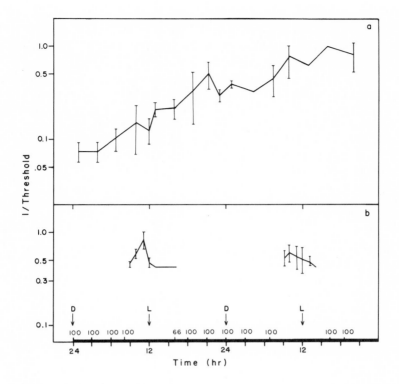

Fig. 8.   Circadian rhythm in responsiveness by *Gyrodinium dorsum*.
(a) Cells were placed in constant darkness and the threshold sensitivity for a positive stop-response to 470 nm
light was determined at times throughout the day as indicated on the abscissa. "D" is time in LD cycle for the
beginning of the dark phase, and "L" for the light phase.
Ordinate is as described for Figure 3.   (b) Ordinate:
inverse of threshold for a positive stop-response after
cells were preirradiated with red light. Abscissa: time
in LD cycle. The numbers directly above the times indicate
the percentage of nonresponsive cultures tested (replotted
from Forward and Davenport, 1970).

sensitivity of the cells was increased to a level sufficient to allow
responsiveness to the lower stimulus intensity. Irradiation of the
phytochrome serves to increase sensitivity only at a particular time
during the day (Fig. 8b). At all other times no responsiveness was
observed. Three important characteristics of *Gyrodinium*'s photo-
response system are evident in this figure. First, a circadian
rhythm in sensitivity is apparent, with maximum responsiveness occurring shortly before the time for the beginning of the light phase.
Second, since blue sensitivity alone does not vary rhythmically (Fig.
8a), this suggests that the circadian rhythmicity resides within the
phytochrome pigment system. Third, by comparison to the light intensity threshold levels seen upon blue stimulation alone (Fig. 8a),
irradiation of the phytochrome with red light serves to greatly

increase the sensitivity of the cells.

Thus, although the internal mechanisms appear to be different, both *Gyrodinium dorsum* and *Gymnodinium splendens* have circadian rhythms in their photoresponses with the time for greatest light sensitivity within a 24-hr period occurring at the beginning of the light phase. Correspondingly, in nature dawn is the time for the beginning of the diurnal migration upward toward the surface. It can be hypothesized that this rise places the organisms in an area of higher light intensity which would be available for photosynthesis.

Sweeney (1960) clearly demonstrated that photosynthetic activity in the dinoflagellate *Gonyaulax polyedra* varied rhythmically over the day with maximum activity during the day phase. A similar pattern for photosynthetic activity is seen for *Gyrodinium* (Fig. 9). In this

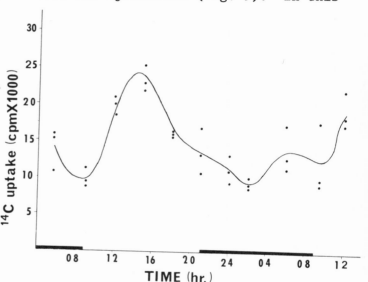

Fig. 9. Photosynthetic activity for *Gyrodinium dorsum* throughout the 12:12 LD cycle. Dark phase indicated by dark bands on abscissa. Each point represents the sample counts/min minus those counts due to dark incorporation as determined by the dark controls.

experiment photosynthetic activity was measured at 3-hr intervals for cells growing on a 12:12 LD cycle (light intensity = 3.2×10^4 ergs $cm^{-2}s^{-1}$). At each time aliquot cell samples were inoculated with 2 microcuries $NaHC^{14}O_3$ and illuminated for 30 min under three cool white fluorescent bulbs (intensity – 1.25 ergs/$cm^{-2}s^{-1}$). The cells were then vacuum-filtered onto a Millipore filter and washed first with seawater saturated with $NaHCO_3$, then by seawater adjusted to a pH of 3.4. The incorporation of C^{14} was then measured as counts per minute by a liquid scintillation counter. The greatest photosynthetic activity occurs in the middle of the light phase (Fig. 9).

CONCLUSIONS

Assuming that the preceding results reflect physiological characteristics of most dinoflagellates, the pattern of events that emerges with respect to diurnal vertical movement, phototaxis, and photosynthesis is as follows. At night, dinoflagellates are observed to descend from 3 to 14 m below their day depths (Hasle, 1950, 1954; Eppley, Holm-Hansen, and Strickland, 1968). The time when the light intensity increases at dawn corresponds to the beginning of the light phase and the time within the circadian rhythm of phototaxis in both *Gyrodinium* and *Gymnodinium* for maximum sensitivity to light. Thus, at dawn the cells should begin their migration upward in the direction of increasing light intensity. Assuming swimming speeds 1 to 2 m/hr (Hand, Collard, and Davenport, 1965; Eppley, Holm-Hansen, and Strickland, 1968) in 6 hrs, i.e., by 12:00 noon, most dinoflagellates would attain the surface area of the ocean, presumably when the maximum amount of sunlight is available underwater and when the greatest photosynthetic activity occurs. The present data do not permit speculation about physiological events occurring at the end of the day phase. However, the descent must be an active process, since the cells sink too slowly to attain the observed night depths (Smayda, 1970).

ACKNOWLEDGMENTS

I thank Meg Forward for her technical assistance. Portions of the study were supported by a Biomedical Science Support Grant from the National Institutes of Health to Duke University and by a Major Grant from the Duke University Research Council.

LITERATURE CITED

Bruce, V. G. 1970. The biological clock in *Chlamydomonas reinhardi*. J. Protozool. 17:328-34.

Butler, W. L., S. B. Hendricks, and H. W. Siegelman. 1964. Action spectra of phytochrome in vitro. Photochem. Photobiol. 3: 521-29.

Diehn, B. 1969. Action spectra of the phototactic response in *Euglena*. Biochim. Biophys. Acta 177:136-43.

Dring, M. J. 1967. Phytochrome in red algae. Nature 215:1411-12.

_____. 1970. Photoperiodic effects in microorganisms. In: Photobiology of Microorganisms (P. Halldal, ed.), pp. 345-68. New York: Wiley-Interscience.

_____. 1971. Light quality and the photomorphogenesis of algae in marine environments. In: Fourth European Marine Biology Symposium (D. J. Crisp, ed.), pp. 375-92. London: Cambridge University Press.

Eppley, R. W., and O. Holm-Hansen, and J. D. H. Strickland. 1968. Some observations on the vertical migration of dinoflagellates. J. Phycol. 4:333-40.

Feinlieb, M. E. H. and G. M. Curry. 1971. The relationship between stimulus intensity and oriented phototactic response (topotaxis) in *Chlamydomonas*. Physiol. Plant. 25:346-57.

Forward, R. B., Jr. 1973. Phototaxis in a dinoflagellate: action spectra as evidence for a two pigment system. Planta (Berl.). In press.

_____ and D. Davenport. 1968. Red and far-red light effects on a short-term behavioral response of a dinoflagellate. Science 161:1028-29.

_____ and _____. 1970. The circadian rhythm of a behavioral photo-response in the dinoflagellate *Gyrodinium dorsum*. Planta (Berl.) 92:259-66.

Halldal, P. 1958. Action spectra of phototaxis and related problems in volvocales, ulva-gametes and Dinophyceae. Physiol. Plant. 11:118-53.

_____. 1961. Ultraviolet action spectra of positive and negative phototaxis in *Platymonas subcordiformis*. Physiol. Plant. 14: 133-39.

Hand, W., P. A. Collard, and D. Davenport. 1965. The effect of temperature and salinity change in swimming rate in dinoflagellates *Gonyaulax* and *Gyrodinium*. Biol. Bull. 128:90-101.

_____, R. Forward, and D. Davenport. 1967. Short-term photic regulation of a receptor mechanism in a dinoflagellate. Biol. Bull. 133:150-65.

Hasle, G. R. 1950. Phototactic vertical migration in marine dino-flagellates. Oikos 2:162-75.

_____. 1954. More on phototactic diurnal migration in marine dino-flagellates. Nytt Magasin for Botanikk 2:139-47.

Hastings, J. and B. Sweeney. 1958. A persistent diurnal rhythm of luminescence in *Gonyaulax polyedra*. Biol. Bull. 115:440-58.

_____ and _____. 1960. The action spectrum of shifting the phase of the rhythm of luminescence in *Gonyaulax polyedra*. J. Gen. Physiol. 43:697-706.

Haupt, W. and E. Schönbohm. 1970. Light-oriented chloroplast movements. In: Photobiology of Microorganisms (P. Halldal, ed.), pp. 283-307. New York: Wiley-Interscience.

Huth, K. 1970. Bewegung und Orientierung bei *Volvox aureus*. Ehrb. Z. Pflanzenphysiol. 62:436-50.

Jerlov, H. G. 1968. Optical Oceanography. New York: Elsevier Publishing Co.

Loeblich. A. R., III and V. E. Smith. 1968. Chloroplast pigments of the marine dinoflagellate *Gyrodinium resplendens*. Lipids 3:5-13.

Pohl, R. Z. 1948. Tagesrhythmus im phototakischen Verhalten der *Euglena gracilis*. Z. Naturforsch. 3b:367-74.

Scheibe, J. 1972. Photoreversible pigment occurrence in a blue-green alga. Science 176:1037-39.

Smayda, T. J. 1970. The suspension and sinking of phytoplankton in the sea. In: Oceanogr. Mar. Biol. Ann. Rev. (H. Barnes, ed.) 8:353-414.

Sweeney, B. M. 1960. The photosynthetic rhythm in single cells of *Gonyaulax polyedra*. CSHSQB 25:145-48.

_____ and J. W. Hastings. 1958. Rhythmic cell divisions in populations of *Gonyaulax polyedra*. J. Protozool. 5:217-24.

Taylor, A. O. and B. A. Bonner. 1967. Isolation of phytochrome from the alga *Mesotaenium* and liverwort *Sphaerocarpos*. Plant Physiol. 42:762-66.

West, J. A. 1968. Morphology and reproduction of the red alga *Acrochaetium pectinatum* in culture. J. Phycol. 4:89.

SUMMARY

An estuary represents a dynamic, demanding environment in which organisms must employ various evolutionary strategies to survive. The papers included in this volume have been divided into five sections dealing with various aspects of physiological adaptations to different ecological stresses, an interdisciplinary field known as physiological ecology. In recent years particular attention has been focused on estuaries as a result of the intensive conflicts between various segments of human society to utilize this natural resource. An understanding of the functional capabilities of the estuarine biota to survive fluctuations in both "normal" environmental factors and man-induced factors has never been more vital than it is now. Such understanding is essential to the wise management of the estuarine ecosystem.

Section 1. <u>Resistance Adaptations</u>.

The first group of papers dealt with the physiological response of algae and animals to extreme environmental stress. In this case the extreme stress would result in the death of the organism. Precht[1] called these physiological responses to extreme conditions resistance adaptations. Rice and Ferguson have reviewed extensively the response of phytoplankton to the extreme ends of various environmental gradients. Not only did they discuss the influence of "natural" parameters, but also included a section on responses of algae to man's perturbations of the estuarine environment. That mercury in excessive concentrations adversely effects the growth and photosynthetic activity of algae was reported by Zingmark and Miller. Inhibition of cellular processes is a function of concentration of mercury, length of exposure, and the phase of the growth cycle.

A characteristic vertebrate inhabitant of estuarine waters, the killifish, *Fundulus heteroclitus*, experiences extreme excursions in their thermal environment throughout the year. Umminger reported that freezing resistance was correlated with chemical composition changes and in terms of enzymatic, hormonal, and neural regulatory mechanisms. He emphasized that interspecific variation is to be expected in estuarine fish because of differences in phylogenetic histories and in ecological and physiological requirements. Another obvious environmental variable in estuaries is salinity. Moreira found a zoogeographical difference in the ability of the copepod, *Euterpina acutifrons*, to survive low salinity. Further, the lethal response to low salinity was influenced by previous thermal acclimation; generally cold-acclimated animals survived low salinity better than warm-acclimated animals. These findings highlight the fact that populations from different regions may have evolved different physiological responses to meet the combination stresses characteristic

---

[1]Precht, H. 1958. In: <u>Physiological Adaptation</u>. C. L. Prosser (ed.). Am. Physiol. Soc., Washington, D. C. pp. 50-77.

of their particular region.

The differential limiting nature of salinity on different stage_ in the life cycle was demonstrated when comparing larval and adult stages of the fiddler crab by the Vernbergs. The larvae are more sensitive than adults, but in contrast, adult fiddler crabs from the tropics are more sensitive than their larvae to low temperature.

Another important point to emphasize is that two or more enviro_ mental factors at a stressful, but sublethal, level may interact to influence the functional response of an organism more markedly than any one factor acting independently. The studies of Moreira and the Vernbergs involving temperature and salinity demonstrate this point, while Rice and Ferguson reported on the influence of multiple factor interaction on algae.

Although the remaining contributions to this volume were concerned with the functional response of organisms to sublethal exposures of various environmental parameters, the limiting nature of the environment for both individual organisms and populations was emphasized in other papers. For example, Queen described the adapti_ mechanisms of halophytes to extreme salt concentrations and Theede emphasized adaptation of individual animals to salinity at the organismic and the cellular level of biological organization. At the population level, Sastry and Bayne discussed reproductive responses as reflecting environmental flux.

Section 2. Respiration and Energetics.

Darnell and Wissing presented extensive data on the food relationships of the common estuarine pinfish. This study is of special interest in that much of their data was derived from natural populations. Thus it has relevance in attempting to correlate laboratory and field studies in an effort to ecologically interpret the adaptive nature of the physiological responses of organisms. Darnell and Wissing suggested that behavior plays an important role in the physiological processes of the pinfish and must be carefully considered when designing future studies. That the metabolic capacity of an intertidal animals is influenced by many extrinsic and intrinsic factors was well-documented by Newell's work on the snail, *Littorina*. He emphasized that only after detailed quantitative comparisons between the metabolism of intact animals and subcellular components can biochemical events be related to those occurring in whole organisms.

The respiratory response of the horseshoe crab, *Limulus polyphemus*, was carefully analyzed by Johansen and Peterson at various level of locomotor activity and in field-simulated condition of burrowing. Their work indicated that the distinction between standard and active metabolism is not clear and presents problems unless the actual work output of the organism can be measured. Further they suggested that the "principle" of low tension transport of oxygen commonly ascribed to invertebrates may not be valid for many crabs. The reversed Bohr shift benefits buried *Limulus*. Oxygen extraction rates for a number of invertebrates representing various phyla were reported by Mangum and Burnett. Passive and an actively generated flow of water over a respiratory surface enhanced respiration of certain estuarine species that experience low oxygen environments. These investigations further examined the respiratory economy of worms that depend on muscular activity for ventilation.

Section 3. <u>Water and Ions</u>.

Polychaete worms occupy a wide range of habitats which vary markedly in terms of salinity. Oglesby developed a theoretical formulation of the maximum amount of water a worm would take up after transfer to a reduced salinity. This formulation permits a quantitative assessment of the efficiency of water content regulation in various species. Results of a comparison of water-control regulation in seven species of polychaetes are consistent with known data on their osmotic behavior.

Queen reviewed the adaptive mechanisms of the few halophytic vascular plants found in estuaries. Included was a discussion of both morphological and physiological mechanisms. In turn Theede reviewed, with special emphasis on work from the University of Kiel, the various adaptive mechanisms to salinity change used by poikilosmotic and homoiosmotic animals. Responses discussed randed from intact organisms to the molecular level of biological organization. Although these two papers dealt with two diverse biological groups, plants and animals, the similarity in some of their adaptive strategies is remarkable. Further, both stressed the need for more detailed studies on the role of enzymes in salinity adaptation.

The remaining two papers in this section emphasized the role of amino acids in osmoregulation by two estuarine animals, a barnacle and a clam. The barnacle, *Balanus improvisus*, is dependent upon cell volume regulation for its characteristic euryhalinic mode of life. This regulation is accomplished with the aid of amino acids and other α-amino compounds. However, below a sea water osmolality of about 100 m - osmoles, Fyhn and Costlow felt that either α-amino compounds may be less effective in cell volume regulation or the hemolymph osmolality may not conform osmotically with the sea water. Anderson found that in an euryhalinic clam, *Rangia cuneata*, cells in their gills were the major site of C^{14}-glycine uptake. In addition the gills have the necessary metabolic machinery to regulate rapidly the accumulated glycine. Uptake of glycine by the gills was not inhibited by low salinity, although the response of the whole organism was the opposite. Further, he found that sodium was necessary for glycine uptake.

Section 4. <u>Reproduction and Development</u>.

Although reproduction and development are of obvious importance to the continuity of populations in estuaries, until recently relatively few detailed studies on this subject have been published. Both Bayne and Sastry reviewed the existing data on reproductive cycles and presented their ideas on needed research. Bayne presented a schematic model of the complex interrelationships between food sources, sites of energy reserve in animals, and the utilization of materials in gametogenesis in *Mytilus edulis*, a mussel. Sastry, using the scallop, *Aequipecten irradians*, as his principal example, emphasized the interaction between environmental and endogenous factors in influencing reproductive cycles. Considerable diversity in reproduction patterns of invertebrates exists, but the causal relationships in determining these patterns need examination. Hopkins and Dean dealt with the effects of thermal shock on the development of the common estuarine fish, *Fundulus heteroclitus*. A thermal shock (40°C for 5 min) applied at various developmental stages resulted in death and/or malformation. The authors suggested that different mechanisms may be influenced at different stages of development. These data were related to the possible environmental impact of locating electrical generating plants on estuaries. That other environmental factors also influence larval development was demonstrated by Dupuy. The growth rate of oyster larvae was greatly influenced by the combination of algal species they were fed and by the size of the culture container. The general feeling that I gained from the papers included in this section is that this is a fertile field for future investigation and one in which data are vitally needed to understand not only natural phenomena, but also, how man and the estuarine biota can live together.

Section 5. Perception of the Environment.

How organisms are able to perceive and respond behaviorally to the ambient environment represents an important step in understanding the total integrated response of an organism to its environment. Hyatt determined that fiddler crabs have diphasic, cornea-negative electroretinograms which are characteristic of "fast" eyes found in other crustaceans. Further he observed the behavior of these crabs to light in order to define the sign of phototaxis and to assess true color discrimination. Light is an important environmental factor and coloration may be an important role as a "close-up" identifier to initiate or perpetuate courtship or other behavior patterns of fiddler crabs. That light is important was demonstrated further by Forward for dinoflagellates. He presented physiological data to support his hypothesis that the greatest sensitivity to light by dinoglagellates occurs around dawn, which can be correlated with the onset of their upward vertical migration. Thus, these responses would bring the dinoglagellates to the surface when the maximum amount of sunlight is available and when greatest photosynthetic activity occurs.

In summary, the twenty papers included in this volume represent a status report of our knowledge concerning some problems in the physiological ecology of estuarine organisms. They represent various viewpoints ranging from the population level of biological organization to the subcellular and from resistance to capacity adaptations. It is evident that more detailed studies are needed not only for the joy of pure scientific discovery but also for the necessity of supplying the vital physiological data essential to wise environmental management of the estuaries. Few present day scientific fields have more relevance to pure and applied aspects of science than does physiological ecology.

INDEX